T0295066

Kuhn's *The Structure of Scientific Revolutions* at 60

Thomas Kuhn's *The Structure of Scientific Revolutions* has sold over a million copies since its publication in 1962, is one of the most cited academic books of all time, and continues to be read and studied today. This volume of new essays evaluates the significance of Kuhn's classic book in its changing historical context, including its initial reception and its lasting effects. The essays explore the range of ideas that Kuhn made popular with his influential philosophy of science, including paradigms, normal science, paradigm changes, scientific revolutions, and incommensurability; and they also look at less-studied themes in his work, including scientific measurement, science education, and science textbooks. Drawing on the latest scholarship as well as unpublished material in the Thomas Kuhn Archives at MIT, this volume offers a comprehensive way into Kuhn's philosophy and demonstrates the continuing relevance of his ideas for our understanding of science.

K. Brad Wray is Associate Professor at the Centre for Science Studies, Aarhus University. His recent publications include *Interpreting Kuhn: Critical Essays* (Cambridge University Press, 2021) and *Kuhn's Intellectual Path: Charting "The Structure of Scientific Revolutions"* (Cambridge University Press, 2021).

Cambridge Philosophical Anniversaries

The volumes in this series reflect on classic philosophy books from the second half of the twentieth century, assessing their achievements, their influence on the field, and their lasting significance.

Titles Published in This Series

Kuhn's *The Structure of Scientific Revolutions* at 60
Edited by K. Brad Wray
Kripke's *Wittgenstein on Rules and Private Language* at 40
Edited by Claudine Verheggen
MacIntyre's *After Virtue* at 40
Edited by Tom Angier
Rawls's *A Theory of Justice* at 50
Edited by Paul Weithman
Cavell's *Must We Mean What We Say?* at 50
Edited by Greg Chase, Juliet Floyd, and Sandra Laugier

This list can also be seen at www.cambridge.org/cpa

Kuhn's *The Structure of Scientific Revolutions* at 60

Edited by

K. Brad Wray

Aarhus University

CAMBRIDGE
UNIVERSITY PRESS

Shaftesbury Road, Cambridge CB2 8EA, United Kingdom

One Liberty Plaza, 20th Floor, New York, NY 10006, USA

477 Williamstown Road, Port Melbourne, VIC 3207, Australia

314–321, 3rd Floor, Plot 3, Splendor Forum, Jasola District Centre, New Delhi – 110025, India

103 Penang Road, #05–06/07, Visioncrest Commercial, Singapore 238467

Cambridge University Press is part of Cambridge University Press & Assessment, a department of the University of Cambridge.

We share the University's mission to contribute to society through the pursuit of education, learning and research at the highest international levels of excellence.

www.cambridge.org
Information on this title: www.cambridge.org/9781009100700

DOI: 10.1017/9781009122696

First published 2024

A catalogue record for this publication is available from the British Library

A Cataloging-in-Publication data record for this book is available from the Library of Congress

ISBN 978-1-009-10070-0 Hardback

This book is dedicated to
Pablo Melogno[†]
1979–2023
Friend and Scholar

Contents

Part III   Incommensurability, Progress, and Revolutions

Part IV   Kuhn's Impact on the Philosophy, Sociology, and History of Science

Figures

Tables

Contributors

HANNE ANDERSEN is a Professor in the Department of Science Education at the University of Copenhagen.

FEDERICA BOCCHI is completing her PhD in the Department of Philosophy at Boston University.

ALISA BOKULICH is a Professor in the Department of Philosophy at Boston University.

ALEXANDRA BRADNER is a Visiting Assistant Professor of Philosophy at Kenyon College.

HASOK CHANG is Hans Rausing Professor of History and Philosophy of Science at the University of Cambridge.

RICHARD CREATH is President's Professor of Life Sciences and of Philosophy, and the director of the program in history and philosophy of science, at Arizona State University.

HAIXIN DANG is an Assistant Professor in the Department of Philosophy at the University of Nebraska Omaha.

CHRIS HAUFE is the Elizabeth M. and William C. Treuhaft Professor of the Humanities at Case Western Reserve University.

VASSO KINDI is a Professor of Philosophy of Science at the National and Kapodistrian University of Athens.

ALEX LEVINE is Professor and Chair in the Department of Philosophy at the University of South Florida.

PABLO MELOGNO[†] was a Professor at the University of the Republic in Uruguay.

MARKUS SEIDEL is a Senior Lecturer in the Center for Philosophy of Science (ZfW) at the Westfälische Wilhelms-Universität Münster, Germany.

JAMIE SHAW is a Postdoctoral Fellow at the Leibniz Universität Hannover.

JONATHAN Y. TSOU is a Professor of Philosophy and the Marvin and Kathleen Stone Distinguished Professor of Humanities in Medicine and Science at the University of Texas at Dallas.

K. BRAD WRAY is an Associate Professor at the Centre for Science Studies at Aarhus University in Denmark. He is also an Associate Fellow at the Aarhus Institute of Advanced Studies.

Acknowledgments

This book is the result of the efforts of many people and a number of organizations. I want to acknowledge my gratitude to them here.

Hilary Gaskin of Cambridge University Press originally approached me with the idea of taking on the project of editing a volume of papers in celebration of Thomas Kuhn's influential book *The Structure of Scientific Revolutions* (SSR) as part of a new series intended to celebrate key books in philosophy on the anniversary of their publication. This seemed like a great idea, and a great opportunity to foster scholarship on Thomas Kuhn and SSR.

After planning the volume, I was encouraged by Paul Hoyningen-Huene to run a conference in 2022, in celebration of the sixtieth anniversary of the publication of SSR, as well as in honor of the hundredth year since Kuhn's birth. A number of happy coincidences occurred that made running such a conference feasible. First, I had the good fortune to be appointed as an Associate Fellow at the Aarhus Institute of Advanced Studies (AIAS) in 2021. AIAS is an exciting interdisciplinary research environment organized to encourage fruitful interactions between scholars across traditional disciplinary lines. Further, AIAS provided some financial support for a conference, as well as wonderful facilities in which to host the conference. I want to thank the director of AIAS, Søren Rud Keiding[†], for his support, as well as the staff at AIAS. In addition, with the help of my colleague Kristian Hvidtfelt Nielsen, I applied for and was awarded a conference grant from the Carlsberg Foundation. I want to thank the director and staff at AIAS for their help in organizing and running the conference, as well as the Carlsberg Foundation for their generous support (Grant No. CF21-0698). The conference allowed me to invite a number of the contributing authors to Aarhus to present their papers, as they worked on them, as well as other Kuhn scholars. The excitement and energy at the conference was wonderful. And the constructive exchanges of ideas are reflected in a number of the pieces in the volume. I want to thank all those who participated in the conference for making the event so successful and enjoyable. Paul deserves a special thanks for encouraging me to organize the event.

I also thank Lori Nash for working with me on preparing the volume for production, preparing an index and a comprehensive bibliography, and for proofreading the proof pages with me. The production of the book, in particular, Lori's work on the book, was supported by a publication grant from the Aarhus Universistets Forskningsfond (AUFF), grant number AUFF-E-2022-4-34. AUFF has been very supportive over the last five years, supporting four of my book projects.

I thank Hilary and Abi Sears at Cambridge University Press for their work on this volume, from its conception to its final production. I also thank Thomas Haynes from CUP for seeing the book through production.

Finally, I thank the contributors for their thoughtful papers on Kuhn's *The Structure of Scientific Revolutions*. It is a testament to the vigor and creativity of scholarship on Kuhn and his remarkable book. It has been an honor and privilege to work with these colleagues on this volume.

Abbreviations

Thomas Kuhn's Books

BBT (1978). *Black-Body Theory and the Quantum Discontinuity, 1894–1912.* Oxford: Oxford University Press.

CR (1957). *The Copernican Revolution: Planetary Astronomy in Development of Western Thought.* Cambridge, MA: Harvard University Press.

ET (1977). *The Essential Tension: Selected Studies in Scientific Tradition and Change.* Chicago: University of Chicago Press.

- "Preface," pp. ix–xxiii (1977a).
- "The Relations between the History and the Philosophy of Science," pp. 3–20 (1976/1977b).
- "Mathematical versus Experimental Traditions in the Development of Physical Science," pp. 31–65 (1976/1977a).
- "The History of Science," pp. 105–126 (1968/1977).
- "The Relations between History and History of Science," pp. 127–161 (1971/1977).
- "The Function of Measurement in Modern Physical Science," pp. 178–224 (1961/1977).
- "The Essential Tension: Tradition and Innovation in Scientific Research," pp. 225–239 (1959/1977).
- "Second Thoughts on Paradigms," pp. 293–319 (1974/1977).
- "Objectivity, Value Judgment, and Theory Choice," pp. 320–339 (1973/1977).

LW (2022). *The Last Writings of Thomas S. Kuhn: Incommensurability in Science,* ed. Bojana Mladenovic. Chicago: University of Chicago Press.

QPT (2021). *The Quest for Physical Theory: Problems in the Methodology of Scientific Research,* ed. George A. Reisch. Boston: MIT Libraries.

RSS   (2000). *The Road since Structure: Philosophical Essays, 1970–1993, with an Autobiographical Interview*, ed. J. Conant and J. Haugeland. Chicago: University of Chicago Press.

- "The Road since *Structure*," pp. 90–104 (1991/2000).
- "The Trouble with the Historical Philosophy of Science," pp. 105–120 (1992/2000).
- "Afterwords," pp. 224–252 (1993/2000).
- "A Discussion with Thomas S. Kuhn" (with Aristides Baltas, Kostas Gavroglu, and Vasso Kindi), pp. 253–323 (1997/2000).

SSR   *The Structure of Scientific Revolutions*, Chicago: University of Chicago Press.

- First edition (1962). SSR-1
- Second edition, includes "Postscript-1969" (1962/1970), pp. 174–210. SSR-2
- Third edition, includes "Postscript-1969" (1962/1996), pp. pp. 174–210. SSR-3
- Fourth edition, 50th anniversary edition, with an introductory essay by Ian Hacking (1962/2012), "Postscript-1969," pp. 173–208. SSR-4

TSK Archives–MC240

The following are documents from the Thomas Kuhn Archives at the Institute Archives and Special Collections, MIT Libraries, Cambridge, MA. They are cited according to date.

(1959).   Lectures, General, 1957–1959, box 3, folder 12.

(1961b).   *Proto-Structure*. Unpublished early manuscript of *The Structure of Scientific Revolutions*.

(1964).   Lectures, General, 1960–1964, box 3, folder 13.

No Abbreviations Used

The following are papers of Kuhn's that do not appear in one of the above books. They are cited according to date.

(1952).   "Robert Boyle and Structural Chemistry in the Seventeenth Century." *Isis* 43, 1: 12–36.

(1961a).   Letter from Thomas S. Kuhn to James B. Conant. Berkeley, CA, June 29. Harvard University Archives.

(1963).   "The Function of Dogma in Scientific Research."
                 In A. C. Crombie (ed.), *Scientific Change: Historical
                 Studies in the Intellectual, Social and Technical Conditions
                 for Scientific Discovery and Technical Innovation, from
                 Antiquity to the Present*, pp. 347–369. London:
                 Heinemann; New York: Basic Books.

(1970a).   "Logic of Discovery or Psychology of Research?"
                 In I. Lakatos and A. Musgrave (eds.), *Criticism and the
                 Growth of Knowledge*, pp. 1–23. London: Cambridge
                 University Press.

(1970b).   "Reflections on My Critics." In Imre Lakatos and Alan
                 Musgrave (eds.), *Criticism and the Growth of Knowledge*,
                 pp. 231–278. Cambridge: Cambridge
                 University Press.

(1971).   "Notes on Lakatos." In R. C. Buck and R. S. Cohen
                 (eds.), *PSA 1970: In Memory of Rudolf Carnap*,
                 pp. 137–146. Dordrecht: Reidel.

(1972).   "Scientific Growth: Reflection on Ben-David's
                 Scientific Role." *Minerva* 10: 166–178.

(1974).   "Discussion." In F. Suppe (ed.), *The Structure of
                 Scientific Theories*, pp. 500–517. Urbana: University of
                 Illinois Press.

(1976).   "Mathematical versus Experimental Traditions in the
                 Development of Physical Science." *Journal of
                 Interdisciplinary History* 7, 1: 1–31.

(1979).   "Foreword." In Ludwik Fleck, *Genesis and Development
                 of a Scientific Fact*, ed. T. J. Trenn and R. K. Merton,
                 trans. F. Bradley and T. J. Trenn, pp. vii–xi. Chicago:
                 University of Chicago Press.

(1980).   "The Halt and the Blind: Philosophy and History of
                 Science." *British Journal for the Philosophy of Science*
                 31: 181–192.

(1983).   "Reflections on Receiving the John Desmond Bernal
                 Award." *4S Review* 1, 4 (Winter): 26–30.

(1984).   "Professionalization Recollected in Tranquility." *Isis*
                 75, 1: 29–32.

(1986).   "The Histories of Science: Diverse Worlds for Diverse
                 Audiences." *Academe* 72, 4: 29–33.

(1989/1993).   "Foreword." In Paul Hoyningen-Huene, *Reconstructing
                 Scientific Revolutions: Thomas S. Kuhn's Philosophy of
                 Science*, pp. xi–xiii. Chicago: University of
                 Chicago Press.

(2016).   "The Nature of Scientific Knowledge: An Interview with Thomas Kuhn (conducted by Skúli Sigurdsson)." In A. Blum et al. (eds.), *Shifting Paradigms. Thomas S. Kuhn and the History of Science*, pp. 17–30. Berlin: Max-Planck-Institute for the History of Science.

Introduction
The Impact of *The Structure of Scientific Revolutions*

K. Brad Wray

The influence of the *Structure of Scientific Revolutions* (SSR) has been remarkably wide-ranging. Thomas Kuhn, the author of the book, was honored by the History of Science Society, the Philosophy of Science Association, and the Society for the Social Studies of Science (see Buchwald and Smith 1997, 361), three very different academic societies.

Given SSR's wide-ranging influence, it is useful to review the impact of SSR, and the changing perceptions of its significance, one discipline at a time. Necessarily this survey is very selective. Nothing approaching a comprehensive literature review is possible for a book that has been cited more than 135,000 times. My focus here will be on book reviews of SSR, some written soon after the book was first published, and others written as much as fifty years after its publication, in response to the publication of the fourth edition of the book. I will also discuss articles that reflect on the impact of the book and eulogies or appreciations of Kuhn marking his death in 1996.[1]

I.1   History of Science

Let us first consider the reception of SSR among historians of science, as Kuhn's professional identity was initially as a historian of science. I will rely heavily on the pages of *Isis*, the journal of the History of Science Society, to provide a window into how historians have responded to the book over the decades, though I will discuss a few reviews from other sources as well.

Mary Hesse wrote a very positive review of the book when it was first published (see Hesse 1963). Her first sentence says it all: "this is an important book" (p. 286). Her second sentence, though, hardly captures the spirit of today. It reads as follows: "it is the kind of book one closes with the feeling that once it has been said, all that has been said is *obvious*,

---

[1] *Structure* has been published in four editions, in the following years: 1962 (SSR-1), 1970 (SSR-2), 1996 (SSR-3), and 2012 (SSR-4).

1

because the author has assembled from various quarters *truisms* which previously did not quite fit and exhibited them in a new pattern in terms of which our whole image of science is transformed" (ibid., emphases added). On the one hand, Hesse is correct to say that the book transformed our whole image of science. But many readers today would object to her claim that Kuhn has assembled various *truisms* and that what he says is *obvious*.

Hesse rightly recognized that Kuhn sought to replace the philosophical view of science associated with the positivists "with a view of science as a historical succession of *paradigms*" (ibid.).[2] Further, she also grasps his "method of argument": to examine historical examples of scientific revolutions (ibid.). She praises Kuhn for his "deft explication of those tricky conceptual tools of the historians trade: 'discovery,' 'priority,' 'anticipation,' and many others" (p. 287).

Despite her praise of the book, she did not think it was a foregone conclusion that the book would be well received by historians of science. She ends the review noting that "the major question for historians of science ... is whether history bears the interpretation here put upon it" (ibid.). In a reflective turn, she claims that "the answer, as in the case of any paradigm shift, will be partly dependent on impressionistic and non-logical factors and will be subject to the kinds of resistance Kuhn finds to paradigm-change within the sciences" (ibid.). Hesse, though, claims that her "own impression is that Kuhn's thesis is amply illustrated by recent historiography of science and will find easier acceptance among historians than among philosophers" (ibid.).[3]

Charles Gillispie reviewed SSR for the journal *Science*. He begins his review noting that "this is a very bold venture" (Gillispie 1962, 1251). Rightly, Gillispie recognizes that Kuhn "is not writing history of science proper. His essay is an argument about the nature of science, drawn in large part from its history but also, in certain essential elements, from considerations of psychology, sociology, philosophy, and physics" (ibid.). From Gillispie's point of view, "Kuhn's critique of the very

---

[2] Hesse ends the review suggesting that "Kuhn has at least outlined a new epistemological paradigm which promises to resolve some of the crises currently troubling empiricist philosophy of science" (Hesse 1963, 287).

[3] The distinguished historian of science Marie Boas Hall reviewed the book for *The American Historical Review*. Though she describes it as a "closely reasoned monograph," one hardly gets the impression from her review that she thought the book would have much of an impact. She suggests that Kuhn was probably influenced by Crane Brinton's *Anatomy of Revolution* and by George Sarton (see Boas Hall 1963). Given the research I have conducted on the influences on Kuhn's intellectual development, I doubt that either of these had much influence on Kuhn, especially not Sarton (see RSS 275 and 281–282; also Wray 2021a, especially chapter 2).

notion of scientific discovery may … be the strongest part of his argument, and is certainly at the heart of it" (ibid.). Gillispie sees it as part of an attack on a wrong-headed theory of discovery, according to which "inventions of theory … were found like hidden treasure or a misplaced hat … wanting mainly to be revealed" (ibid.).

Though Gillispie expresses some minor concerns, he ends his review noting that "there can be only admiration for the erudition, the scholarship, the fidelity, and the seriousness that the enterprise reflects on every page" (p. 1253). Further, he remarks that "every historian … will surely applaud one recurrent and fundamental emphasis, which is that the development of science must be set into the context of a Darwinian historiography and treated as a circumstantial evolution from primitive beginnings rather than the ever closer approach to the telos of a right and perfect science" (ibid.). Oddly, this dimension of Kuhn's view was not discussed much at all, at least not until the last decade or so (see Renzi 2009; Reydon and Hoyningen-Huene 2010; and Wray 2011b).

It is worth contrasting these early reviews of the first edition of SSR with Joel Isaac's review of the fourth edition, published in 2012 to mark the fiftieth anniversary of the book. Isaac focuses mainly on the features that distinguish the fourth edition from earlier editions, specifically:

(i) the fact that it has been newly typeset, which has the consequence of shifting some passages from one page in the first, second, and third edition, to a different page in the fourth edition;
(ii) the new and expanded index; and
(iii) the Introductory essay by Ian Hacking, a long-time sympathetic reader of Kuhn's SSR (see Isaac 2013).

The latter two features, Isaac notes, are "much more unambiguously goods" than the first (p. 658).

In praising the new index, Isaac notes that the older index, prepared for the third edition, was merely two pages long, and erroneously listed an entry for "Clarant" intended to guide readers to a brief mention of Alexis Claude Clairaut (ibid.). Oddly, Isaac does not mention that the index in the new edition includes an entry for "Foucault, Michel," which is intended to guide readers to the Foucault of "Foucault's pendulum," that is, Jean Bernard Léon Foucault (see SSR-4 211).

Isaac ends his review noting that "of course I shall make no attempt to review the book itself. I can do no better than repeat Hacking's opening comments: 'Great books are rare. This is one. Read it and you will see' (p. vii)" (p. 659). Despite the high praise, Isaac does note a few shortcomings of Kuhn's analysis of science. Specifically, he notes that "Kuhn's conception of science was indelibly shaped by his own training as a physicist and by the ascendency of physics among the sciences during the early years of the Cold War" (ibid.).[4] He also notes that "Kuhn's claims about theory change and experimentation do not obviously have purchase on the contemporary scientific world of biotechnology, information science, and computer simulation" (ibid.).

Peter Dear also provides some reflections on the fiftieth anniversary of the publication of SSR. He claims that "any historian of science who sits down to reread *SSR* will be struck by its almost archaic historiographical sensibilities" (Dear 2012, 426). In light of this assessment, it is not surprising that Dear claims that Kuhn "was never really a guide to historical research except by association" (p. 425).

Between these two dates, that is, 1962 and 2012, specifically, in 1982, the History of Science Society honored Kuhn with the Sarton Medal. The president of the Society, Frederic L. Holmes, provides some reflections on SSR and its impact on the history of science, though the prize was not awarded specifically for SSR but rather for Kuhn's contributions to the profession as a whole.[5]

Holmes notes the wide appeal of the book. "Ever since [its publication in 1962], that book has remained the focal point for passionate debate among historians, scientists, social scientists, and even those in the arts and in political movements to whom the author had not imagined his ideas were relevant" (Holmes 1983, 247, in Hannaway et al. 1983). Writing in the early 1980s, Holmes reports that "the influence of the book shows no signs of having run its course" (ibid.).

Already, though, only twenty years after its publication, historians were more or less finished with the book. In fact, as Holmes notes, "historians of science have, curiously, been on the whole the most reluctant to enter into the discussions evoked by *The Structure of Scientific Revolutions*" (ibid.). Elaborating, Holmes explains that "some have admired it, others have tried to ignore it, still others have asserted that what Kuhn had to say was merely a codification of the common practices of historians of science" (ibid.). Holmes, though, insists that "whatever the personal reactions of individual members of our field may have been ... the history of science has not been the same since 1962" (ibid.).

Holmes then suggests that "all of us, whether we wished to or not, have had to locate ourselves with reference to Kuhn's framework. Whenever we have described particular historical events, we have had to ask

---

[4] A number of studies have explored the influence of the Cold War on Kuhn as he wrote SSR (see, especially, Fuller 2000b and Reisch 2019).

[5] As John Heilbron explains, the Sarton Medal is "the Society's highest award" (Heilbron 1998, 514).

ourselves whether they fell within some phase of his cycle of pre-paradigm, paradigm, normal, crisis, or revolutionary science" (ibid.). Whether this is how historians of science *felt* in the early 1980s, I cannot say, but one sees little evidence that between 1962 and 1982 historians of science located their own work with reference to Kuhn's framework.

Incidentally, it is worth noting that in 1985 Paul Josephson reported on the influence of Kuhn's SSR on Soviet historians of science (Josephson 1985). SSR was "translated into Russian in 1975" and "has been the subject of many articles in Soviet journals" (p. 76). Josephson attributes some of the appeal of the book to Soviet historians of science to the fact that "Kuhn's postulated sequence of 'normal science – anomalies – crisis/ revolution – normal science' ... fits the dialectical explanation of revolutions" (p. 551).

It is fitting to end this quick tour of the responses of historians of science to SSR by looking at remarks in Kuhn's eulogy in *Isis*. Kuhn's former student, John Heilbron, wrote the memorial notice. I will limit my analysis to Heilbron's remarks on SSR. Heilbron describes SSR as an "enduring book" (Heilbron 1998, 505). As Heilbron explains, "it made 'paradigm shift' as common and misused a metaphor as 'quantum leap' and 'critical mass.' It achieved what few philosophical books have done. It simultaneously instructed a wide academic public and a specialist community" (ibid.). I think it is telling that Heilbron, who knew Kuhn well, and worked closely with him, describes the book as a philosophical book.

Heilbron summarizes the book's effects:

the book comforted social scientists who wanted to assimilate their discipline to physics, Luddites who blamed social problems on scientists and engineers, and everyone who rejected authority. It repelled the philosophers of science at which it was aimed for the good reason that it undercut their belief that scientific knowledge advances by application of rational criteria to the products of observation and experiment. (ibid.)

Indeed, Heilbron captures well the wide range of people to whom the book spoke.

I.2   Philosophy of Science

The reactions of philosophers of science were somewhat hostile right from the beginning. Dudley Shapere's review of SSR, published in *The Philosophical Review*, set the stage for the book's reception among philosophers of science. Shapere begins the review noting that "this important book is a sustained attack on the prevailing image of scientific change as a linear process of ever-increasing knowledge, and an attempt to make us see that process of change in a different and ... more enlightening way" (Shapere 1964, 383). Shapere rightly anticipates the book's place in the history of philosophy of science. He notes that Kuhn's "view, while original and richly suggestive, has much in common with some recent antipositivistic reactions among philosophers of science – most notably, Feyerabend, Hanson, and Toulmin" (ibid.). Thus, already, only two years after its publication, Kuhn's book is characterized as a typical contribution to what we now often refer to as "the historical school in philosophy of science."[6]

The more lasting impact of Shapere's review is his critique of Kuhn's use of the term "paradigm." In Shapere's assessment,

[Kuhn's] view is made to appear convincing only by inflating the definition of 'paradigm' until that term becomes so vague and ambiguous that it cannot easily be withheld, so general that it cannot easily be applied, so mysterious that it cannot help explain, and so misleading that it is a positive hindrance to the understanding of some central aspects of science. (p. 393)

As I have discussed in detail elsewhere, Kuhn would spend the next ten years working out what he meant by the term "paradigm" (see Wray 2011b, chapter 3). Ultimately, Kuhn restricted its application to the exemplars that scientists appeal to in their research to solve research problems in the normal course of conducting research.

Knowing that Kuhn intended to write an expanded version of the book, Shapere ends the review suggesting that "the difficulties that have been discussed here indicate clearly that the expanded version of this book which Kuhn contemplates will require not so much further historical evidence (p. xi) as ... more careful scrutiny of his tools of analysis" (Shapere 1964, 394).

In light of the critical nature of Shapere's assessment, it is quite surprising that the book went on to have the impact it had.

Indeed, one would have been left with the same impression if one had read Harry Stopes-Roe's review of SSR in the *British Journal for the Philosophy of Science*. Stopes-Roe remarks that "one's first impression is of enthusiasm and vitality. The author clearly feels himself to be opening up a new world of appreciation and understanding" (Stopes-Roe 1964, 158). Stopes-Roe continues:

---

[6] On the historical school, see Kuhn's presidential address to the PSA, "The Road since *Structure*" (see RSS 91).

In the face of such force and charm, it seems mean to question the lasting value of the work; but it must be said that many of its features are already well established ...; and the author's enthusiasm leads him to over-state his novelties in a way that prejudices the appreciation of those things of value he has to say. (ibid.)

Stopes-Roe also registered a complaint about "the ubiquitous use of the odd word 'paradigm'" (p. 159). He even goes so far as to "suggest ... that if the reader wishes to bring out the real content of what Kuhn is saying, he may find it advantageous to try substituting 'basic theory' for every occurrence of 'paradigm' in the book" (Stopes-Roe 1964, 159).

This suggestion betrays the fact that Stopes-Roe has missed the importance of the paradigm concept for Kuhn's analysis. In many instances where Kuhn uses the term he is drawing attention to the reasoning by analogy that scientists engage in when solving research problems. Kepler's mathematical model of the orbit of Mars provided a template for modeling the orbits of other planets, the orbit of the Moon, and the orbits of other satellites, and even, ultimately, the paths of comets. Similarly, Planck saw similarities between Boltzmann's modeling of gasses and the black-body problem that led him to develop a hitherto unimagined solution to the latter. This is an aspect of paradigms that Margaret Masterman saw and appreciated, though scholars tend to cite Masterman's remarks about the many different ways that Kuhn used the term "paradigm" without acknowledging that she was extremely supportive of Kuhn's project (see Masterman 1970).

Alexander Bird wrote an essay review of the fiftieth anniversary edition of SSR. Bird, though, does not approach the task as Isaac did in his review in *Isis*, assessing the new features of the fourth edition. Instead, Bird consciously limits himself to assessing the content and impact of the first edition (see Bird 2012b, 860 n. 1).

Unlike the early reviews by philosophers of science, discussed above, Bird claims that "Kuhn's *The Structure of Scientific Revolutions* ([1962]) is in many ways an unusual and remarkable book" (p. 859). With the benefit of hindsight, Bird rightly notes, "it has a strong claim to be the most significant book in the philosophy of science in the twentieth century" (ibid.). Bird further describes it as "an original, wide-ranging, interdisciplinary, and bold book" (p. 878).

Perhaps one of the most noteworthy parts of Bird's review is his final remark on exemplars. According to Bird, "the exemplar idea is ripe for renewed investigation and development with the tools of current psychology and cognitive science, in a climate that is more receptive than that which Kuhn himself faced" (Bird 2012b, 880). So the concept that caused Kuhn so much grief in the initial years after its publication turns out to be the concept that seems most relevant to contemporary philosophy of science.

Upon Kuhn's death, Jed Buchwald and George Smith wrote a memorial notice for *Philosophy of Science*. Though it provides a useful overview of Kuhn's whole career, in both philosophy of science and history of science, I will focus narrowly on their remarks on SSR. They note that "in remarkably few words, Kuhn advanced the argument that the development of science cannot be understood simply as a process in which more accurate conceptions gradually replace less accurate ones under the impetus of experiment" (Buchwald and Smith 1997, 365).

Buchwald and Smith rightly note that "Kuhn's claims provoked strong resistance, particularly within the philosophic community" (p. 368). Elaborating, they note that "many felt, and continue to feel, that *SSR* did not fit well with claims to rationality and objectivity ... for scientific knowledge" (ibid.). And "others saw SSR as advancing theses about science that seemed to be paradoxical" (ibid.). Kuhn's remarks about "world changes" were often singled out as especially problematic.

Both Buchwald and Smith knew Kuhn, and knew him for a long time. Consequently, they felt many of his critics were uncharitable. In their words, "the picture of how science develops that Kuhn had formed came not out of philosophical reasoning, but from personal encounters with episodes in the history of science. The seemingly conflicting positions that his critics accused him of trying to maintain were merely artifacts of the way he communicated this picture" (ibid.).

Interestingly, and importantly, Buchwald and Smith also rightly note that "opposition to *SSR* did not prevent its impact on the philosophy of science. The 'problem of conceptual change' – i.e., the problem of incorporating something akin to Kuhn's conceptual readjustments into an account of the cumulative growth of scientific knowledge – took center stage in the wake of *SSR*" (ibid.). Consequently, Buchwald and Smith note, "philosophers of science began to look more closely and in much greater detail at the historical development of science, and they became increasingly attentive to the complexities of scientific practice" (ibid.).

Indeed, in highlighting normal science, Kuhn gave birth to the philosophy of science in practice, a development and movement in philosophy of science that generally eschews the more traditional focus on the logic of science, the traditional focus of the logical positivists and their heirs. Rather, those working in this new tradition are more inclined to examine laboratory practices, developments in techniques and instruments, and their impact on the advance of scientific knowledge than the logical relations between data and theory.

And as I have argued in detail elsewhere, the focus on the problem of conceptual change has had a profound impact on the realism/antirealism debates since the mid-1970s (see Wray 2021a, chapter 10). No longer are these debates concerned with understanding the meaning of theoretical propositions or whether our theoretical vocabulary in science is reducible to and expressible in observation terms. Rather, central to the contemporary debates is a concern for understanding how, or if, we can reconcile radical theory change with a central tenet of scientific realism, that our theoretical knowledge is increasing with the development of science. If successive theories are incommensurable, as Kuhn suggests, it is challenging to understand how to ground the traditional realist assumption of convergence on the truth.

Buchwald and Smith provide a useful analysis of Kuhn's later work, especially the work that was meant to clarify and develop the general theory of science presented in SSR. On their reading, incommensurability figured importantly (see Buchwald and Smith 1997, 375).

David Hull wrote a brief commentary for the journal *Nature*, reflecting on Kuhn's career after his death. Hull notes that "professional philosophers of science were put off by Kuhn's views, especially his principle of incommensurability" (Hull 1996, 204). As Hull explains, "Kuhn was deeply frustrated by the philosophical responses to his views – so much so that he claimed that, among all the readers of his work, philosophers were uniquely unable to understand him" (ibid.).

Hull also claims that philosophers of science failed to appreciate Kuhn's philosophy of science. Indicative of this "is the fact that many younger, less influential philosophers … were elected president of the Philosophy of Science Association before Kuhn was elected in 1988" (ibid.). Further, Hull suspects that Kuhn will have a more lasting impact. In Hull's words, "I suspect that a hundred years from now, Kuhn will be one of the few philosophers of science who will be looked back upon as having radically changed our understanding of science" (ibid.).

Also following Kuhn's death, Richard Rorty wrote a short reflective piece on Kuhn's impact. What is particularly interesting about Rorty's perspective is that he draws attention to the wide-ranging significance of SSR. Unlike many philosophers, Rorty welcomed the appropriation of Kuhn's work throughout the academic world. Rorty begins by noting that "the death … of Thomas S. Kuhn, the most influential philosopher to write in English since the Second World War, produced many long, respectful obituaries. Most of these obituaries referred to him as a historian of science rather than as a philosopher" (Rorty 1997/1999, 175). Rorty then remarks that

if I had written an obituary, I should have made a point of calling Kuhn a great philosopher, for two reasons. First, I think that 'philosopher' is the most appropriate description for someone who remaps culture – who suggests a new and promising way to think about the relation among various large areas of human activity.... My second reason for calling Kuhn a great philosopher is resentment over the fact that Kuhn was constantly being treated, by my fellow professors of philosophy, as at best a second-rate citizen of the philosophical community. (ibid.)

Rorty's second point is interesting because he too was an outsider of sorts in mainstream American philosophy.

The first point is more substantive. And Rorty makes it clear exactly what Kuhn did in writing SSR. According to Rorty, "Kuhn's major contribution to remapping culture was to help us see that the natural scientists do not have a special access to reality or to truth. He helped dismantle the traditional hierarchy of the disciplines" (p. 176).

On a more personal note, Rorty explains, "Kuhn was one of my idols, because reading his *The Structure of Scientific Revolutions* (1962) had given me the sense of scales falling from my eyes" (p. 175). Many readers, no doubt, have had a similar experience with the book.

Despite the fact that philosophers of science were so displeased with the image of science represented in the book, it has become, without a doubt, a canonical text in the philosophy of science, and the history of the philosophy of science.

I.3   The Sociology of Science

Bernard Barber provides valuable insight into how the book was received by sociologists of science. Reading his review, though, one could easily get the impression that Barber had read an entirely different book from the book that Shapere and Stopes-Roe read and reviewed. Barber claims that "Kuhn's book is offered as an essay in the sociology of scientific discovery" (Barber 1963, 298). Barber notes that "Kuhn's subtle, rigorous analysis of the social process of scientific discovery is ... different from that presented in the reports of 'normal science' ... and especially in the textbooks of the reigning 'normal science'" (ibid.).

Whereas Stopes-Roe questioned the lasting value of SSR, Barber expresses unrestrained enthusiasm. In fact, Barber makes two prescient observations. First, noting that "Kuhn has limited himself to examples chiefly from the physical sciences," Barber suggests that the book "has obvious and important relevance to the social sciences" (ibid.). In fact, as I have argued elsewhere, social scientists have found the book a rich source for reflecting on their own fields (see Wray 2021a, chapter 5). Across the disciplines – economics, political science, sociology, and anthropology – social scientists reflected on (1) whether their own fields had paradigms, (2) whether their own fields followed the developmental cycle Kuhn described, and (3) whether specific changes in their own fields can be accurately characterized as paradigm changes.

Second, Barber notes that Kuhn's analysis does not "include some sociological factors that would improve his analysis by enlarging it," specifically, a consideration of the effects of "external factors" (Barber 1963, 298). Barber does not fault Kuhn for this. Rather, he suggests that "when we are given so much … we should not ask for more. We can instead, take it as a challenge to try and give it ourselves" (p. 299). Indeed, this is exactly what sociologists of science did. First, the Strong Programme in the Sociology of Scientific Knowledge emerged, displacing the Mertonians. Their studies began by blurring the boundary between internal and external factors. Then the sociology of science underwent a metamorphosis into science studies. Now, in science studies, it is regarded as quaint or antiquated to speak of external factors.

In 2012, the journal *Social Studies of Science* ran a series on the fiftieth anniversary of SSR, which coincided with the twenty-fifth anniversary of Bruno Latour's *Science in Action*. There is quite extensive disagreement about the value of SSR among sociologists of science. As well, there is significant disagreement among those who value the book about what Kuhn's most important insight was.

On the one hand, Michael Lynch argues that "Kuhn's book effected a revolution in history and philosophy of science, and set the stage for a 'paradigm shift' in the sociology of scientific knowledge during the 1970s which is widely regarded as a crucial turn in the establishment of STS" (Lynch 2012, 450). In this respect, he sees it as foundational to the development of the sociology of scientific knowledge and STS. But Lynch rightly notes that "given the half-century that has passed since its publication, it is not surprising that Kuhn's *SSR* is not often treated anymore as a source of novel insight into the history and social study of science. *SSR* is mentioned much more than it is used, but the uses and abuses of Kuhn's ideas in the past half-century have been legion" (p. 452).

Lynch notes that "Kuhn … famously denounced the Edinburgh School's Strong Programme by linking it to 'deconstruction'" (p. 454). But Lynch astutely observes that "this remark seemed to be designed less as a characterization of the research done by the Science Studies Unit in Edinburgh, and more as an attempt to deter his critics in philosophy and history of science from linking his philosophy to relativism and subjectivism" (ibid.). Lynch also suggests that Kuhn benefited from the "misuses" of the ideas in SSR that he objected to (ibid.).

On the other hand, Harry Collins seems far less impressed by SSR. He suggests that many of the ideas in Kuhn's SSR "were largely already in place," specifically in the work of Ludwik Fleck, Ludwig Wittgenstein, and Peter Winch (see Collins 2012, 420 and 421). But Collins also suggests that "we would never have had the idea of incommensurability without Kuhn, and it is a vastly under-exploited idea" (p. 422).

Andrew Pickering also provides a more positive assessment of SSR. On his assessment, "Kuhn assembled provocative lines of contemporary thought in history, philosophy and psychology into a coherent, seductive and unforgettable, neo-Hegelian vision of what science is and how it changes" (Pickering 2012, 467). But, for Pickering, the most fertile notion in SSR was "[Kuhn's] theme of 'different worlds'" (ibid.). On Pickering's reading, "different worlds … can be produced, not by anything linguistic, but by different 'material grips' on nature. Grab hold of nature this way, with this material set-up, and you get this world; and another world can … be produced by a different set-up" (p. 468).

I.4   Social Scientists

Sociologists of science were not the only social scientists to read SSR. In fact, social scientists were some of the most enthusiastic early readers of SSR (see Wray 2021a, chapter 5). Consequently, it is not surprising that they also reviewed the book when it was initially published, and then reflected on its influence when Kuhn died.

The anthropologist, Marshall Sahlins reviewed the book for *Scientific American*. Sahlins raises the same concern about the paradigm concept that Shapere did. In Sahlins' words,

the term 'paradigm,' which carries the burden of [Kuhn's] argument, is so loosely defined that it covers not only well-articulated theories and vague conceptual schemes but also single scientific laws and specialized instrumental techniques. As a result almost any discovery or innovation can be considered the beginning of a new period of normal science or the occasion for a scientific revolution. (Sahlins 1964, 144)

Sahlins also raises the concern that "it is difficult to see how, in view of Kuhn's claim that a paradigm determines the way scientists perceive their subject matter and thereby completely controls their interpretations of observed data, paradigm-based expectations can ever be defeated or periods of crisis ever arise" (ibid.). The idea here is that paradigms have such a firm grip on scientists that it becomes inexplicable how any scientist could possibly see outside a paradigm, and thus bring about a paradigm change. Though Kuhn makes ample remarks that address this concern, and thus makes clear why a paradigm will inevitably fail and be replaced by a new paradigm, this reading of Kuhn became quite widespread among philosophers of science.

Interestingly, Larry Laudan would later raise the opposite concern about Kuhn's account of science. As Laudan explains, "Kuhn's model correctly predicts that dissensus should be a common feature of scientific life. What it cannot explain so readily, if at all, is how ... scientific disagreements are ever brought to closure" (Laudan 1984, 16). In fact, Laudan claims that "Kuhn's analysis has several features built into it which seem to foreclose any possibility of accounting for the emergence of consensus formation" (p. 17).

Sahlins ends his review remarking that "the book succeeds in presenting sound but familiar reflections on the nature of science; it is also much ado about very little" (Sahlins 1964, 144). What is striking about Sahlins' assessment is that he completely fails to anticipate the impact the book would have generally, but especially among social scientists. Where Sahlins sees "much ado about very little," other social scientists, including leaders in a number of fields, saw a set of concepts and a general framework that proved to be a rich source for reflections on key methodological and epistemological issues in their own disciplines (see Wray 2021a, chapter 5). Most significantly, many social scientists thought about the natural sciences in quite a different way after reading SSR. They no longer thought of them as shielded from the impact of social factors external to science.

The anthropologist Clifford Geertz provides another interesting perspective, writing one year after Kuhn's death and from the perspective of someone who knew Kuhn, worked with him, and liked him. Kuhn was affiliated with the Institute for Advanced Studies at Princeton between 1972 and 1979, which overlapped with Geertz's time at the Institute (see Wray 2021a, 8).[7] On Geertz's assessment, "*Structure* was the right text at the right time" (Geertz 1997/2000, 161). Interestingly, Geertz claims that "what remains of Kuhn's legacy, what enrages his most intransigent opponents and befuddles his most uncritical followers, is his passionate insistence that the history of science is the history of the growth and replacement of self-recruiting, normatively defined, variously directed, and often *sharply competitive scientific communities*" (p. 163; emphasis added). In this way, Geertz argues, "*Structure* opened the door to the eruption of the sociology of knowledge into the study of those sciences about as wide as it could be opened" (p. 164). Before the publication of SSR, sociological studies of science were confined narrowly to "external" factors, "concerned with the social effects of science, the institutional

[7] The historian William Sewell, Jr., recounts his involvement in a seminar in 1975–1976 led by Geertz at the Institute for Advanced Studies, in which Kuhn participated (see Sewell 2005, 42–43).

norms which govern it, or the social origin of scientists" (p. 162). The new sociological studies, in contrast, were wholly unconstrained. The Strong Programme sociologists of science, for example, were committed to finding the social causes of both false and *true* beliefs. Geertz notes that "this particular genie, once out of the bottle, can't be stuffed back in" (p. 164).

On the overall impact of the book, Geertz summarizes things quite well: "he had prayed for rain and got a flood" (p. 165). The astounding impact of SSR continues to perplex people.

I.5   Natural Scientists

The physicist David Bohm reviewed SSR in *The Philosophical Quarterly*. Bohm claims that "this book is without a doubt one of the most interesting and significant contributions that has been made in recent years in the field of the history and philosophy of science" (Bohm 1964, 377). Bohm was especially struck by its relevance to working scientists. And, though he acknowledges the wide application of the 'paradigm' concept, he does not see any difficulties with it (p. 378). Rather, he ends his review suggesting that "it may be worthwhile for scientists to try to become aware of the role that paradigms actually play in the life of scientific research in order that they shall be able more easily to realize the need for a change of Gestalt, when a particular field of study has been characterized by general confusion for some time" (p. 379). Thus, Bohm had no doubt that there are paradigms in science, as Kuhn describes. I do not think Kuhn imagined the book aiding scientists in the way that Bohm suggests. In fact, Kuhn even suggests that scientists' ignorance of the history of their disciplines may serve an important functional role.

It is striking that a physicist of Bohm's caliber would appreciate the book, given that many philosophers of science would come to regard the book as representing science in a degrading manner (see, e.g., Scheffler 1967; Popper 1970; and, later, Fuller 2004). Many philosophical critics were concerned that Kuhn made scientists look irrational, or at least nonrational. Clearly, what Kuhn was degrading was the philosophers' image of science, not science itself. Kuhn took the success of science for granted. His aim was to explain how it could be so successful, but to do so in a manner that was consistent with the historical record. Many of Kuhn's readers, though, missed this.

Reflecting on SSR two years after Kuhn's death, the distinguished physicist Steven Weinberg gave a mixed assessment. On the one hand, Weinberg was confident that something like paradigms characterize normal science. As Weinberg explains, "there is more to a scientific consensus than just a set of explicit theories" (Weinberg 1998, 48). Indeed, reflecting on his own initial reactions to the book, Weinberg claims that he was especially impressed by Kuhn's "treatment of normal science." Specifically, Weinberg claims that "Kuhn showed that a period of normal science is not a time of stagnation, but an essential phase of scientific progress" (pp. 50–51). On the other hand, Weinberg found Kuhn's characterization of scientific revolutions quite unsettling, and described it as "seriously misleading" (p. 51). Like many philosophical readers, Weinberg thought that "Kuhn made the shift from one paradigm to another seem more like a religious conversion than an exercise of reason" (p. 48).

Weinberg's appraisal illustrates that every reader can find something in the book that speaks to their experiences or preconceptions. No doubt, that is why the book has sold so well, and continues to influence how we think about science and scientific knowledge.

I.6   The Birth of Kuhn Scholarship

As we have seen above, right from the beginning, many scholars across the disciplines drew on or criticized Kuhn's SSR, but a new phenomenon emerged in the late 1980s. What emerged was a new subfield in the history of philosophy of science: Kuhn studies. These are proper scholarly studies of Kuhn's views, with the aim of understanding the development of his views and assessing the influences on, and subsequent impact of, his work. Much of this research is less concerned with ensuring that philosophers know what is and is not true about Kuhn's theory of science and more concerned with ensuring that philosophers know the facts about what Kuhn *really* believed, the influences on him, and the development of Kuhn's theory of science.

The pioneering study of this sort is, without a doubt, Paul Hoyningen-Huene's *Reconstructing Scientific Revolutions: Thomas S. Kuhn's Philosophy of Science*. It was first published in German in 1989, but it was quickly translated and published in English in 1993. James Marcum describes it as "the first full length analysis of Kuhn's philosophy of science" (Marcum 2015, 183).

The book was partly written while Hoyningen-Huene was a visiting scholar at the Massachusetts Institute of Technology in the 1984–1985 academic year (see Hoyningen-Huene 1989/1993, xvii). It was published with a very supportive endorsement from Kuhn: "No one, myself included, speaks with as much authority about the nature and development of my ideas" (Kuhn in Hoyningen-Huene 1989/1993, xi). The central conceptual innovations developed and applied by Hoyningen-

Huene were the distinction between genetically subjected-sided and genetically object-sided aspects of perception and cognition more generally,[8] and the distinction between phenomenal worlds and the world-in-itself. In invoking these distinctions, Hoyningen-Huene was able to clarify many of the apparently paradoxical aspects of Kuhn's view, aspects that had led to many uncharitable readings of SSR.

Any new book in this subfield must come to terms, to some extent, with the interpretation advanced by Hoyningen-Huene. And a perusal of many of the influential books in Kuhn studies that have been published since illustrate this clearly. In fact, "Hoyningen-Huene" is typically an entry in the book indexes (see, e.g., Bird 2000b; Sharrock and Read 2002; Nickles 2003a; Andersen et al. 2006; Wray 2011b; Kindi and Arabatzis 2012; Marcum 2015; Mayoral 2017; Mizrahi 2018).[9]

Alexander Bird has raised doubts about Hoyningen-Huene's idealist, Kantian reading of SSR. In fact, Bird makes the provocative suggestion that "Kuhn was himself influenced by Hoyningen-Huene's" reading (Bird 2012b, 869). More precisely, Bird claims that Hoyningen-Huene gave Kuhn's "earlier thought a ... philosophical sophistication that it did not really have" (Bird 2012b, 869). Whether or not one agrees with Hoyningen-Huene's reading of SSR, it is undeniable that he continues to make important and insightful contributions to Kuhn scholarship.

Another early influential book in Kuhn studies was Steve Fuller's *Thomas Kuhn: A Philosophical History for Our Times* (Fuller 2000b). In marked contrast to Hoyningen-Huene's book, which is aptly characterized as an "internalist" analysis, as it traces the development of Kuhn's ideas, Fuller's is decidedly an "externalist" analysis. Fuller argues that Kuhn's book is the product of Cold War American culture, produced under the influence of James B. Conant, who was himself involved in the Manhattan Project, and rationalizes a particular relationship between science and society, one that shields science from public scrutiny. On Fuller's reading, Kuhn's self-policing scientific specialty communities are answerable only to themselves, and are, at least in their own minds, not affected by broader societal concerns.

These two books, representing the two poles of research in Kuhn studies, speak to quite different audiences. Hoyningen-Huene is concerned with the development of Kuhn's thought, specifically, with the intellectual sources that influenced him as he wrote SSR. Fuller, on the other hand, is concerned with showing how Kuhn's view is merely

---

[8] "Genetically" here refers to their origins.
[9] With respect to collections, I only list those to which Hoyningen-Huene did not contribute a piece.

reflecting recent developments in the relationship between science and society, developments that were a consequence of the First World War.

Another major development in Kuhn studies is the use of the archival materials at the Massachusetts Institute of Technology, deposited there by Kuhn upon his retirement. Gradually, some of this material has been published thanks to the efforts of George Reisch and Pablo Melogno†, among others. For example, readers can now access Kuhn's Lowell Lectures, *The Quest for Physical Theory*, which Kuhn describes as his first attempt to write SSR (see QPT). It is now commonplace for those working in Kuhn studies to draw on the material in the archives (TSK Archives–MC240).

And no serious Kuhn scholar can ignore the candid and insightful interview that Kuhn gave in October 1995, the year before he died. The interview was conducted by Aristrides Baltas, Kostas Gavroglu, and Vasso Kindi, while Kuhn was in Greece, on the occasion of his being awarded an honorary doctorate from the department of philosophy and history of science, from the National and Kapodistrian University of Athens (see RSS 253–323). Kuhn remarked that he rarely agreed to give interviews (see RSS 321). But in Athens, Kuhn speaks frankly about his upbringing and education, his career successes and disappointments, as well as his legacy and the impact of SSR, as he saw it in 1995. Indeed, during the interview, Kuhn claims, "Look, I'm saying some things that I'm glad to think will be around somewhere" (RSS 321).

The contributions to this volume build on the vast body of literature on Kuhn's philosophy written over the past three decades, as well the material in the Kuhn Archives and the interview. Some of the contributors to the volume have already made significant contributions to our understanding of Kuhn, while others provide new perspectives on Kuhn, the development of his ideas, and the value of the Kuhnian account of science. Together the essays in this volume constitute a rich contribution to our understanding of Kuhn, SSR, and science.

The essays are organized into four thematic parts. Part I contains essays that address influences on Kuhn as he wrote SSR. Richard Creath revisits the topic of the relationship between Rudolf Carnap's philosophy of science and Kuhn's. And Jamie Shaw examines the influence of the post–Second World War science policy debates on Kuhn, given that Kuhn was a protégé of James B. Conant, a key player in these debates. Part II contains essays on normal science and science education. Alisa Bokulich and Federica Bocchi examine Kuhn's so-called Fifth Law of Thermodynamics, which states that "no experiment gives quite the expected numerical results" (see ET 184). Pablo Melogno† examines the role of normal science in Kuhn's cycle of scientific change. And

Alexandra Bradner and Hasok Chang both examine aspects of science education, in the light of Kuhn's remarks on science education and science textbooks. Part III is concerned with the concepts of incommensurability, scientific progress, and revolutions. Alex Levine examines Kuhn's appeal to the notion of translation in explaining the relationship between one theory and its successor. Haixin Dang examines the social dimensions of group belief change during scientific revolutions. Chris Haufe examines Kuhn's appeal to fruitfulness as a value affecting theory choice. K. Brad Wray examines Kuhn's appeal to George Orwell's *1984* as a means to illuminate the nature of theory change. And Hanne Andersen raises concerns about the applicability of Kuhn's account of scientific change to contemporary science, given the dramatic ways in which the practices of science have changed since 1962. The essays in Part IV are concerned with the impact of SSR on the philosophy of science, the sociology of science, and the history of science. Jonathan Tsou examines Kuhn's ambiguous legacy in philosophy of science. Markus Seidel examines the appropriation of Kuhn's views by the Strong Programme in the Sociology of Scientific Knowledge. And Vasso Kindi provides a comprehensive overview and assessment of Kuhn's impact in the history of science.

Part I

Writing *Structure*

1 "... I probably would never have written *Structure*"

Richard Creath

1.1 Introduction

In 1995, less than a year before his death, Thomas Kuhn engaged in a long autobiographical interview (RSS 253–323). Among the topics covered was some of the then new literature, specifically "Carnap and Kuhn: Arch Enemies or Close Allies" (Irzik and Grünberg 1995), showing that on a wide variety of important topics, Rudolf Carnap and Kuhn had strikingly similar views. This was part of a wider reappraisal of Carnap that had been developing for about ten years. In the interview Kuhn made an astonishing claim: "You know this article that recently appeared.[1] It's a very good article. I have confessed to a good deal of embarrassment that I didn't know it [the Carnap]. On the other hand, it is also the case that if I'd known about it, if I'd been into the literature at that level, I probably would never have written *Structure*." Kuhn was fully aware that his and Carnap's views were not exactly the same. But he obviously believed that the similarities were enormously important and went to the heart of his motivating ideas in writing *The Structure of Scientific Revolutions* (SSR) in the first place.

Kuhn must have been astonished on reading Irzik and Grünberg (1995). It is difficult to appreciate now, some sixty plus years after the publication of SSR, what a sensation it had caused in the late 1960s and early 1970s. The book has sold over a million copies and was required reading in virtually every department in most major universities. But Kuhn had gotten a lot of abuse in those early years, especially from philosophers. And some of those wounds remained quite raw. Moreover, he must have thought that he was fighting an uphill battle against the very ideas that he thought Carnap represented. Of course, the similarities that Kuhn suddenly saw between himself and Carnap diminish neither the importance nor the originality of SSR. But the dislocation

---

[1] Kuhn cites G. Irzik and T. Grunberg, "Carnap and Kuhn: Arch Enemies or Close Allies?," *British Journal for the Philosophy of Science* 46 (1995): 285–307 (RSS 306).

that those similarities provoked for the then-standard narrative of the recent history of the philosophy of science and Kuhn's place in it must have been shocking. And that Kuhn more or less instantaneously accepted that dislocation shows enormous largeness of mind and generosity of spirit.

First, in Section 1.2, I will highlight some of the similarities and differences that Irzik and Grünberg discuss. My aim is not to reargue the case that they made quite well but to pick out some themes for further discussion. Kuhn's remark quoted above shows that he took some of the similarities quite seriously. As it happens, Carnap saw and took them seriously as well. Carnap and Kuhn saw differences as well. I am not concerned to say whether the similarities or the differences are more important. Instead, in Section 1.3, I shall then argue that, given the similarities, the two views could be and were attacked in similar ways. My aim is not to give a detailed history here but to indicate, in broad terms, what two of those challenges were. Finally, in Section 1.4, I shall argue that, given their differences, the two views can be allies in an even deeper sense, that is, help each other meet their respective challenges.

1.2   Similarities and Differences

Let us begin with some obvious differences. Kuhn was writing as a historian of science focused on scientific change as a social phenomenon. And what he offered was more or less a theory of punctuated equilibrium in which periods of rapid change, revolutions, alternated with other periods in which the changes are more incremental and somehow more orderly, normal science. The difference between these two sorts of episodes is that in normal science a scientific community is governed by a single paradigm, whereas revolutions begin with a weakening of that paradigm, proceed through the strife of alternative paradigms, and end by the establishment of a new paradigm and a new period of normal science. The notion of "paradigm" is thus central, and it has a number of components[2] that were delineated more clearly as Kuhn's career progressed. Paradigms involved examples of important scientific achievements that could guide and thus help evaluate further research in the area. These evaluations express the epistemic values of the community

---

[2] Kuhn recognized that his talk of paradigms in SSR ran together a number of considerations that are usefully separated. I will use the term "paradigm" for the collection of factors without itemizing them. As long as one such collection dominates periods of normal science, what I need to say about them here is undisturbed (SSR-2/ SSR-3, 174–210; RSS).

and affect the meanings of its entire vocabulary, including those terms used in the most basic of observational reports. During normal science, paradigms and the standards or values of evaluation that they include are conveyed to each new generation of researchers by an educational strategy of textbooks that stress examples. But paradigms are not eternal, for they change in revolutions. As a consequence, there would seem to be no theory-neutral observations that could be used to compare two paradigms and choose between them. Paradigms must therefore be a different kind of commitment and evaluated differently from the more usual kind of theory. Such themes as community, education, values, and, above all, change are at the center of Kuhn's approach.

Carnap, by contrast, was a logician and philosopher of science who explored formal and artificial languages as a way of understanding both language in general and its role in science considered as a body of knowledge, that is, a body of claims taken to be true. He developed detailed accounts of the logical structure of scientific claims, gave formal theories of confirmation, and rarely invoked notions of community or values and said little about how scientific languages were learned. Logic, broadly understood, was to be at the center of philosophy, and all that could be said about rational belief and theory choice was somehow to proceed from that. The objects that such formal systems, called linguistic frameworks, seemed to be about were abstract and thus not themselves subject to change. What language we use, however, can and does change, often under pressure from changes in what we believe about the world.

Even so, Carnap's particular concern was that there are alternative logics that are not just verbal variants of one another. Since the logics embody the standards for evaluating claims, the choice among them is problematic. A search for the uniquely correct one had no standards of a theoretical sort to guide it. So Carnap rejected the idea that there was such a uniquely correct logic and treated the choice as one between alternative languages or linguistic frameworks to which the notion of correctness does not apply. Once the linguistic framework was identified, there would be sentences, the analytic ones, the truth of which would be guaranteed by the linguistic framework alone. But these sentences do not describe the world but collectively give the meanings of all the expressions of the language and determine the rules for assessing the nonanalytic, that is, synthetic, sentences. Instead of looking for the correct linguistic framework or language, Carnap took the choice as a practical choice of a tool rather than a theoretical one. There were theoretical choices to be made, namely, among the synthetic sentences considered as genuine descriptions of the independent world.

Carnap and Kuhn seem therefore to be miles apart: in what problems they take up, in their manners of treating them, in the themes they stress, and, it may appear, in the results they get. Given these differences of approach and vocabulary, it is all the more surprising that the similarities that Irzik and Grünberg point out are so central and so extensive. But it is not surprising that the similarities went unnoticed for a long time.

Kuhn had help in missing Carnap's point. Kuhn says in the Preface to SSR: "W. V. O. Quine opened for me the philosophical puzzles of the analytic-synthetic distinction" (SSR-4 xli). But Quine viewed Carnap's talk of analyticity as an attempt to find logical certainty, which Quine rendered as our never having to give up such sentences. This is misleading at best. Quine did, however, have important objections to the analytic/synthetic distinction, to some aspects of which we will return later. Quine was not the only one of Carnap's detractors who was eager to explain what was wrong with Carnap's views. And the technical aspects of Carnap's writing make it difficult for those not versed in the details of alternative logics to see for themselves the general thrust of what Carnap was doing.

I will not try to reargue the case that Irzik and Grünberg have made, but I do want to highlight a few important similarities. Perhaps the most important of these similarities is that, for both, our scientific commitments sort into two tiers. The commitments of Tier 1 are evaluated according to something like the common picture of how theories meet experience. I will call the Tier 1 commitments *ordinary theories*. By this I mean those commitments that can change without changing the meanings of their constituent terms. For Kuhn, these are claims that are not part of the paradigm itself, and for Carnap, these claims are synthetic. For example, when we change the maps showing magnetic declination because we have remeasured and have a new value, this does not change the meanings of any of the terms involved. The commitments of Tier 2 are evaluated in some other way. For Kuhn this tier is the paradigm, and it is to be evaluated in terms of how well it guides the normal science done under its aegis. For Carnap the second tier is the linguistic framework. And this is to be evaluated in terms of our convenience in using the language thus specified, for example, in furthering the simplicity, fruitfulness, or convenience of the theories that are confirmed according to its directions. Thus, for both writers the choice among alternatives at this level is a practical one.

The similarities do not end with just having two tiers of scientific commitments. Their accounts of what Tier 2 contributes are also strikingly similar. The commitments of this second level determine the meanings of all expressions in the language and the standards of evaluation of theories at the first level. It follows that changes in the linguistic framework or paradigm involve changes of meaning, that is, conceptual change, throughout the language, including at not only the theoretical level but also the observational level, should we choose to distinguish those. There is thus no independent, theory-neutral (paradigm- or linguistic framework–independent) observation language by which to evaluate a linguistic framework or paradigm or even to compare theories drawn from different frameworks or paradigms. This in turn precludes a development-by-accumulation picture of scientific change across revolutions or changes in framework. Moreover, the long-run changes in science are not toward some predetermined truth that could even be expressed in our current language. Instead, such changes represent the refinement of our conceptual structure for current purposes of getting on with science in the most convenient and productive way as measured by our current understanding.

The remark by Kuhn as quoted at the start of this chapter shows that he took these similarities to be very important and as going to the very heart of his motivations in writing SSR. There is reason to think that Carnap took them seriously as well. As George Reisch (1991) pointed out, Carnap was the editor for SSR, and in this capacity, wrote Kuhn two revealing letters. In the first Carnap is commenting on some manuscripts that Kuhn had sent in preparation for writing SSR. Carnap says in part:

I am myself very much interested in the problems that you intend to deal with, even though my knowledge of the history of science is rather fragmentary. Among many other items I liked your emphasis on the new conceptual frameworks which are proposed in revolutions in science, and, on their basis, the posing of new questions, not only the answers to old problems. (Carnap 1960, quoted in Reisch 1991, 266)

In the second letter Carnap officially accepts SSR for publication but includes a long paragraph on what he likes about the book. It is worth quoting in full.

I am convinced that your ideas will be very stimulating for all those who are interested in the nature of scientific theories and especially the causes and forms of their changes. I found very illuminating the parallel you draw with Darwinian evolution: just as Darwin gave up the earlier idea that evolution was directed toward a predetermined goal, men as the perfect organism, and saw it as a process of improvement by natural selection, you emphasize that the development of theories is not directed toward the perfect true theory, but is a process of an improvement of an instrument. In my own work on inductive logic in recent years I have come to a similar idea: that my work and that of a few friends in the step for step solution of problems should not be regarded as leading to "the ideal system," but rather as a step for step improvement of an instrument. Before I read your manuscript I would not have put it in just those words. But your formulations and clarifications by examples and also your analogy with Darwin's theory helped me to see clearer what I had in mind. (Carnap 1962, quoted in ibid., 266–267)

Interestingly, Kuhn says in the autobiographical interview, "I would now argue very strongly that the Darwinian metaphor at the end of the book is right, and should have been taken more seriously than it was; and *nobody* took it seriously. People passed it right by" (RSS 307).

Well, Carnap took it seriously, and he thought that the similarities between his own and Kuhn's work were serious indeed. Of course, there is more to it than that. From "On Protocol Sentences" (1932/1987) onward Carnap held that the meanings of the most basic evidential reports changed as the language changed. In *The Logical Syntax of Language* (1934/1937) he held that (what we take to be) basic theoretical laws were best understood as P-rules and hence as among the determiners of meaning throughout the language. This continued right on through the end of his life. When challenged to provide the analytic sentences that would endow novel theoretical terms their meanings, Carnap suggested what is now sometimes called the "Carnap sentence" for the theory. The technical details do not matter here, but the upshot is that the statement of the theory itself is not analytic but is involved in the statement of the relevant analytic sentences in such a way that when the theory changes, so does the (analytic) Carnap sentence. This means that when basic theory changes, so do all meanings of the expressions of the language, whether those expressions are at the theoretical or observational level.

There is one of Kuhn's themes, values, that it might seem at first that Carnap would resist. The values in question are epistemic ones governing what puzzles are most worth solving and what counts as a better solution or even an adequate one. Carnap, by contrast, says little about values in the early part of his career, and what he does say sounds negative: ethics is a branch of metaphysics and hence without cognitive content (cf. Carnap 1935). Set aside the issue of whether for Carnap there are other kinds of content – there are. Since Carnap is known to have strongly held moral and political views, his comments on this score can hardly amount to a rejection of ethics or evaluative topics tout court. In fact, what Carnap is really rejecting is the idea that there is a uniquely correct evaluative system and especially the idea that philosophers have a special nonempirical means of identifying that uniquely correct system of values. This, then, is just the same sort of pluralism that Carnap readily applied to logic combined with a denial that philosophers have some insight into a reality that is deeper than or lies behind what empirical scientists can know.

In his logical writings Carnap also resisted formulating logic as an account of how people *ought* to reason. This may seem to be a rejection of the evaluative. But this was because he thought the formulation in terms of how people ought to reason did not add anything and hence was equivalent to his own usual formulation. So Carnap is not really rejecting those apparently evaluative formulations as wrong.

In this section we have highlighted several similarities and differences between Carnap and Kuhn. Both are important and were recognized as such by the two writers. As we shall see in the next section some of the similarities suggest that the same or similar criticisms can be lodged against both views. As we shall also see in the final section, some of the apparent differences can help each to be a deeper ally to the other by helping him to address the criticisms presented in the next section.

1.3   Two Criticisms

In the section that follows I shall consider two broad families of criticism that have been lodged against the views we have been considering. I will avoid here the detailed histories of specific texts. This is because there are many minor variations of the complaints and neither the criticisms nor the responses that can be made depend on the details.

The first criticism has been lodged more often and more vocally against Kuhn, so I will frame it that way first. But because the criticism depends only on features that Kuhn and Carnap share, it would apply equally well to the latter. The criticism comes in two parts:

Criticism 1.a: If paradigms are the bearers of the standards of rationality, of the standards of theory choice, then those standards cannot be invoked in that transition. Therefore, paradigm change cannot be rational. Moreover, we can never have any basis for saying that an ordinary theory from one paradigm is rationally better than another ordinary theory from another paradigm. (As indicated earlier, an ordinary theory is just a commitment that can change without affecting the meanings of its constituent terms.)

Admittedly, Kuhn's talk that in paradigm change "the world has changed" and that the process is one of "conversion" seems to suggest just this outcome. But it does not have to. Those could be just phenomenological descriptions of what it feels like at the time of paradigm transition.

Criticism 1.b: Without neutral observational claims there is no hope of giving an objective answer to the question of whether a proposed new paradigm makes significant, or indeed any, empirical progress over its rivals. Since paradigms do not meet experience in anything like the traditional ways that ordinary theories are thought to, what else are we to think?

Kuhn speaks of observation as "theory bound," and some have interpreted this to mean that theories in general simply generate data that uniformly supports that theory. This, obviously, would not be a good thing. But presumably it is not what Kuhn had in mind. Instead, his idea is that paradigms (and not just any claims at all) will affect the meanings of observational claims. So I have framed Criticism 1.b in terms of whether a paradigm-induced change in the meanings of claims at the observational level precludes a reasonable assessment of the paradigms themselves. Kuhn does say that the new paradigm will offer striking new empirical successes and be chosen on that basis. But again, one does not have to say this. But one does have to say what makes one paradigm better than another.

I have framed this criticism as though it were directed toward Kuhn rather than against both him and Carnap. And indeed, Kuhn has more often been charged with irrationality. Perhaps this is because he emphasized more explicitly that observation is not paradigm neutral and possibly because his writing style is more dramatic. He is, after all, talking about revolutions, and his audience, especially in the 1960s and 1970s, might be tempted to view him as a radical. In some ways he was. Nonetheless, the features of Kuhn's view that the above criticism responds to are shared by Carnap. So the criticism, if it is valid, ought to be equally valid against Carnap. By the 1960s he was a grandfatherly figure who came across as ever the careful logician with a writing style that is measured and even ponderous. Carnap was a radical too, but people did not see it.

I shall argue in the third section that Carnap stresses the themes and distinctions that allow him to respond to this objection. And there is no reason why that machinery could not be mobilized by Kuhn as well.

The second criticism has been lodged more forcefully and more persistently against Carnap. This criticism is directed at Carnap's two-tier account of our scientific commitments. The linguistic framework, or language, by itself guarantees that certain sentences are true. Those are the analytic sentences. To abandon one or more of those sentences would be to change languages. The remaining sentences, the synthetic ones, are given their meaning by the linguistic framework, but their truth is to be adjudicated, according to the rules laid down by the framework, by comparing them with observational judgments. As is well known, W. V. Quine has challenged Carnap on this, saying that the distinction between analytic and synthetic sentences – that is, between the commitments of the linguistic framework and other more ordinary commitments – simply cannot be drawn. The argument is roughly this:

Criticism 2: Any two-tier account must have a clear way to distinguish the tiers and to apply that distinction empirically to natural languages and to real languages used in science. Such an empirical distinction would amount in Carnap's case to clear behavioral criteria that would mark the difference between analytic and synthetic sentences. According to Quine, such behavioral criteria have not been found and thus probably cannot be. If so, a two-tier account must be defective.

Carnap would like to sidestep Quine's demand for empirical/behavioral criteria (Quine 1951) on the grounds that he is not concerned with natural languages but considering only abstract proposals for constructing the language of science in a certain way. Quine can argue, however, that unless a clear behavioral difference can be specified there can be no way to tell whether one of Carnap's proposals has been adopted, and hence, there is no clear sense to 'making a proposal' (cf. Creath 2004 and Creath 2007). Moreover, without such criteria there would be no way to compare proposals for practical utility. Thus, it seems that Carnap cannot dodge Quine's demand. The question is whether a demand such as Quine's can be met even in principle. Carnap would not have to show that current scientific languages or current natural languages have the structure he proposes, only that we can tell empirically whether they have it or not.

It may not be obvious, but Kuhn faces a version of this objection as well. We saw above that he does distinguish two tiers of scientific commitments: paradigms and more ordinary theories. And he uses that distinction to formulate an account of the development of advanced sciences according to which a field oscillates between periods dominated by a single paradigm, normal science, and other periods, revolutions, in which no one paradigm is dominant or in which the dominant paradigm is replaced by another.

Kuhn's critics have often wanted to deny, say, that any period is governed by a single paradigm. The problem is not that Kuhn may be wrong and his critics right. Rather, the worry in this second criticism is that without a clear empirical criterion of what it is to be a paradigm and what it is for work to be done under the aegis of a paradigm, there is nothing for Kuhn or his critics to be right or wrong about. If Kuhn does have such a criterion, then it should be a fairly straightforward empirical matter to sort this out.

What I shall argue in the next section of this chapter is that the themes that Kuhn stresses do allow for a criterion that can turn Kuhn's model into an empirically decidable historical claim. I am not claiming that Kuhn has given an explicitly formalized and idealized statement of these behavioral criteria or that he should. Rather, it is that his themes and practices suggest that this *can be done* to the required degree of specificity. They suggest, moreover, where to find them. I shall suggest in the next section that these same Kuhnian themes and practices will show that the demands that Quine raises against the analytic/synthetic distinction can be met as well.

1.4   Allies in Meeting the Criticisms

In this section I want to reflect on the criticisms just outlined and consider the prospects for avoiding them. I shall argue that, while much remains to be done, there is reason to think that the criticisms can be deflected and that it is the dissimilarities in their views and approaches that will allow Carnap and Kuhn to help each other do this.

Criticism 1.a is lodged primarily at Kuhn, though it can readily be transformed into a criticism directed at Carnap. Roughly, the question is: If paradigms provide the standards of rational theory choice in science, how can they be appealed to when the choice is between competing paradigms? Such choices must be irrational.

We said that Carnap's background is different from Kuhn's. Carnap had a long and deep involvement with and developed detailed accounts of logic, language, and meaning. From the 1930s on, he treated philosophical "theories" that seem to be about the world, for example, 'There are numbers' and 'Theoretical entities are real,' as better understood as proposals for structuring the language of science. Because he is a pluralist even about logic, he has explored not only the details of alternative logics and conceptual systems but also what kinds of reasons are available for choosing among them. Logics are to be construed as languages, and languages are not true or false, correct or incorrect. Some may be easier to use or allow for simpler descriptions of the world. The choice is a practical rather than a theoretical one. Carnap explored, as well, the options we have for fundamental concepts and how we might choose among them. Shortly, we will discuss some specific examples, namely, choices among time metrics and among alternative logics.

In any case, this long experience exploring logic, language, and meaning puts Carnap in a good position to respond to Criticism 1.a. Carnap did not tend to use the phrase 'standards of rational theory choice,' but it is plain that he would consider them as embedded in the inference and confirmation rules that constitute languages. So the choice among such standards, like a choice among languages, is a practical rather than a theoretical choice; it is a choice among tools, as he suggested in the second letter to Kuhn quoted above. So what do we need to know to make that choice? We need to know what the tools are like and what the effects of using them would be. This is a straightforward empirical question (insofar as any such questions are straightforward), and Kuhn would treat it as such. And we would have to evaluate those outcomes to decide which we would prefer. There is absolutely no problem, absolutely no question begging, in using our current standards of rational theory choice in these deliberations. Besides, there is no danger of choosing falsely and no contradiction if you choose differently than I. Of course, this does not eliminate the possibility of bias and error in scientific deliberations. It is not intended to. But it does show that one can use one's current standards of rational theory choice in these deliberations and on that basis quite possibly decide to change those standards.

Carnap gives many examples. One is in the choice of basic units of temporal length. One needs for this a periodic process that one then declares by definition to measure out units of equal length. There is a great deal of freedom in this. I could choose my own heartbeats, and by this definition they would be of equal length. This would have the undesirable practical consequence that when I am resting, the world speeds up, and when I drink strong coffee that world slows down. And that would then have to be built into temporal laws of nature, making them complex indeed. There is no contradiction here, just an unwise practical choice. Or I could choose the ticks of a mechanical clock as my definition of equal temporal length. This would be better, but the world would still speed up as the wheels and cogs wore down. Suppose this were my current choice. I could still deliberate about whether to use this system and decide to change it by adopting the oscillations of a cesium atom as my unit of choice (Carnap 1966, 78–85).

The second example is that of a choice between our now standard elementary logic and a more limited one in which our current quantifiers cannot be expressed. In the former we can express classical mathematics, and that is very convenient for expressing powerful physical theories. The more limited logic cannot do this and so is less convenient. But our standard logic is more likely to be inconsistent than its weaker sibling. As a result, we have to choose between one logic that is convenient and another that is safer. The choice you make will depend on what your values are. Criticism 1.a provides no argument against the rationality of such a choice.

These Carnapian considerations generalize to a defense of Kuhn against this criticism as well. Carnap's very different starting point allows him to be Kuhn's protector as well as his own.

Carnap's careful discussions of languages and meaning are also helpful in dealing with Criticism 1.b. The issue is this: Kuhn's claim is that changes of paradigm induce changes throughout the language, including at the observational level. This implies that observation is not utterly theory neutral. Does this preclude the possibility of a new paradigm being a dramatic empirical success in any interesting sense or even an ordinary theory being tested observationally? Does it render them untestable? No.

Of course, it is possible to invent a nontestable theory, one that no claims at the observational level could ever contradict or even disconfirm. But such theories would be empty, telling us nothing about the world around us. But having observational claims that do not change their meanings as high-level commitments change will not prevent that. It is true, of course, that people sometimes see what they want to see or fail to notice something that threatens a cherished belief or fail to recognize its relevance even when they do notice. Again, meaning change has little or nothing to do with this.

The mere fact that observational claims change their meanings with a new paradigm does not say what those changes are or whether those changes have the consequences that many of Kuhn's critics seemed to fear, namely, that, once the meanings of observational claims can change, those observational claims will always agree with the theory that they are used to test.

The way to defend Kuhn here against such fears is to spell out *how* changes at one level induce changes at another. This is precisely what Carnap's delineation of the operative inference rules is designed to do. The inference rules spell out what is logically and evidentially relevant to what is within the body of scientific claims. And by doing so those rules spell out what the public norms are for challenging or defending those scientific claims. We will return to this idea of public norms shortly.

When we use the explicit detailed inference rules to examine specific examples, the fears of the critic have not been realized. The historical evidence is that meaning change has not held theories immune from disconfirmation. And Carnap's abstract examples of changing definitions of scientific terms has not had that result either. If the critic is to make a serious case along the lines of Criticism 1.b, that critic will have to show that both the historical and abstract cases that Kuhn and Carnap discuss are exceptions rather than the general rule. So far, no such general argument has been forthcoming, and it is doubtful that it ever could be given.

For Kuhn much of the virtue of a given paradigm is indirect. It consists in its ability to guide research and successful puzzle-solving. But this does not change the situation regarding Criticism 1.b.

Let us then turn to the second criticism. This is lodged primarily at Carnap and stems most famously from Quine. Carnap had distinguished analytic from synthetic sentences, and drawing this distinction was essential for giving a special status to linguistic frameworks and for talking in a precise way about meaning, synonymy, logical implication, and the like. Quine challenged Carnap to provide behavioral criteria that would allow the distinction to be applied to natural or even actual scientific languages. This is not a problem just for Carnap, as it would seem that if it is essential to give special status to linguistic frameworks, it would be equally necessary to do that for paradigms.

Here I think that the themes and historical practices that Kuhn has stressed can be useful in deflecting Criticism 2 from himself but also from Carnap. Kuhn can thus be an important ally to Carnap. There is no intent on my part to state a precise set of behavioral criteria or to find in Kuhn's writings any such precise statement. But I do think that Kuhn's writings help us understand the empirical factors involved and where the relevant empirical data is to be found. This is not entirely surprising, for history is an empirical discipline.

The Kuhnian themes that I want to highlight are *education, community,* and *values.* Education, by which I mean in this context scientific education, is for Kuhn the process by which the paradigm is conveyed from one generation to the next. Kuhn's discussion of education in SSR is designed to explain why paradigms are largely invisible. His explanation, using a Wittgensteinian distinction, argues that paradigms are not stated but shown in examples. The textbooks give the examples that embody the paradigm. And then at the end of each chapter are problem sets where the student is expected to apply the paradigm in ever more complicated ways. At no point along the way is it necessary to explicitly state the paradigm. We learn our native languages initially without being given explicit rules. There is no reason to think that learning scientific languages could not proceed in the same way.

I have no wish to deny any of what I have just recounted. But I would add what I take to be a Kuhn-friendly addendum. In the cases of both textbook education and learning our native language there is more going on than just the structured series of examples with the hope that the student catches on. In response to the problem sets, the students try out answers, and those answers are evaluated and graded. Often those grades are accompanied by comments about what went wrong. In a natural language we also have more than examples to go on. We try to use the language for ourselves. And our attempts are rewarded with approval or else suppressed in some way. Of course, for any finite number of corrected examples it is logically possible to carry on in infinitely many different ways. While this is logically possible, after a remarkably short number of trials different students from different backgrounds will carry on in largely the same way. As social creatures we are remarkably good at recognizing the public norms and conforming to them. In this way each paradigm serves as a set of public norms, as a *community wide set of values* about how to proceed in science, what the important problems are, how to evaluate scientific claims, and how one's own claims are properly evaluated. Corresponding statements can be made about the community-wide sets of values or norms that constitute a natural language. These public norms are empirically accessible. They have to be in order to be learned. Norms can and do change, and those changes in norms are empirically accessible as well. Spelling and grammar are usually in flux, but the norms are evident in every proofreading or copy-editing. That there can be disagreements about what the rules are is evidence for rather than against the idea that there are such rules.

Norms are not facts about universal behavior that everyone speaks or thinks in a certain way. They are, rather, second-order facts about how in a given community various kinds of behavior are widely evaluated. And as with most social phenomena, community-wide norms are likely to be vague. They might be made more precise and even codified as in the law. But their vagueness does not imply that there are not such norms, or that the norms have no structure, or that they are empirically inaccessible. Such norms can be conveyed by explicit instruction. But no doubt they are often conveyed by seeing examples and learning what gets praised or scorned.

Where would we find evidence of community-wide values or norms? In the scientific case they can be found in the process of education and grading, in referee reports, in grant applications, in promotion letters, in review articles, and in the journal articles themselves. Every scientific writer puts down on paper what they expect the reader to accept as evidence and argument. Many such papers begin with a brief literature review in which specific work is picked out for praise or blame. All this evidence is in the historical record. Sometimes the current norms and values are more available to historians. But they are also available to us as scientists in real time because we have had that education, had our work reviewed, and written our own evaluations, which often themselves will be reviewed. I won't say that all historians of science try to reconstruct the community-wide norms of some former time or try to assess whether an earlier writer meant the same as we might mean by a given term. But Kuhn did. He aimed both (1) to reconstruct community-wide norms of former times and (2) to assess whether earlier writers (scientists) meant the same as we might by a given term. And it is that practice that gives reason to think that the various distinctions he needs are sufficiently clear and evidence based. Nothing in the issue at hand requires us to say that Kuhn was right in all his historical judgments. The issue is whether there is something to be right or wrong about on the basis of the empirical evidence. Clearly, there is.

Criticism 2 would have it that a distinction between the two tiers of the sort of accounts that Carnap and Kuhn give is unintelligible because there are no empirical criteria for drawing that distinction. I do not claim that Kuhn has explicitly stated any such criteria. He doesn't need to. But it is clear that there is an abundance of empirical evidence that is relevant to the kind of distinction he wants to draw between a paradigm and the other scientific commitments we make. His various themes and his historical practice point us in the direction of that evidence and reassure us that criteria can be found that will allow that practice to go forward.

Much the same can be said for Carnap. He is not a historian or sociologist. Nor does he need to be. What he does need is for the historian or sociologist to be able to determine empirically what the community-wide norms of evidence gathering, argumentation, and theory evaluation are in a particular time and place. That Kuhn can do what he does and that his themes and practices indicate where the appropriate empirical evidence is to be found is sufficient reassurance that there are in principle the behavioral criteria to apply Carnap's distinctions to natural and scientific languages. In this way Kuhn can be a particularly close ally to Carnap.

In Section 1.2 we highlighted various similarities and differences between Carnap and Kuhn. Among the similarities was a two-tiered system of scientific commitments. Even the characters of the two tiers were analogous. Importantly, the commitments of the broader tier could change, and such changes brought with them changes in meaning throughout the language, including at the observational level. This meant that there could be no fully development-by-accumulation picture of scientific development and no utterly theory-neutral observational basis for our theories. There were important differences too, especially in background. Kuhn was a historian who wanted to say how science did in fact develop. Carnap was a logician and philosopher of language whose abstract structures were more akin to mathematics than to history or sociology or even to empirical linguistics.

In Section 1.3 we explored some of the similarities and found that this meant that the same or similar objections could be raised against the two accounts. Criticism 1.a argued that changes from one set of commitments at the broader tier to another such set must be irrational. And Criticism 1.b argued that the lack of completely theory-neutral observational claims precluded rational theory choice even for ordinary theories. Criticism 2 argued that even drawing the distinction between the two tiers was likely to be impossible because of the lack of empirical/behavioral criteria for doing so.

In Section 1.4 we argued that the two criticisms above could be met and that it was the very difference between Carnap and Kuhn that allowed each to help the other to do so.

I do not know how much of this Irzik and Grünberg had in mind when they wrote "Carnap and Kuhn: Arch Enemies or Close Allies," but they certainly saw the parallel between Carnap and Kuhn. That insight began with the then recent historical scholarship revealing the underappreciated complexity and depth of Carnap's thought – indeed, its radical character (cf. Friedman 1987; Creath 1990; and Richardson 1998). Kuhn was a radical too and often in ways like Carnap. That they should have reached so nearly the same conclusions from such different points of departure is perhaps the second most surprising result of all. What is the most surprising? Surely, it is that their real differences in starting points allow each to help the other where they have often been thought to be most vulnerable.

2   The Influence of Science Funding Policy on Kuhn's *Structure of Scientific Revolutions*

Jamie Shaw

2.1   Introduction

In 1983, Kuhn received the J. D. Bernal Award. The namesake of the award was infamous for his contention that science itself is inherently a practical endeavor, geared toward useful goods and services rather than pure knowledge of the natural world. So-called "basic" or "pure" sciences, Bernal argued, are mere fictions – science itself is nothing but an economically driven endeavor. Bernal was one of the most vocal advocates of organizing scientific research to ensure that science is driven toward the *right* kinds of practical endeavors and liberated from the false consciousness of producing pure, apolitical scientific knowledge (Bernal 1939). Given this legacy, there is an irony of Kuhn receiving this award. Along these lines, Kuhn writes,

But when I wrote [*The Structure of Scientific Revolutions* (SSR)] I thought of it as exclusively internalist, a limitation by which I was actually a bit embarrassed. In the introduction, I took pains to document my awareness that conditions in the larger society did in fact impinge on scientific advance. I had elsewhere, I pointed out, written on Copernicus and the calendar as well as on Carnot and the steam engine. But considerations of the social setting of science had, I insisted, no place in the present book. It was based entirely on the technical writings of scientists.... My principal efforts have, that is, been directed to what I have sometimes called "the dynamic interrelationships of pure ideas." (Kuhn 1983, 26–27)

The ways in which scientific developments were affected, for better or worse, by factors external to science was intentionally marginalized. The political dimensions of science funding policy surely fall in this category (see ET 127–161), and so one may think that an analysis of Kuhn's positions on such matters would make little difference for understanding the central tenets of SSR. This inference, however, would be mistaken. Kuhn does not marginalize external forces because he thinks they are unimportant. Rather, he writes that they would "surely add an analytic dimension of first-rate importance for the understanding of scientific advance" but that "explicit consideration of effects like these would not ... modify the main theses developed in this essay" (SSR-4 xliv). This implies that considerations of external factors, science funding policy included, will not *modify* Kuhn's claims but add to their interpretative depth. Indeed, after SSR, Kuhn is explicit that external and internal considerations are complementary and his focus on internalist explanations is a biographical artifact rather than a philosophical conviction. Bringing considerations regarding science funding policy into a more complete account of the material presented in SSR is the primary goal of this chapter.

Since these considerations revolve around *policy*, it should be clear that a fuller appreciation of the claims in SSR have a *political* dimension. Kuhn is not merely describing or explaining the place of scientific institutions in society, as a sociologist would. He has views about the *proper* place science has in society. That is, Kuhn was not merely attempting to argue how science would function if science funding policy were this or that, but he has a view about how science funding policy *should* be organized. I am not the first scholar to attempt to appreciate how political considerations pervade SSR. Juan Mayoral documents Kuhn's early investments in politics and how they trickled into his interest in science (Mayoral 2009). George Reisch (2019) elaborates on Kuhn's use of political metaphors such as "revolution", and how they were shaped by the surrounding Cold War political scene. Others have focused on Kuhn's substantive views about the place of science in society. Mary Jo Nye (2014), for example, claims that Kuhn's politics were nearly identical to Conant's; specifically, Conant's fierce advocacy for the freedom of science – the view that scientists should be allowed to direct their own research without interference from the state. Nowhere was this debate more important or pronounced than during the intense debates surrounding the foundation of the National Science Foundation from 1945 to 1950. While Conant's influence on Kuhn is palpable on this issue,[1] Kuhn did not merely mimic the views of his mentor. Rather, I argue, Kuhn came to a distinct view. Steve Fuller (2000b) also insists that Kuhn merely reappropriates a view of science that is in servitude to the military-industrial complex. As others have hinted at, this speculation[2] will also be shown to be off the mark as Kuhn was consciously developing his own, unique position on science funding policy (see also Wray 2021a, 26–28).

---

[1] See Wray (2016b) for a broader overview of Conant's influence on Kuhn.

[2] I call this view "speculative" because Fuller provides no strong textual evidence for his interpretation (see Andersen 2001a).

2.2   Some Intellectual Background

During the Second World War, the United States made unprecedented advances on various fronts that were essential for winning the war. The development of penicillin, radar, proximity fuses, and the atomic bomb were a credit to the ingenuity of wartime science policy wherein the practical power of the sciences showed their value. After the Second World War, Franklin Roosevelt requested a reflection on these successes from the administrative head of wartime science policy, Vannevar Bush. Bush, along with several committees, constructed a vision for a national science funding policy for peace times in a now widely read pamphlet, *Science: The Endless Frontier* (Bush 1945).

This document launched five years of debate about national postwar science policies (see Kelves 1977). Many things happened in the interim: Roosevelt died and was replaced by Harry Truman, who had less sympathy for Bush's aspirations (see Reingold 1987); congressional debates about a wide variety of issues about the place of science in society wore on with ferocious divisiveness; and many compromises and changes of mind made the debates extremely difficult to chart. The result of these debates was the establishment of the National Science Foundation (NSF). The proximity of the NSF to Bush's vision has been a source of a great deal of controversy (see Edgerton 2004). However, the topics Bush tabled dominated deliberations and were instrumental in framing the NSF's mission statements and subsequent discussions about the political functioning of the NSF. These topics, among others, included the following:

- The scope of the autonomy of scientists to direct funds toward their own priorities
- The relationship between basic/pure science and the applied sciences
- The relationship between democracy and science
- The public value of basic science
- The relationship between science and national education programs.

For decades, one of Bush's primary allies was Kuhn's mentor James B. Conant. Conant had a wealth of professional and political experience. In addition to being the president of Harvard University and the American Association for the Advancement of the Sciences, Conant was a chemist by training who aided in developing poison gases during the First World War, and the chairman of the National Defense Research Committee since 1941 where he oversaw the development of synthetic rubbers, DDT, antimalarial medicine, as well as the atomic bomb. This inside experience made his perspective on science funding policy especially important.

While Conant's view complemented Bush's in several ways, it had its own idiosyncrasies. Conant argued that applied sciences were an extension of ordinary empirical reasoning and that basic sciences were extensions of applied sciences. The difference was the "degree of empiricism," or the degree to which they engage with material objects and processes. What science provides, Conant claims, are conceptual schemes with low degrees of empiricism – highly abstract, lawlike statements – that act as guides for applied sciences. To support basic sciences, then, is to create better guides for progress in the applied sciences. Applied sciences are less likely to be "fruitful" or show their value when extended to novel applications. The opposite is true for basic sciences that unearth significant facts about nature that will be relevant for many, unforeseeable uses. In Conant's words, "in general, the more we understand the fundamentals, the more likely we are to succeed in a new scientific or technical endeavor; in short, the lower the degree of empiricism involved, the better" (Conant 1951, 40).

Kuhn was likely well aware of Bush's and Conant's views and at least some of the criticisms of their views. While Kuhn never explicitly mentions them, there is plenty of circumstantial evidence that he not only knew such debates existed but knew some of the important details as well. First, Kuhn published a reconstruction of Bush's and Conant's argument that well-educated civilians would be able to harness the powers of science that had been demonstrated during the war (Kuhn 1984, 30). More acutely, Kuhn calls himself a "product" of this postwar, intellectual environment, suggesting that he knew these arguments before SSR. Kuhn also demonstrates intimate familiarity with arguments Bush makes about the Americanization of basic science in his own course notes before SSR (TSK Archives–MC240 1959). More poignantly was Kuhn's attachment to Conant's program of creating history of science courses to educate laypeople. This program was partially motivated by Conant's egalitarian views about access to education. However, its primary purpose was explicitly connected to issues closely related to science funding policy. In Conant's manifesto, *Understanding Science*, he lists three reasons for the importance of scientific education for laypeople. First,

we need a widespread understanding of science in this country, for only thus can science be assimilated into our secular cultural pattern. When that has been achieved, we shall be one step nearer the goal which we now desire so earnestly, a unified, coherent culture suitable for our American democracy in this new age of machines and experts. (Conant 1947, 19)

Second,

Because of the fact that the applications of science play so important a part in our daily lives, matters of public policy are profoundly influenced by highly technical scientific considerations. Some understanding of science by those in positions of authority and responsibility as well as by those who shape opinion is therefore of importance for the national welfare. (ibid.)

And third,

My ... final reason ... for urging an improvement in the scientific education of the layman rests on the need for clarification of popular thinking about the methods of science. We need to lay the basis for a better discussion of the ways in which rational methods may be applied to the study and solution of human problems. (p. 21)

Each of these purposes is tied to debates surrounding science funding policy. Broadly speaking, Conant thought that the education of laypeople was necessary so that science can be harnessed for the good. In Kuhn's course notes from the time period just before the completion of proto-*Structure*, he repeats this vision to the letter. He claims that

science has become a creative force of great magnitude, constantly reshaping the structure of modern society, and altering the relations among nations ... AND it's just in the last portion of this century that the educated layman has for the most part lost touch with science. This is dangerous, because it means that one of the most effective forces shaping our society is for the first time understood only by a tiny minority of the citizenship and because there is no reason to suppose that this small minority ... are any better equipped than anyone else to make the decisions about the way in which their discoveries are to be employed. (TSK Archives–MC240 1959)

This is Conant's view. Educated laypeople have become unable to understand the recent technical findings in the sciences not because of shifting interests, but because of the increased specialization of the sciences. As Kuhn declares in SSR, "a paradigm ... transforms a group previously interested merely in the study of nature into a profession" (SSR-4 19). As a profession, the scientists who work with the same paradigm have their own journals, specialized vocabulary, and require extensive training to become members (SSR-4 131; see also Patton 2018). This makes public education all the more difficult and presents obstacles for incorporating detailed scientific knowledge into common knowledge. But, Kuhn continues, "Science itself must remain the task of the professional scientist. It's a specialists pursuit. But the harnessing of science, the integration of its new products into society is a job which must concern all those who are affected by the new products which science provides" (TSK Archives–MC240 1959). This is Conant's view as well, but it also clarifies the scope of the freedom of science for Kuhn. Science may be free in the laboratory, or when it comes to deciding what research is a priority to pursue, but it is not free to be left unchecked once it becomes incorporated into society at large. This is what makes educating laypeople so important. The discoveries of science, once they become socially relevant, must be safe for everyone according to democratic standards. Without programs that make this goal possible, it would be dangerous to leave science alone.

This view had important implications for science funding policy on two fronts. First, and more obviously, it allowed the distribution of funds to be decided on by practicing scientists. It was commonly held that public scientific institutions should be democratically accountable, so leaving scientists *entirely* free would be a non-starter. Second, it holds that this freedom from outside interference should only be held when there are other ways of holding science in check democratically. Thus Kuhn had his eyes wide open about the political value of teaching the history of science, and it was seen as essential for providing freedom to scientists.

2.3   Kuhn's Turn away from Conant

This suggests that Nye is correct in her assertion that Kuhn's "*Structure of Scientific Revolutions* is a book *fully resonant* with the aims of Conant's general education program and with the liberal humanist politics of American intellectuals in the postwar period" (Nye 2014, 214, emphasis added). As time went on, however, Kuhn's views shifted in an important respect that distanced him from Conant and Bush. The view that science has "products" for society implies the linear model of innovation, where science can have downstream consequences for practical applications. While this implication leaves it vague as to what the linear model entails, it is clear that Kuhn thinks that applied progress at least partially depends on discoveries in the basic sciences.[3] Conant's presumption of the linear model is readily apparent, and its truth seemed so obvious that it needed no argument. Take one passage as an illustrative example:

A society bent on material prosperity and involved in an armament race (at least temporarily) had better have the maximum amount of applied science. But in a given segment, the application of science sooner or later comes to an end unless new concepts are developed ... [T]he applied scientist reaches the dead end of a road and calls to his colleague in the university laboratory for new supplies of scientific knowledge. Whence it follows that the nation must speed up the advance of science itself. This means that large sums of money must be spent

[3] To be clear, neither Bush nor Conant held the view that progress in the applied sciences *only* or even *mostly* came from basic sciences. Both recognized that progress on both fronts could come from any discipline (see Zachary 1997, 218–19; and Conant 1951, 38).

on another speculation ... adventures in pure science. The implementation of this conclusion through a National Science Foundation unfortunately still awaits Congressional action. (Conant 1950, 198)

Here we see Conant echo Bush's argument for investment into basic research to stockpile the reservoir from which the applied sciences may draw (see also Conant 1951, 39ff.). While Kuhn accepted the linear model early on in his career, he came to reject it. It is unclear when this transition took place, but his course notes at Berkeley announce this change of mind:

Today we are all accustomed to thinking of science as an enterprise that constantly alters our way of life. Radio and television, plastics and miracle drugs, atomic bombs and atomic power. As a result of this impressive indoctrination we are, I think, usually quite unprepared to discover how useless and impractical science has been during most of its long development.... [T]here were many many inventions of immense social import [e.g.,] gunpowder, compass, and printing ... metallurgy, power sources, and many others besides. But with a few isolated exceptions these were produced by craftsmen – men who knew little or none of the science available. (TSK Archives–MC240 1964)

This point becomes repeated several times in his post-SSR publications. According to Kuhn, science and technology are separate ventures that rarely interact. Far from being applied *sciences*, many inventions were developed independently of then-current developments in basic sciences. Kuhn writes that

during most of recorded history science and technology, including technological innovation, have been quite separate pursuits, involving, except occasionally in the military sphere, different people drawn largely from different social groups. While that was the case, however, the relation between the enterprises was anything but reciprocal. Only in minor instances did technological advance occur by the application of knowledge created by science. (Kuhn 1972, 176)

This passage is particularly interesting due to the caveat "except occasionally in the military sphere," which could easily be read as a wink to Conant wherein Kuhn is claiming that Conant's wartime experiences are an exception rather than the rule. But Kuhn also allows for a principled exception – namely, that the linear model may hold more commonly for pre-paradigmatic or immature sciences: "early in the development of a new field ... social needs and values are a major determinant of the problems on which its practitioners concentrate" (ET 118). Kuhn explains how professionalization separates social groups, leading to less ability to have fruitful cross-disciplinary collaborations (ET 134ff.). As paradigms become established, they become detached from social practices with which they were once entangled. So Kuhn's strong wording should be tempered: the linear model is only rejected in cases of *mature sciences*, the kinds of sciences Kuhn is concerned with in the first place.

But this caveat introduces further restrictions on the contexts in which Kuhn takes the linear model to hold. Kuhn argues that this pattern of professionalization is a fairly recent phenomenon that appears in most scientific fields only during the late eighteenth and early nineteenth century (ET 60; ET 135). This is when the scientific journal becomes commonplace (Csiszar 2018), and the era of amateur scientists became replaced by one of specialists (Shapin 2008). Thus the failure of the linear model for mature sciences is a recent phenomenon. Kuhn is committed to the view that the pattern described in SSR pushes us away from the linear model. Confusingly, though, at times Kuhn argues for the opposite claim:

The emergence of science as a prime mover in socioeconomic development was a … sudden phenomenon, first significantly foreshadowed in the organic-chemical dye industry in the 1870s, continued in the electric power industry from the 1890s, and rapidly accelerated since the 1920s. To treat these developments as the emergent consequences of the Scientific Revolution is to miss one of the most radical historical transformations constitutive of the contemporary scene. Many debates over science policy would be more fruitful if the nature of this change were better understood. (ET 142)[4]

I am not sure how to reconcile this tension in Kuhn's thought. On the one hand, he claims that the linear model presents a false image of progress in the applied sciences since the nineteenth century. On the other hand, he claims there was a "radical transformation" where the linear model became salient in the late nineteenth century. Whether or not Kuhn agrees with Conant in the end, he certainly did not appropriate or even come to accept the linear model in his mature works because of Conant. Kuhn was consciously, even if awkwardly, developing his own line of thought.

If we read Kuhn as rejecting the linear model, which seems more plausible given how central this rejection fits into his broader views, then he not only rejects part of Conant's purpose of scientific education for laypeople, but also undercuts Conant's and Bush's primary argument for the *value* of basic science. Bush's and Conant's linear model had the happy consequence that one need not settle on the seemingly intractable question of whether basic science is intrinsically valuable. Regardless of whether scientific knowledge is valuable in itself, it can be pursued *as if* it

---

[4] The final sentence in this passage is likely intended for scholars arguing over the linear model.

is. Bush needed the linear model to convince his opponents in Congress (e.g., Harvey Kilgore). An argument from an appeal to the intrinsic value of basic science would have been politically inert. Kuhn's position removes this argument in favor of Bush's and Conant's view. While I will return to what this leaves us with in the final section, this further demonstrates that Kuhn was no mere mouthpiece for Conant in his mature work.

2.4   Detour on Unpredictability

Before moving on to the implications all this has for reinterpreting SSR, it is worth noting another stance Kuhn takes on a central issue discussed in the science funding policy debates. Kuhn did not, so far as I can tell, come to this stance *from* science funding policy but it is relevant nonetheless and deserves mention. According to Bush, the possible applications of basic science are unpredictable in advance – especially in the long term:

Discoveries pertinent to medical progress have often come from remote and unexpected sources, and it is certain that this will be true in the future. It is wholly probable that progress in the treatment of cardiovascular disease, renal disease, cancer, and similar refractory diseases will be made as the result of fundamental discoveries in subjects unrelated to those diseases, and perhaps entirely unexpected by the investigator. (Bush 1945, 14)

Because of this, Bush thought we should allow a fair amount of latitude to scientists to pursue research freely as we will not know where applications may come from. Conant shares this view and often speaks of unpredictability as the sine qua non of scientific inquiry: "Investigations in science represent a leap into the unknown; betting on science is betting, it is not buying a sure thing. All trials of new procedures are uncertain" (Conant 1950, 198). Indeed, the more fruitful a conceptual scheme is for Conant, the more unpredictable progress will be.

For Kuhn, this might be true during periods of revolution. He claims:

A decision between alternate ways of practicing science is called for, and in the circumstances that decision must be based less on past achievement than on future promise. The man who embraces a new paradigm at an early stage must often do so in defiance of the evidence provided by puzzle-solving. He must, that is, have faith that the new paradigm will succeed with the many large problems that confront it.... A decision of that kind can only be made on faith. (SSR-4 156–157)

The decision can only be made on faith because there is no scientific backing for promise when a new paradigm has not yet had a chance to prove its mettle. So progress cannot be reliably anticipated when choosing a paradigm while closing a revolution. However, once a new paradigm has settled in, then progress becomes more assured and the expectation that a puzzle has a solution within some well-defined boundaries becomes justified in the track record of the paradigm. That is, "the insulation of science from society permits the individual scientist to concentrate his attention upon problems that *he has good reason to believe he will be able to solve*" (SSR-4 163, emphasis added).[5]

Put in other words, puzzle-solving is a much more predictable mode of research than Bush or Conant would have us believe. Bush and Conant valorized a conception of science as an open-minded, self-critical, and free intellectual pursuit. Given this view of science, it is no small wonder that science is unpredictable in advance given the indefinite number of possible directions scientists may take. But Kuhn saw the closing of the scientific mind as characteristic of science in its most mature phases; normal science is a time where dogmatism prevails and scientists trade dozens of possible routes of inquiry for the puzzles that they are most confident they can solve with the tools at hand. A bird in the hand is worth far more than two in the bush. If Kuhn's view of normal science came with a version of the linear model, then he would surely not have the grounds to justify as much freedom as Bush or Conant allotted to scientists. This is because scientists are not free to follow their curiosity and are constrained in their choice of topics and how they study those topics. However, if Kuhn rejects the linear model for normal science, and the public value of basic science is left up in the air for Kuhnians, it remains to be seen what consequences this holds for a Kuhnian science funding policy.

2.5   Implications for Interpreting *Structure*

On its surface, SSR appears to be an attempt to understand the development of scientific knowledge. However, there are many passages where this interpretation is difficult to defend. Kuhn regularly makes claims or uses language that suggests a *normative appraisal* of particular scientific practices. This ambiguity was noted immediately by Paul Feyerabend in a note to Kuhn on an earlier draft of SSR (see Hoyningen-Huene 1995, 366ff.). Evidently, Feyerabend did not think that Kuhn made the necessary changes in the published text as Feyerabend repeats this complaint almost a decade later:

---

[5] For related discussion, see Šešelja and Straßer (2013).

Whenever I read Kuhn, I am troubled by the following question: are we here presented with *methodological prescriptions* which tell the scientist how to proceed; or are we given a *description*, void of any evaluative element, of those activities which are generally called 'scientific'? Kuhn's writings, it seems to me, do not lead to a straightforward answer. (Feyerabend 1970, 198)

Kuhn's response to Feyerabend, and others who launched the same criticism, is perplexing. Kuhn claims that SSR

should be read in both ways at once. If I have a theory of how and why science works, it must necessarily have implications for the way in which scientists should behave if their enterprise is to flourish. The structure of my argument is simple and, I think, unexceptionable: scientists behave in the following ways; those modes of behavior have ... the following essential functions; in the absence of an alternative mode *that would service similar functions*, scientists should behave essentially as they do if their concern is to improve scientific knowledge. (SSR-2 237)

What Kuhn claims to offer is a characterization of how scientists *must* behave. If scientists were to abandon the development of science in the way described in SSR, then they would likely fail at improving scientific knowledge. For Kuhn to show this, he would need to engage in some counterfactual reasoning about alternative ways of organizing science and why they are unlikely to fulfill the same goals.[6] Otherwise, Kuhn would have no grounds to insist that the way science is practiced *now* is inherently the way it *should* be practiced. This reveals a normative element in Kuhn's thinking. Science could be otherwise, but it should not be other than it has been for the previous couple of centuries. As we look into the background of SSR, as we have, we can see that this argument involves political premises.

The first of which is more explicit in SSR: paradigms have a political function to isolate scientific communities from the rest of society. They play not just a sociological role in explaining the organization of science, but a normative role in placing science in its *proper* place in society. Scientists need not pander to its outside critics, reflect on the social importance of their trade, or worry about whether their knowledge will be practically fruitful. In Kuhn's words, "some of these [benefits of normal science] are consequences of the unparalleled insulation of mature scientific communities from the demands of the laity and of everyday life" (SSR-4 163). While Kuhn arrives at the same conclusion as Conant on this point, his argument is obviously unique as Conant did

---

[6] Kuhn goes on to say that Feyerabend's "plea for hedonism is ... irrelevant" (SSR-2 237). This is entirely mistaken as Feyerabend is claiming that the goal of developing scientific knowledge, if done in Kuhn's way, would run against the goal of developing enlightened human beings.

48 Jamie Shaw

not accept anything like Kuhn's conception of normal science. Paradigms, therefore, are not only helpful for expediting scientific discovery, but when enforced by scientists, they also play the role of gatekeeper and make certain kinds of democratic demands illegitimate.

This can also be gleaned from Kuhn's brief, otherwise cryptic remarks on the social sciences. Kuhn claims that the social sciences, unlike the natural sciences, must "defend their choice of a research problems – e.g., the effects of racial discrimination or the causes of the business cycle – chiefly in terms of their social importance of achieving a solution" (SSR-4 164). Kuhn then suggests that this explains their lack of progress. When scientists are held to standards other than their own, they are unable to focus their attention toward theoretical problems that arise from trying to understand more pervasive features of nature. Again, the appraisal of scientific progress with little practical benefit justifies the isolation of science from society via the acceptance of a paradigm.

The second, hidden political premise concerns Kuhn's *internalist historiography*. According to Kuhn, the historian of science largely reconstructs the development of scientific knowledge by explaining how particular ideas or theories influenced subsequent intellectual developments. This is distinct from *externalist* approaches that explain scientific changes by appeal to happenings outside science (e.g., political, aesthetic, and/or cultural movements or sociological explanations). Lakatos (1971) separates the two to preserve scientific rationality from the "mob psychology" of externalist explanations. Kuhn, on the other hand, is (mostly) an internalist because it is appropriate to the subject matter (ET 117). Science is best explained internally because scientific discussions are largely segregated from other kinds of external features that might incentivize or disincentivize particular kinds of inquiry. But now we know that this is only the case because science *should* be separated from other parts of society. Clearly, events like the Lysenko affair demonstrate that externalist explanations are sometimes correct, even if science *at its best* is guided by its own inner logic. While internalism is not made false by any political argument against scientific freedom, it can be made *irrelevant* if freedom were to be removed. Without freedom, then science may not be the kind of enterprise that can be explained internally. While discussions of Kuhn's historiographical approach are rich, this political foundation has been hitherto unnoted (e.g., Hoyningen-Huene 2012; Bird 2015; Melogno 2022).

Finally, it is worth mentioning that the analysis in SSR is built to become irrelevant. This point has been mentioned before. Ian Hacking, in his introduction to the fiftieth anniversary edition of SSR, suggests that "*The Structure of Scientific Revolutions* may be – I do not say is – more relevant to a past epoch in the history of science than it is to the sciences as they are practiced today" (Hacking 2012, viii). Hacking grounds this hypothesis in the increased digitization of science communication and the role of computers in simulations as well as the increased presence of the biomedical sciences. Hanne Andersen (2013) similarly considers how the increased presence of interdisciplinary research recasts Kuhn's "essential tension" in new ways. We now have more to add to this story. As the social and political situations change, so too does the argument for the freedom of science. Because of this, we should not hear anyone parroting Kuhn's views as an appropriate framework to understand or evaluate science without close attention to the social and political circumstances that made them sensible (or, at least, seem sensible) to Kuhn in the contexts he was looking at. Figuring out what those circumstances are, however, is made difficult by the fact that Kuhn is not explicit in SSR about what those circumstances are. However, with a clearer appreciation of the conditions under which science should be free, we are now better positioned to assess the plausibility of Kuhn's position in the twenty-first century. This is a pressing question as science has become more applied. The NSF, for example, began to employ social impact metrics during the 1990s where scientists began to have to anticipate the social value of their research and not just its value to a paradigm (Holbrook 2005). Should this trend be reversed? Can a Kuhnian analysis shed light on what resistance or consequences there have been and might be toward the breakdown of the political function of paradigms (see, e.g., Porter and Rossini 1985)? These questions are opened once we appreciate the contingency of the position outlined in SSR. As Kuhn writes in his course notes: "science is not an inevitable concomitant of advanced civilizations … [This] raises a host of questions: What are the characteristics of a civilization that produces science? How is it to be promoted and preserved? Could we lose it?" (TSK Archives–MC240 1959). The answer to the final question is obvious: yes, we could lose it.

2.6   Open Questions for a Kuhnian

One disadvantage of leaving much of the influence of science funding policy behind closed doors is the added difficulty of assessing the merits of Kuhn's position. Before concluding, I would like to pose two questions for a Kuhnian and motivate their importance. My own sense is these questions will lead to an abandonment of Kuhn's view, but I will not argue this here. Rather, I would like to invite self-identified Kuhnians to answer these questions in such a way that Kuhn's position, or something close to it, can be advanced in a plausible manner.

First, I want to suggest that Kuhn's rejection of the linear model seems to rest on a needlessly narrow definition of the linear model. Kuhn thinks of the linear model as a description of engineers or craftspeople using *then current* science to develop their trade. Or, conversely, the linear model could involve scientists also being innovators of concrete deliverables. The linear model need not be so quick. The more common understanding, accepted by Bush and Conant, is that *knowledge* from basic sciences can eventually have uptake in the applied sciences. Indeed, Bush was famous for arguing that the benefits of basic science take a long time before they are manifested in concrete artifacts: "Basic research is a long-term process – it ceases to be basic if immediate results are expected on short-term support. Methods should therefore be found which will permit the agency to make commitments of funds from current appropriations for programs of five years duration or longer" (Bush 1945, 33).

Conant makes similar remarks when speaking historically: "After the birth of the experimental philosophy in the seventeenth century, considerable time elapsed before advances in science began appreciably to influence progress in the practical arts" (Conant 1951, 34). Indeed, Conant even makes similar historical claims as Kuhn does, writing that "the men engaged in both enterprises [science and industry during the industrial revolution] were in touch with each other, yet the progress of the iron industry and even the development of the steam engine owed but little to the advance of science" (Conant 1951, 34). Yet Conant recognizes that this does not refute the linear model. Kuhn narrowly focuses on *collaborations* between science and more practical trades rather than seeing long, drawn-out processes by which scientific knowledge becomes a part of a "pipeline" toward downstream applications. The broader understanding of the linear model, which Kuhn originally accepts, allows science and engineering or crafts to be separate ventures while allowing scientific discoveries to positively impact the development of the applied sciences. Indeed, Kuhn writes:

Until late in the nineteenth century, significant technological innovations almost never came from the men, the institutions, or the social groups that contributed to the sciences. Though scientists sometimes tried and though their spokesmen often claimed success, the effective improvers of technology were predominantly craftsmen, foremen, and ingenious contrivers, a group often in sharp conflict with their contemporaries in the sciences. (ET 142)

This is a much narrower conception of the linear model than the version we saw Kuhn reject earlier. Thus Kuhn seems to have overgeneralized from a version of the linear model, where applications come from *scientists*, to a broader formulation of the linear model where scientific *knowledge* can aid in applications.

The conflict between Kuhn's narrow conception of the linear model, which he rejects, and the broader understanding of the linear model, which he also rejects, can be illustrated by a brief history of the development of radar. Kuhn may have had some inside knowledge with the development of radar given his position at the Radio Research Laboratory. He describes the main research there concerned triangulating radar sites, constructing radar profiles as a function of distance, and designing rotating antennas, which are feats of engineering first and foremost (RSS 269; see also Galison 2016, 44–47). Moreover, the (then) recent history of the development of radar seems to conform to Kuhn's view. Early developments of radar sprang from technical problems observed by Alexander Popov, who noticed interference caused by a passing ship while he was transmitting wireless signals (see Guarnieri 2010). However, a look at the longer history shows that these developments were made possible by Maxwell's equations, which were realized in the first wireless devices designed by Heinrich Hertz (James 1989). Thus Kuhn's experiences and, perhaps, his knowledge of the local history of radar would have confirmed his contention that radar was not developed by scientists, but the larger history reveals that this does not refute the linear model. Rather, the linear model explains parts of the development of radar, though Kuhn was probably right that for a part of its history, radar was developed independently of ongoing scientific research.

The development of the atomic bomb points in the same direction. It is true that many engineers contributed to the art of containment and passage of nuclear material and the ability to create stable fusion reactions within the confines of a bomb. However, much of the basic theoretical material that is mentioned by Oppenheimer and repeated by Bush came from discoveries in basic science. This includes, but is not limited to, Thompson's early isolation of the electron, early atomic theories of Bohr and Rutherford, the liquid drop model of the nucleus, and so on. Since Kuhn explicitly calls nuclear theory a "mature science" (SSR-4 64), he recognizes the bomb as an example of the linear model, although, as we have seen, he thinks this historical episode is an exception rather than the rule.

More foundationally, Kuhn seems vulnerable to the question Feyerabend asks in *Science in a Free Society*: "What's so great about science?" (Feyerabend 1978b, 73). It is not clear why Kuhn thought science was worth funding. Bush and Conant have an answer and their answer requires the linear model to be correct, to some degree. Kuhn seems trapped in a corner where he must defend the intrinsic value of scientific knowledge or else there would be no reason to fund science.

There are passages where Kuhn seems to lean toward this view. The argument that it satisfies curiosity might work to secure funding in the same way that governments fund basketball courts for those who happen to enjoy basketball, but hardly the billions upon billions of dollars that are currently invested in scientific research. The Kuhnians seems to have their backs against the wall and I, for one, am curious to see if they could find their way out.

2.7   Concluding Remarks

Kuhn was a complicated thinker, and much of what informed his judgments in SSR cannot be lifted out of the text itself. The many roles and contexts in which he found himself coalesce to produce such a rich, multidimensional text, and his relationships with debates about science funding policy are no different. While Kuhn was hesitant to discuss politics much in his more famous works, this should not lead us to think that he was an apolitical thinker. It is my hope that closer attention to Kuhn's reflections on issues surrounding science funding policy will only add to clarifying and expanding the views contained in one of the most important texts in the history of philosophy of science.

Acknowledgments

I thank Jon Tsou and K. Brad Wray and the audience at the Aarhus Institute of Advanced Studies conference in Aarhus, Denmark, for their helpful feedback.

Part II

Normal Science and Science Education

3 Kuhn's "Fifth Law of Thermodynamics"
Measurement, Data, and Anomalies

Alisa Bokulich and Federica Bocchi

3.1 Introduction

A central component of Thomas Kuhn's philosophy of measurement is what he calls the *fifth law of thermodynamics*. According to this "law," there will always be discrepancies between experimental results and scientists' prior expectations, whether those expectations arise from theory or from other experimental data. These discrepancies often take the form of what Kuhn calls *quantitative anomalies*, and they play a central role in both normal and revolutionary science. Whether the effort to resolve these anomalies is taken to be a part of normal or revolutionary science depends in part on the ever-evolving and context-dependent standards of what Kuhn calls *reasonable agreement*. In *The Structure of Scientific Revolutions* (SSR), Kuhn identifies as one of the most important types of experiments those aimed at determining the values of the fundamental physical constants. Why would he emphasize this seemingly obscure class of experiments? The answer, we argue, requires paying closer attention to, first, the historical context of a prominent research program in the physics department at Berkeley when Kuhn arrived in the 1950s and, second, Kuhn's broader philosophy of measurement and data. As we show, the *fifth law of thermodynamics* and the failure of *reasonable agreement* played a fundamental role in both.

In Section 3.2, we reconstruct Kuhn's philosophy of measurement and philosophy of data, as laid out primarily in his 1961 paper "The Function of Measurement in Modern Physical Science" (ET 178–224), where he introduces this fifth law. We discuss the important role of quantitative anomalies in Kuhn's philosophy, noting his emphasis on the iterative process of improving reasonable agreement. Section 3.3 turns to the historical context at Berkeley and the research program initiated by the longtime physics chair, Raymond T. Birge, who first called attention to the widespread discrepancies and inconsistencies in the experimental data on fundamental constants. We illustrate the quantitative anomalies uncovered in this research on constants, using the example of the speed of light (*c*), for which there were many different (and inconsistent!) experimentally determined values measured during the Birge-Kuhn era.

We follow this important research program forward in time in Section 3.4, highlighting Kuhnian elements taken up by the metrology institution subsequently charged with periodically adjusting the values of the fundamental constants, known as the Committee on Data for Science and Technology (CODATA). In particular, we identify three striking points of similarity. First, like Kuhn, these metrologists emphasize the iterative and ever-changing standards of reasonable agreement, prioritizing the identification of quantitative anomalies. Second, the metrology community also expresses a fundamental skepticism about scientists' ability to ever know the "true value" of a fundamental constant. Third, in the absence of any access to the true values of the constants, these metrologists emphasize the values of consistency and coherence as the only arbiters in deciding what numerical value to adopt. We connect these points to the ongoing effort to determine the value of the gravitational constant (*G*), which is the fundamental constant that Kuhn emphasizes as being particularly problematic in SSR.

In Section 3.5, we discuss Kuhn's later reflections on the formative role that his earlier work on the philosophy of measurement and data had for the development of his views in SSR. By paying closer attention to Kuhn's work on the philosophy of measurement we are also able to recover a key notion of scientific progress in Kuhn's thinking that goes beyond the increase in puzzle-solving ability later identified in the Postscript to SSR. We conclude by reflecting on the continuing relevance of Kuhn's views for the philosophy of metrology and philosophy of data today.

3.2   Kuhn's Philosophy of Measurement and Data

In SSR, Kuhn highlights as one of the "most important of all" classes of experiments in normal science the "determination of physical constants" (SSR-3 27). He notes, for example, that the "improved values of the gravitational constant [*G*] have been the object of repeated efforts ever since [the 1790s] by a number of outstanding experimentalists" (SSR-3 27–28). Kuhn goes on to mention the ongoing work to improve the values of several other fundamental constants, such as "the astronomical unit [AU], Avogadro's number [N_A], Joule's coefficient [μ_{JT}], the electronic charge [e], and so on" (SSR-3 28). This emphasis on the iterative determination of the values of fundamental constants might prima facie be surprising given the paucity of attention this topic has received from philosophers of science. However, when Kuhn arrived at Berkeley in the mid-1950s he was surrounded by discussions of the adjustment of fundamental physical constants, which was one of the towering achievements of his colleague Raymond T. Birge, who had served as chair of the Berkeley physics department for over twenty years.[1] While the "true values" of the fundamental physical constants are assumed to be constant, their values can only be known empirically through experimentation and measurement. And as Kuhn and his Berkeley colleagues were acutely aware, the empirically determined values of these constants over time are anything but constant.

In order to better understand Kuhn's views on this work, and its possible influences, we must look to Kuhn's writings on the philosophy of measurement and philosophy of data, which are worked out in detail in a paper he published just one year before SSR titled "The Function of Measurement in Modern Physical Science."[2] It is in this paper that Kuhn introduces the eponym for the title of our paper: *the fifth law of thermodynamics*. In physics, of course, there are only the three traditional laws of thermodynamics, related to the conservation of energy, the increase of entropy, and the limit of absolute zero.[3] Both the fourth and fifth laws of thermodynamics are humorous aphorisms of the experimental sciences. The more familiar fourth law (which Kuhn relegates to a passing footnote) states that "no piece of experimental apparatus works the first time it is set up." However, Kuhn's primary interest is in articulating what he calls the *fifth law of thermodynamics*, which states that "no experiment gives quite the expected numerical result" (ET 184).

These prior "expected numerical results" can come from two kinds of sources. On the one hand, they can be based on data from other experiments that have measured the same quantity, either by the same experimental method (an attempted replication) or via a different experimental approach. On the other hand, expected numerical results can also come from predictions that were calculated on the basis of theory. For example, in a collision experiment involving rigid bodies in a low friction environment, one would theoretically expect momentum to be conserved. As anyone who has been in a high school physics lab knows, however, it can be extremely difficult to get the experimental data to come out exactly the way the theory predicts. Indeed, Kuhn rightly

---

[1] We will return to discuss Birge's program of the periodic adjustment of physical constants and the legacy of its influence on the philosophy of metrology in the next section.

[2] This paper, originally published in 1961 in the journal *Isis*, is reprinted in Kuhn's (1977) collection of essays *The Essential Tension* (ET 178–224), which is the source for our page numbers cited here.

[3] To complicate the numbering, there is also a "zeroth law of thermodynamics," which establishes a kind of transitivity relation for systems in thermal equilibrium.

argues that the numbers will never agree exactly, even in the best-case scenario, due to the resolution limits of the measuring instruments employed, which will always yield fewer decimal places than one could in principle calculate from theory. More importantly, Kuhn argues that the theoretical expectations will involve various idealizations, and the experimental setup various "approximations," such that the concrete experiment will never be a perfect realization of the theoretical schema of the experiment. There will be any number of interfering influences, some perhaps known or guessed at, but others unknown. For all these reasons, experimenters are constrained by the fifth law of thermodynamics and can at best hope for what Kuhn calls "reasonable agreement." This notion of reasonable agreement is a crucial one for Kuhn, which we will discuss in more detail shortly, but before doing so it may be helpful to provide some background and terminology in the philosophy of measurement and data to help understand Kuhn's ideas.

In trying to better understand the discrepancy that Kuhn identifies in his fifth law of thermodynamics, it is helpful to turn to Pierre Duhem's chapter "Experiment in Physics" in his book *The Aim and Structure of Physical Theory*.[4] In this chapter, Duhem introduces the following distinction: "When a physicist does an experiment, two very distinct representations of the instrument ... fill his mind: one is the image of the concrete instrument that he manipulates in reality; the other is a schematic model of the same instrument, constructed with the aid of symbols supplied by theories" (Duhem 1914/1954, 155–156). Duhem goes on to explain that "the schematic instrument is not and cannot be the exact equivalent of the real instrument" (157). He continues: "the physicist, after reasoning on a schematic instrument that is too simple ... will seek to substitute for it a more complicated scheme that resembles reality more. This passage from a certain schematic instrument to another which better symbolizes the concrete instrument is essentially the operation that the word *correction* designates in physics" (ibid., emphasis added). Kuhn describes this very process using the example of Newton's *theoretical* description of a pendulum and a *concrete* pendulum in the lab:

The suspensions of laboratory pendula are neither weightless nor perfectly elastic; air resistance damps the motion of the bob; besides, the bob itself is of finite size. ... If these three aspects of the experimental situation are neglected, only the roughest sort of quantitative agreement between theory and observation can be

---

[4] Duhem's book was originally published in 1906, and his work influenced many of the authors that Kuhn cites. Here we quote from the second edition of Duhem's book, published in 1914.

expected. But determining how to reduce them ... and what allowance to make for the residue are themselves problems of the utmost difficulty. Since Newton's day much brilliant research has been devoted to their challenge. (ET 191)

As Kuhn emphasizes, this gap between theory and data is not one that is initially directed at the theory, or yet attempting any sort of confirmation or falsification. Instead, the theory is initially held "fixed," while scientists attempt to bridge the gap through a complexification of our understanding of the experimental apparatus and an extended process of wrestling with various "corrections" to the data.

The process that Kuhn describes here is similar to what metrologists today would refer to as creating a model of the measurement process and an associated uncertainty budget. The international organization responsible for coordinating the vocabulary and standards of measurements in metrology (i.e., the science of measurement) is the International Bureau of Weights and Measures, known by the acronym of its French title BIPM, which was established by the Metre Convention in 1875. In discussing measurements in physics, it is helpful to briefly review some of the basic terminology in metrology to bring philosophical clarity to the discussion. The quantity one is trying to measure in the world is known as the *measurand*, which is assumed to have some definite, but unknown (and arguably unknowable) *true value*. When a measurement is performed using some apparatus, the scientist will obtain what is called a measurement *indication* – a reading on the dial, for example. A measurement result requires turning this measurement indication into a measurement *outcome*, which is the scientist's considered *estimate* of the measurand's "true value." As Eran Tal explains

'indication' ... does not presuppose reliability or success ... but only an intention to use such outputs for reliable indication of some property of the sample being measured ... A measurement outcome, by contrast, is an estimate of the quantity value associated with the object being measured ... inferred from one or more indications ... and include[s] either implicitly or explicitly, an estimate of uncertainty. (2011, 143–144).

Turning measurement indications into measurement outcomes (or "results") can involve all sorts of theoretical calculations, such as equations that allow one to convert the measured quantity (e.g., travel time of a light signal measured in seconds) into the quantity of interest (e.g., distance measured in kilometers), a process that is known as *data conversion* (see, e.g., Bokulich 2020a). More broadly, it depends on having a detailed understanding, or model, of the measurement process, as well as an estimate of the various sources of error that may be affecting the measurement indications given by the measurement apparatus.

These sources of error, or uncertainty, can be detailed in an uncertainty budget, which tries to estimate how these various sources of error likely influenced the measurement. Kuhn is calling attention to these many possible sources of error in his discussion of examples such as the "vacuum is not perfect" and "the 'linearity' of vacuum tube characteristics" (ET 184). The data that is obtained from a measurement indication thus often needs to be corrected or processed (i.e., undergo *data correction*) in order to turn it into a measurement outcome.

This metrological distinction between indications and outcomes helps us understand Kuhn's philosophy of data: as he notes in SSR, "the measurements that a scientist undertakes in the laboratory are not 'the given' of experience but rather 'the collected with difficulty'" (SSR-3 126). This key insight is also elaborated in his earlier article on the function of measurement: "the scientist often seems rather to be struggling with the facts, trying to force them into conformity with a theory he does not doubt. Quantitative facts cease to seem simply 'the given.' *They must be fought for and with*"(ET 193; emphasis added).

These uses of the term "given" are of course references to the Latin roots of the word datum (plural data, from the past tense of the verb *dare*, to give). Not only must data be *fought for*, with the design and execution of careful experiments, but as Kuhn says they must also be *fought with*, through various forms of data correction and data processing. How does the scientist judge when the data corrections have been adequately completed? Kuhn explains that "the tests for reliability of existing instruments and manipulative techniques must inevitably be their ability to give results that compare favorably with existing theory" (ET 194). In other words, when the data is in "reasonable agreement" with theory.

What counts as a "reasonable agreement," and when is a discrepancy a problematic anomaly? As Kuhn explains, there is no single, universal criterion; it is instead highly context dependent. "'Reasonable agreement' varies from one part of science to another, and within any part of science it varies with time" (ET 185). First, reasonable agreement is usually field-dependent: the standard of accuracy for an experimental result to be in agreement with theory varies widely across research fields. For example, Kuhn notices that in "spectroscopy 'reasonable agreement' means agreement in the first six or eight left-hand digits in the numbers of a table of wave lengths" (ET 185). In contrast, "there are parts of astronomy in which any search for even so limited an agreement must seem utopian. In the theoretical study of stellar magnitudes agreement to a multiplicative factor of ten is often taken to be 'reasonable'" (ET 185). Evolving norms for reasonable agreement – as established by a given subfield, for a particular quantity, and at a moment in time – set the standards and expectations that practitioners in that field must abide by in their research.

As scientists come to better understand the measurement process and the various sources of error over time – resulting in improved experimental design and protocols, a more detailed model of the measurement process, and improved uncertainty budgeting and data correction – the gap between theory and measurement data should gradually close. George Smith (2014) describes this as a process of "closing the loop," noting its critical role in the Newtonian research program, and spectacularly illustrated by the discovery of Neptune from the no-longer-reasonable agreement between Newton's theory and the observed orbit of Uranus.[5]

Similarly, Kuhn notes that "in overwhelming proportion, these discrepancies disappear upon closer scrutiny. They may prove to be instrumental effects, or they may result from previously unnoticed approximations in the theory, or they may, simply and mysteriously, cease to occur when the experiment is repeated under slightly different conditions" (ET 202). Even if the mismatch between the measurements and theoretical prediction tends to shrink as measurements get more and more refined, the gap is never completely eliminated. In the most troubling manifestation of the fifth law of thermodynamics, however, an anomaly can prove recalcitrant: "a quantitative anomaly [can resist] all the usual efforts at reconciliation. Once the relevant measurements have been stabilized and the theoretical approximations fully investigated, a quantitative discrepancy proves persistently obtrusive" (ET 209).

It is in the course of this "mop-up" work (as Kuhn, in an overly disparagingly way, describes it) of stabilizing the measurements and theoretical approximations that the most productive quantitative anomalies are revealed. At this point, the disagreement between theory and data is no longer reasonable, and the quantitative anomaly could precipitate a crisis.

3.3   Kuhn, Birge, and the Adjustment of Physical Constants at Berkeley

With this background on Kuhn's philosophy of measurement, data, and quantitative anomalies in place, we have one of two key components

---

[5] The Neptune case illustrates the *theory-data discrepancy* and is discussed many places, including in Smith (2014). For a fuller discussion of a case involving *data-data discrepancy*, or more specifically how the iterative comparison between two different experimental approaches to measuring the same quantity can help identify and resolve anomalies in each experimental method, gradually closing the loop, see Bokulich (2020b).

needed to understand Kuhn's puzzling remarks in SSR about one of the most important classes of experiments in normal science being the iterative improvement of the values of the fundamental constants. However, the complete story also requires paying attention to the historical context in which Kuhn found himself at UC Berkeley when he was working out these ideas and writing SSR.

When Kuhn arrived at Berkeley in 1956, one of the most prominent figures in the physics department was Raymond T. Birge – a figure who has largely been forgotten today. Birge served as the chair of the physics department from 1932 to 1955 and is credited with being the architect behind the department's rise to international prominence, helping to recruit the likes of E. O. Lawrence and J. R. Oppenheimer (Helmholz 1980). Birge's impact on Berkeley was so great that when the new physics building was completed in 1964, it was named Birge Hall. More importantly for our purposes here, Birge was also the architect behind the project to coordinate a set of best current values for all the major fundamental physical constants to be prescribed for the entire physics community – a Herculean feat that made him few friends.

When Birge started this project in the late 1920s, the physics community was in disarray, using many different – and often inconsistent – values for the fundamental constants. Birge reports a "surprising lack of consistency, both in regard to the actually adopted values and to the origin of such values" (Birge 1929, 2), not least because of "some peculiar national flavor" (1957, 40) in the choice of values. This lack of standardization undermined any attempt to compare measurement results from different researchers across different labs and in different countries.

Birge's task was to comb through the entire physics literature for the various experimental determinations of the value of some physical constant – which were often arrived at via quite different experimental methods – and then decide which values should be combined to produce an averaged best estimate for the constant, and which values should be discarded as outliers. This project often required that Birge go back to the raw data (measurement indications) from these experiments and reprocess the data himself using improved data correction methods to achieve a better value (or measurement outcome) for the constant. This project of estimating a best value for a given fundamental constant is complicated by the fact that many fundamental constants are part of an interdependent web, and hence cannot have their values fixed in isolation from the values of other fundamental constants. For example, the Rydberg constant, R_∞, is defined by Bohr's formula in terms of electron mass (m_e), electric charge (e), Planck's constant (h), and the speed of

Table 3.1 *Values of the speed of light based on different experimental determinations discussed in Birge (1929)*

Experimental method	Value speed of light (c) (km/s)	Uncertainty/error (km/s)
Michelson (1927)	299,796	± 4
Mercier (1924)	299,700	± 30
Rosa and Dorsey (1907)	299,710 (reported)	± 30
	299,790 (Birge corrected)	± 10

light (c), and hence the adopted values for all these fundamental constants must "form a self-consistent system, as judged by the Bohr formula for R_∞" (Birge 1929, 71). In order to see how this project of determining the values of fundamental constants relates to Kuhn's fifth law of thermodynamics, it is helpful to examine some concrete examples.

The first fundamental constant that Birge discusses in his seminal 1929 article is the speed of light, c. Birge notes that there are three different cutting-edge experimental methods for determining the speed of light. First, there is Albert Michelson's (1927) method using a rotating mirror. Second, there is Jean Mercier's (1924) method measuring the velocity of stationary electric waves along a wire. And a third method is Edward Rosa and Noah Dorsey's (1907) indirect determination from the measured ratio of electrostatic to electromagnetic units for an electric charge. The values for c obtained from these three different measurement approaches are listed in Table 3.1.

The first thing to note is that the experimental values for the speed of light in the center column are not all identical. While at one point in the history of physics they may have been considered to be in "reasonable agreement," by Birge's time, the demands of high-precision measurement physics required a consistent value to higher resolution. At first glance, Rosa and Dorsey's *reported* value appears to be in closer agreement with Mercier's value, suggesting that Michelson's result is the outlier. However, as Birge notes, in calculating their value for the speed of light Rosa and Dorsey used the *international ohm*, and this needs to be converted to *absolute ohms* in order to obtain a proper value for the speed of light; hence, their result needs to undergo data correction. As Birge further emphasizes, it is essential that the "probable error" or uncertainty of a measurement value also be taken into consideration when determining whether or not results are in "reasonable agreement."[6] Birge goes on

---

[6] There are subtle differences between "uncertainty" and "probable error," which we do not engage here.

to reassesses the uncertainties associated with the Rosa-Dorsey measurement values by changing their *maximum* uncertainty value of ± 30 km/s down to a *probable* error value of ± 10 km/s. When these data corrections are taken into account, it turns out their measured value for the speed of light is in closer agreement with Michelson's, and it is instead the Mercier value that is the outlier. Birge concludes that the Rosa and Dorsey value is in fact in "beautiful agreement with Michelson's recent value" (Birge 1929, 10) and given that the Rosa-Dorsey probable error is over twice that of Michelson's, Birge recommends adopting Michelson's result as the recommended value for *c*. Unfortunately, this "beautiful agreement" would not last, and the estimate for the speed of light would drop precipitously in the next decade, before rising again, as seen in Figure 3.2 (from Henrion and Fischhoff 1986, 793), which we discuss below.

Notice that in the case of the speed of light, Birge opted to use the other values of *c* only as a kind of coherence test to help pick out one "best" value.[7] Alternatively, he could have combined the various experimental values for the fundamental constant, such as by weighting and averaging the values in some way, or even just by expanding the associated uncertainty estimate. In his excellent article on the history and philosophy of the adjustment of fundamental physical constants, Fabien Grégis (2019a) describes these two approaches as *arbitrage* and *compromise*, respectively, noting that Birge adopts different methods for different constants at different times, depending on the nature of the discordant data.[8] We will come back to discuss the compromise approach in a moment, but first let us take a closer look at Birge's philosophical views underlying the process of adjusting fundamental physical constants.

In this same 1929 article, Birge lays out a remarkably Peircean[9] view on the process of determining the values of fundamental constants. In particular, Birge sees this as an ongoing, iterative project: "The need is continuous since the most probable value of to-day is not that of to-morrow, because of the never-ending progress of scientific research" (Birge 1929, 2). In addition, Birge highlights the social dimensions of this iterative project of determining the value of physical constants: "there is required the unbiased cooperation of many persons situated in

---

[7] For a discussion of coherence testing of multiple measurement methods, see, for example, Bokulich (2020b) on radiometric dating.

[8] As we discuss below, in his later reassessment of the speed of light in 1941, Birge does opt for the "compromise" over "arbitrage" approach.

[9] The influence of Charles S. Peirce's pragmatic philosophy on Birge is explored by one of us in another work (Bokulich, unpublished manuscript).

scientific laboratories throughout the world" (2). Although the true value of the constant may not be determinable at any given moment of history, it should be a probable value, and more importantly the same value must be coordinated and adopted across the entire scientific community. Though grounded in the best available scientific evidence, the particular value chosen at any given time will, as Birge acknowledges, involve an irreducible element of subjective judgment. For example, the adjuster must make a judgment about which data is "good" and hence will be incorporated into the adjustment, and which data is "bad," and hence may be discarded.

Another noteworthy aspect of Birge's philosophy of data is the careful attention he paid not only to the value of the fundamental constant but also to its associated uncertainty or probable error.[10] In metrology, measurement uncertainty provides the boundaries to the likely range of values within which the "true" measurement value is supposed to lie. He writes that "some estimate of the probable error is ... as important as the constant itself" (Birge 1929, 4). Such a strong emphasis on uncertainty marks an epochal change in precision measurement physics (see Cohen and DuMond 1957; Grégis 2019a). Birge reflects on this important turning point in the physics community in 1943:

Previous to the time that I began publishing critical values of the general physical constants [in 1929], it was rather exceptional to attach a probable error, or other measure of reliability, to suggested "most probable" values. It seemed to me, however, desirable to attempt such an estimate, even in cases where the estimate was admittedly almost a pure guess. (Birge 1943, 213)

Acknowledging the necessity of uncertainty estimates, even if highly speculative, is connected to the recognition, encoded in the fifth law of thermodynamics, that measurements are never a perfect revelation of the true value of any quantity.

Indeed, Birge saw this as one of the strongest arguments for using the now widely adopted least squares method[11] for combining the different

---

[10] As we noted before, although "probable error" was the quantity and term used in Birge's time, today the metrology community prefers to use "uncertainty" rather than "error." See Grégis (2019b) for a more detailed discussion of the shift in the meaning of uncertainty.

[11] Birge in his 1932 expository article defines the least squares method as a statistical method for "(1) the calculation of the 'most probable' values of certain quantities, from a given set of experimental data, (2) the calculation of the 'probable error' of each of the quantities just evaluated, (3) the calculation of the reliability, or probable error, of the probable errors so evaluated" (Birge 1932, 207). The method goes back to the eighteenth-century work of Adrien Marie Legendre and Carl Friedrich Gauss in geodesy (see Struik 1967 for a history).

measured values of a particular constant: "Now in order to evaluate the probable error it is necessary to use the method of least squares. One great objection, it appears to me, to certain methods which have been proposed as substitutes for least squares, is that they give no objective criterion for the error" (1929, 4). In a later work, Birge recalls his seminal role in spreading the use of the least squares method from astronomy and geodesy to the rest of physics: "When I started my work on the calculation of the general constants, the method of least squares was in common use by surveyors and by astronomers. But it was seldom used by anyone else" (Birge 1957, 42). Birge himself made many contributions to the refinement of the least squares method, which had an impact far beyond determining the values of constants. Birge also notes, however, the limits of the least squares method:

one must use some judgement in applying the method of least squares. Otherwise the results may well be absurd. Such a solution applies *only* to observations which are affected merely by accidental errors of observation. If a particular observation deviates too widely from a smooth curve, it should be rejected before attempting to treat the data by least squares. (Birge 1929, 6)

He recognizes here that although the least squares method is well suited for handling *random errors*, it does not address the possibility of *systematic errors* that will skew the result in a particular direction. Another problem with least squares, noted by subsequent researchers (Taylor et al. 1969, 379) and discussed by Grégis (2019a, 49), is that by discarding outliers, this method leads to an underestimation of uncertainty and an overconfidence in the convergence of results.

Later, in his 1957 article, Birge grapples with a further problem, now often referred to as the *bandwagon effect* (or intellectual phase locking), which he defines as follows: the "tendency of a series of experimental results, at a certain epoch, to group themselves around a certain value" – even when that value turns out not to be correct. Convergence, or an agreement of measurement values, is often seen as a hallmark of their accuracy or truth. But, of course, there can be other explanations for such a convergence, apart from the measurements correctly hitting upon the "true value." Birge recounts a conversation he had with his Berkeley colleague, the Nobel Prize–winning experimental physicist Ernest Lawrence, who suggested an alternative explanation based on his own experience as an experimentalist:

In any highly precise experimental arrangement there are initially many instrumental difficulties that lead to numerical results far from the accepted value of the quantity being measured.... [T]he investigator searches for the source or sources of such errors, and continues to search until he gets a result close to the accepted value. *Then he stops!...* In this way one can account for the close agreement of several different results and also for the possibility that all of them are in error by an unexpectedly large amount. (Birge 1957, 51)

Interestingly, Kuhn draws attention to this same phenomenon in his "Function of Measurement" paper when he notes that measurements can be "self-fulfilling prophecies" in the sense that they are adjusted to conform with an expected standard (ET 196). Birge offers as a partial solution to this problem the use of multiple *different* experimental approaches to measuring a quantity, which insofar as they involve very different experimental paths are less likely to suffer from the same systematic errors. Given the common target quantity and much of the common (potentially erroneous) background knowledge, it is unlikely that all systematic errors can be avoided in this way. As we will see next, Birge's own work on the adjustment of fundamental physical constants would also turn out to be plagued by the bandwagon effect.

The experimental work on the speed of light in the decades following Birge's 1929 recommended value turned out to be an illustration of the bandwagon effect in action, as Max Henrion and Barauch Fischhoff (1986) have argued. In 1941 Birge published a reassessment of the value of the speed of light, prompted by new experimental determinations of c that some in the physics community took as evidence that the speed of light was not in fact a constant, but rather was either steadily decreasing or varying sinusoidally.[12] Birge rejected this as nonsense – to echo Kuhn, the theory of the constancy of the speed of light was, for Birge, never in doubt; rather, the anomaly was to be located in the data, which had to be forced into conformity with the theory. Birge then turned back to Michelson's (1927) value to identify previously unrecognized systematic errors, such as Michelson's mistaken use of the *wave* index of refraction instead of the correct *group* index of refraction in correcting from the measurement in air to the needed speed in a vacuum, applying the needed corrections to Michelson's original data, and even exploring possible tectonic changes that might have affected the measured distances of the baseline used in the experiment (Birge 1941, 93). Birge similarly revises the Rosa and Dorsey (1907) value, removing a rounding error and substituting an updated value for one of the quantities in the calculation, again illustrating the ongoing, iterative process of *data correction* (e.g., Bokulich 2020a; Bokulich and Parker 2021). To these two determinations he adds six more recent experimental measures,

---

[12] For a discussion and references see Birge (1941, 92); Birge (1957, 50); and Henrion and Fischhoff (1986, 793–794).

concluding that these eight values for the speed of light are all that need to be taken into account in generating the most probable value for *c*.

Unlike in his 1929 determination of the speed of light where he opted for *arbitrage*, selecting one best value, in this 1941 reassessment of the constants he adopts the *compromise* approach, weighting the different values according to the reciprocal of the square of the probable error to obtain the value 299,776 ± 4 km/s. This value is considerably smaller than Birge's 1929 adopted value of 299,796 ± 4 km/s – and notably well outside the uncertainty bounds of the previous estimate. To address the claim that the value of the speed of light might actually be changing, Birge calculates the weighted average of five measurements of the speed light carried out before the Rosa and Dorsey (1907) experiments. Birge concludes that these

five older results [yielding by a weighted average 299,873 km/s] ... are entirely consistent *among themselves*, but their average is nearly 100 km/sec greater than that given by the eight more recent results. The cause of the sudden change in the experimentally determined values of **c**, at the opening of the 20th century might be an interesting subject for investigation, but I would hesitate to believe this 100 km/sec change is real. (Birge 1941, 100)

The fact that the five older values for *c* agreed well with each other, and that the eight new values also agreed well with each other – but not with the five older values – suggests two possible hypotheses. First, this robust result across multiple experiments from these two different time periods could be interpreted as evidence that the speed of light was in fact physically changing over time – a possibility that, as we saw earlier, Birge rejected. Second, this differential clustering of experimental values for the speed of light at two different time periods could alternatively be interpreted as a *social* phenomenon – the bandwagon effect.

Focusing on the eight more recent determinations of the speed of light (listed in Figure 3.1), Birge writes that "these eight results, obtained by *six* different investigators, using *four* completely different experimental methods, agreed ... one would scarcely anticipate that the several final systematic errors should all be in the *same* direction and of roughly the *same* magnitude" (Birge 1957, 50). But as Birge had realized by 1957, this is in fact just what had happened! Although Birge had hoped in 1941 that "after a long and, at times, hectic history, the value of *c* [had] at last settled down" (1941, 101), the consensus value for *c* was about to jump again, up to around 299,792.4 km/s. Henrion and Fischhoff (1986) provide a helpful graph shown in Figure 3.2 illustrating how estimates of the speed of light changed dramatically in this period of 1929–1973.

These remarkably large swings in the "consensus" recommended values of *c* within a space of just thirty years sent ripples throughout the

Table 1. Velocity of light

Author	Method	Epoch	Corrected result	Adopted probable error r	Adopted weight $100/r^2$	Original published result
Cornu-Helmert	TW	1874·8	299990	200	0·0025	299990
Michelson	RM	1879·5	299910	50	0·0400	299910
Newcomb	RM	1882·7	299860	30	0·1111	299860
Michelson	RM	1882·8	299853	60	0·0278	299853
Perrotin	TW	1902·4	299901	84	0·0142	299901
Rosa-Dorsey	EU	1906·0	299784	10	1·000	299710
Mercier	WW	1923·0	299782	30	0·111	299700
Michelson	RM	1926·5	299798	15	0·444	299796
Mittelstaedt	KC	1928·0	299786	10	1·000	299778
Michelson, Pease and Pearson	RM	1932·5	299774	4	6·250	299774
Anderson	KC	1936·8	299771	10	1·000	299764
Hüttel	KC	1937·0	299771	10	1·000	299768
Anderson	KC	1940·0	299776	6	2·778	299776

TW = toothed wheel ; RM = rotating mirror ; EU = electric units ;
WW = waves on wires ; KC = Kerr cell.

Figure 3.1 Birge's 1941 table showing the clustering of the five earlier
experimentally measured values for the speed of light around a
significantly higher value than the eight later measurements that cluster
around a much lower value for c – a difference he interpreted as a sort of
bandwagon effect, rather than evidence that the constant was
really changing. (Reprinted with permission of IOP Publishing.)

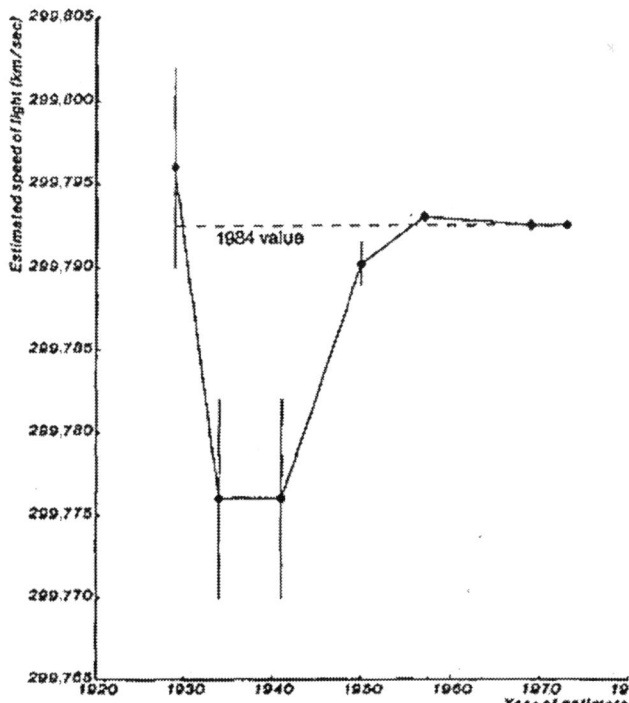

Figure 3.2 Recommended values for the speed of light from 1929
to 1973. (From Henrion and Fischhoff 1986, p. 793, with permission.)

physics community. In his 1957 paper describing this surprising reversal, Birge remarks,

This may well be the last paper I will ever write on the subject of the general physical constants. For that reason I should like to take the opportunity to consider briefly some aspects of the human side of the subject.... For if I have, to any degree, succeeded in calling attention to the numerous pitfalls that menace every research worker in science, and that lead so often to false results and conclusions, I consider that to be a far more valuable accomplishment than any specific scientific advance. (Birge 1957, 39)

This growing mood of crisis and turn to philosophy by one of the most influential members of the physics department marked Kuhn's first year at Berkeley. It is no wonder that Kuhn decided to formulate the fifth law of thermodynamics as the centerpiece of his philosophy of measurement and data, just a few years later.

The year 1957 also marked a symbolic passing of the baton from Birge to another Berkeley physics colleague, Kenneth Crowe, and two other California physicists, E. Richard Cohen and Jesse DuMond, who took up Birge's adjustment project with the publication of their book *The Fundamental Constants of Physics*, which they dedicated to Birge that same year. It should be emphasized that the speed of light was not the only fundamental constant whose experimentally determined values were misbehaving.[13] Similar problems plagued the value for electron charge, e, in the wake of R. A. Millikan's oil-drop experiments, which, as Cohen, Crowe, and DuMond discuss in their book, had a "systematic error [that] remained completely unsuspected for a period of about 15 years ... Because of the great importance of e and its close relationship to many other atomic constants this error had quite far-reaching effects" (Cohen et al. 1957, 116). There were similarly problems with Newton's universal constant of gravitation, G – which was one of the fundamental constants whose determination Kuhn singles out in SSR as being the object of

---

[13] In 1983 it was decided that c would no longer be empirically determined, but instead would become conventionally defined through the redefinition of the meter, tying both to the standard second, which is given in terms of a cesium atom transition (see Tal 2011 for discussion of standard second and see Quinn 2011 for conventional stipulation of c through redefinition of the meter). This is why when you look up the value for c today it says 299,792,458 m/s "exact" – it is because the value became one that was conventionally stipulated, not because of any change in experimental methodology that eliminated all uncertainty – something that would violate Kuhn's *fifth law of thermodynamics*! As W. Rowley further clarifies, "in making the speed of light a fixed constant, we are not attempting to dictate the laws of nature, but merely changing the viewpoint. We are not stating that the speed of light can never change; rather that, if it does, then the size of the metric length unit will change in sympathy so that the numerical value is preserved" (Rowley 1984, 284).

repeated effort by experimentalists ever since the 1790s (SSR-3 27).[14] It was becoming increasingly clear that the project of updating and coordinating a consistent set of values for all the fundamental physical constants, with an ever-growing influx of new experimental data, was neither a project for just one individual nor a project that was ever truly finished. By the late 1960s, CODATA was formed to oversee the project for the entire physics community, and the readjustment of fundamental physical constants would eventually come to be regularized to every four years.

3.4   Data, Anomalies, and the CODATA Philosophy of Metrology

The subsequent history of the process of adjusting the fundamental physical constants is an important one that is only now beginning to receive attention from philosophers of science (e.g., Grégis 2019a; Smith 2010). Relevant for our project here are the number of striking points of similarity between the emerging philosophy of metrology espoused by the CODATA metrologists and Kuhn's own philosophy of measurement.[15] Cohen and DuMond articulate their philosophical approach to the adjustment of physical constants more clearly in a subsequent review paper, where they open with a discussion of the moving goalpost of what Kuhn describes as "reasonable agreement," and how the discovery of anomalies propels this process forward: "the very process of improvement in accuracy and reliability (which the specialists in reviewing the constants themselves stimulate by calling attention to the discrepancies and troubles) whets the appetite for increasing precision, so that the discrepancies, which would have been of negligible magnitude a few years before, become of increasing importance" (Cohen and DuMond 1965, 538).

What counts as reasonable agreement is not fixed once and for all, but rather evolves with the increasing standards of precision as the program to measure a fundamental constant unfolds. To put it in Kuhnian terms, it is only by knowing in detail what to expect the measurement values to be that scientists can recognize a *quantitative anomaly* in the numbers not

---

[14] We will return to briefly discuss ongoing efforts to determine the gravitational constant, G, in the following section.

[15] Due to the lack of citations between Kuhn, on the one hand, and his colleagues Birge, Crowe, Cohen, and DuMond, on the other hand, we make no strong pronouncements about arrows of causation, noting only points of similarity and possible synergy. Our primary interest is in the continuing value of these two threads of ideas for understanding the philosophy of measurement and data today.

turning out as expected. This leads to corrections in either the measurement process, data, or background theory, which in turn yields more precise expectations. As we saw with the case of the value of the speed of light, this is often not a linear process of convergence, but it is one where, gradually over time, the expectation of number of decimal places to which results should agree increases.

Within this CODATA community, the search for anomalies, or what they call discrepancies, is one of the most important parts of the readjustment process – perhaps even more important than the new value of the constant itself. Cohen and DuMond write that "it should be clear that the prime object of these re-evaluations of the constants must always be to *look for discrepancies* and to resolve them by finding errors in either theory or experiment which account for them" (1965, 540). One implication of this approach is the recommendation *not* to expand the uncertainty estimates attached to values of the fundamental constants in the hope that the "true value" will be contained within that expanded uncertainty, as the "safety" approach would recommend. Grégis (2019a) has described this as the dilemma of *safety* versus *precision* in the philosophy of measurement, noting that for the CODATA scientists, *precision* was to be favored over safety, because of its ability to more readily reveal *anomalies*. Although the values of the constants are less likely to be revised outside the previous uncertainty bounds on the "safety" approach, doing so makes the measurements a less sensitive instrument for detecting anomalies. Since it is the disagreement and discrepancies that drive science and lead to new discoveries, the narrower uncertainty estimations of the "precision" approach are to be preferred.

The dismissal of "safety" by the CODATA group working on the revisions of fundamental physical constants also relates to another point of overlap with Kuhn's philosophy, namely, a fundamental skepticism about scientists' ability to ever come to know the true value of a fundamental constant. Cohen and DuMond express this antirealism when they write,

No one can guarantee that an evaluation of the fundamental constants at a given epoch yields the 'true' values. Absolute truth, if these words have any meaning, is beyond the realm of physics. All we can do at each time of re-evaluation is to try to determine a set of values which, in the sense of least squares, and in the light of accepted theory at that time, does least violence to a chosen budget of observational data then believed to be the 'best.' (Cohen and DuMond 1965, 540)

Like Kuhn, these physicists take absolute truth to be unattainable, and so outside of scientific practice, instead adopting a more pragmatic or instrumentalist view. The antirealism of the CODATA metrologists seems to be shaped by the same general considerations that influenced Kuhn: first, a kind of skeptical induction arising from the surprising twists and turns of the history of readjustments and, second, a kind of two-world metaphysics, where true values are something forever beyond our reach.[16]

A third point of similarity between Kuhn and these metrologists involved in the readjustment of fundamental physical constants is an emphasis on the values of consistency and coherence as key arbiters in a conflict between different experimental results, coupled with a commitment to scientific holism. Cohen and DuMond write that

few physicists or chemists fully realize in what a complicated, intricate way the fundamental constants, together with the measurements from which they are derived, are interconnected and interrelated. Everything depends upon everything else ... and one flaw in the picture propagates its defect, to a greater or lesser extent, throughout all the numerical values of the fundamental constants and conversion factors we seek. (1965, 538)

This is reminiscent of Kuhn's idea of the knowledge within a paradigm forming a strongly interconnected web that if modified has to be all together "shifted and laid down again on nature whole" (SSR-3 149). The fact that the fundamental constants of physics are interconnected, exhibiting this holism, also means that they must be evaluated together, and since there is no external arbiter, they must be assessed by the values of consistency and coherence:

the greatest merit in a re-evaluation of the constants resides not in the numerical output values ... but in the fact that the reevaluation constitutes a new test of the validity of all our theoretical preconceptions and their experimental verification over the widest possible domain. The only test of such validity we have is the consistency of the data, and this is indeed all we ask for. (Cohen and DuMond 1965, 540)

This emphasis on holism and the search for quantitative anomalies has come to be a central part of CODATA's philosophy of metrology, as we also see in the writings of Barry Taylor, a metrologist who joined the US National Bureau of Standards (now NIST) in 1970 and who is still involved in the most recent readjustment (Tiesinga et al. 2021). To emphasize the fallible and ever-iterative project of determining the values for fundamental constants, Taylor recommends that all reported values be accompanied by a warning label:

---

[16] For an introduction to these arguments for antirealism and some realist responses, see, for example, McMullin (1984).

[I]n order to bring home to the average worker that a set of recommended constants is not inviolate and handed down on stone tablets, every table of constants, whether original or reprinted, should probably start off with some type of warning label (preferably in large bright red letters) such as:

Warning!

Because of the intimate relationships which exist among least-squares adjusted values of the fundamental constants, a significant shift in the numerical value of one will generally cause significant shifts in others. (Taylor 1971, 497)

As for Kuhn, there is a delicate balance to be struck by the scientific community between, on the one hand, typically suppressing anomalies by excluding outliers from the least-squares adjustment and, on the other hand, periodically allowing these anomalies to trigger an adjustment that, because of this tightly interconnected web, can end up having far-reaching consequences, even for other accepted values that had been thought secure.

The periodic adjustment of fundamental physical constants is still an ongoing project today, and since 1998 has been institutionalized by CODATA to be undertaken every four years.[17] Figure 3.3 shows the results of the most recent (2018) adjustment for twenty-seven fundamental physical constants in relation to the previous 2014 values.

Particularly noteworthy is how many constants have updated values that fall outside the uncertainty bounds of their previously recommended values. The constants without error bars that are listed as "exact" are ones that are no longer empirically determined values, but rather are conventionally defined values due to revisions in the International System of units (SI).

Let us return to the fundamental physical constant that was the primary focus of Kuhn's discussion of constants in SSR – the gravitational constant, G – and assess the status of current efforts to determine its value. As Kuhn notes, G does not appear in Newton's *Principia*, and it was only introduced later when his universal law of gravitation was formulated as the following equation:

$$F = G \frac{m_1 m_2}{r^2}.$$

---

[17] The next adjustment slated for 2022 has not been released yet, and typically lags by a year or so from the official date. More information about the CODATA Task Group on Fundamental Physical Constants and the periodic readjustments can be found at https://codata.org/initiatives/data-science-and-stewardship/fundamental-physical-constants/.

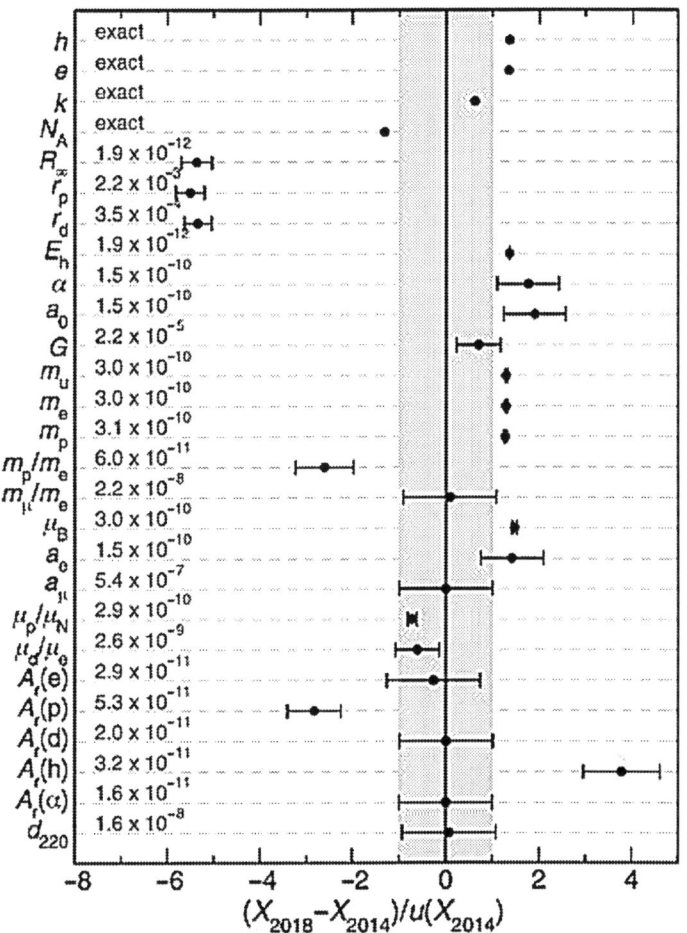

Figure 3.3 Comparison of the new 2018 values for twenty-seven fundamental physical constants (listed on y-axis) with the previously recommended 2014 values for those constants. The vertical solid line and gray band around it represent the 2014 values and their standard uncertainties, respectively. The black circles with error bars show the difference between the 2018 and 2014 values divided by standard uncertainty of 2014 value. (From Tiesinga et al. 2021, 55, with permission.)

The gravitational constant, or "big G" as it is sometimes called, has been the subject of the longest program of experimental investigation of any of the fundamental constants, but at the same time has also had the most problematic history in terms of resisting a reduction in the uncertainty of

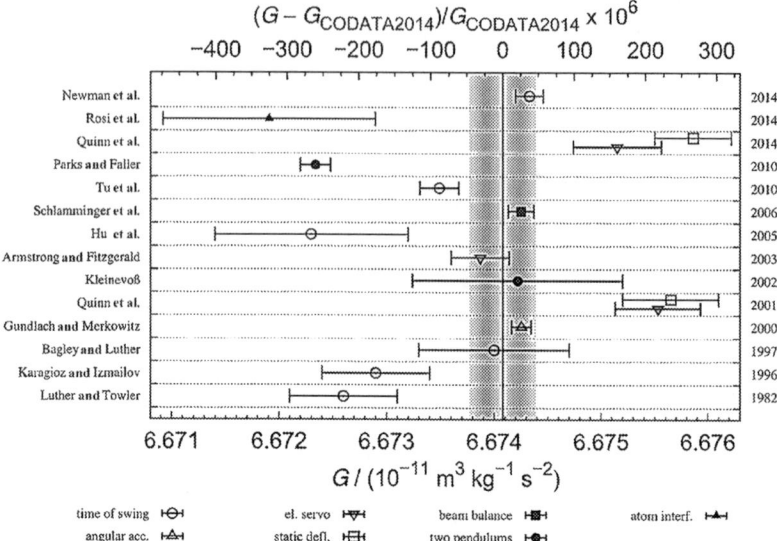

Figure 3.4 Measurement of the gravitational constant, *G*, from 1982 through 2014, involving seven different experimental methods, with the 2014 CODATA recommended value given by the vertical black line. (From Rothleitner and Schlamminger 2017, 22; with permission from AIP Publishing.)

its value. Kuhn cites a classic review article by J. H. Poynting that "reviews some two dozen measurements of the gravitational constant between 1741 and 1901" (SSR-3 28). A more recent review of experimental determinations of *G* that also begins with Poynting's review, but brings it up to the present, is that of Christian Rothleitner and Stephan Schlamminger (2017). As shown in Figure 3.4, they provide a summary of recent measures of *G* from 1982 through 2014.

One of the continuing problems with the gravitational constant (*G*) is that there is a large spread in the data arising from different experimental methods, far more than the uncertainties identified in any individual measurement. As Rothleitner and Schlamminger note, there are three possible explanations for the inconsistent data:

1. Some or all of the experiments suffer from an unknown bias ... [or] systematic effect that shifts the measured result from the true value by a predictable amount.

2. Some or all of the experiments underestimate the relative uncertainty of the measurement. Hypothetically, all of the reported values of the measurements may be correct, but the uncertainties reported may be too small. If the true uncertainty were five times larger, the data set would be perfectly consistent.
3. The most exciting, yet least probable explanation is that new unknown forms of physics can explain the variation in the data. (2017, 22)

The experimental measurements of G are thus a very dramatic illustration of Kuhn's fifth law of thermodynamics that no experiment yields quite the expected value. Even more troubling is that the different experimental determinations of G are not in "reasonable agreement" – their scatter is far outside the uncertainty bounds attached to each experiment. As in Kuhn's day, the value of the gravitational constant remains a puzzle, and it is not yet known whether this quantitative anomaly will someday be resolved within the current physical paradigm or will turn out to someday require fundamentally new physics, hence precipitating a scientific revolution.

3.5   Conclusion: Kuhn's Account of Progress in Measurement

The history of experiments to determine the values of the fundamental constants offers a striking illustration of Kuhn's *fifth law of thermodynamics*, with no experiment giving quite the expected result. Determining the values of the constants involves a long process of data wrangling and remeasurement in an effort to iteratively improve their *reasonable agreement* with both theory and other experimental data. Kuhn knew well that there was a long history of troubling anomalies for even the most central of constants, such as the speed of light (c) and the gravitational constant (G). While Kuhn was working out his views on the "fifth law" and the philosophy of measurement at Berkeley in the years leading up to the SSR, the experimentalists in the physics department were simultaneously struggling with the realization that social phenomena, such as the bandwagon effect or "intellectual phase locking," could influence experimental data. As we saw, this was coupled with an emerging philosophy of metrology that viewed any talk of "true values" as beyond the realm of physics. Instead, the adequacy of an interdependent web of fundamental constants was to be determined by the values of consistency and coherence. There was thus a remarkable synchronicity – if not synergy – between these two intellectual developments concerning the history and philosophy of physics at Berkeley in the late 1950s and early 1960s.

In his later reflections, Kuhn remarked on what a central role this paper on the philosophy of measurement had for his thinking in SSR: "Earlier at Berkeley I was asked to do a command performance ... on 'the role of measurement in xyz.'... The paper that ultimately emerges is *The Function of Measurement in Physical Science*, and that really was extremely important ... that's where the notion of normal science enters my thinking" (RSS 295). Despite the central role of this work in Kuhn's own thinking, his views about the philosophy of measurement and philosophy of data have been largely eclipsed by the theory-centric focus of SSR. This is unfortunate because it obscures a key notion of scientific progress that Kuhn identifies in this 1961 work (ET 178–224) – one that goes beyond the instrumentalist increase in puzzle-solving ability identified in the Postscript to SSR (SSR-3 206). In this pre-SSR paper he writes

I know of no case in the development of science which exhibits a loss of quantitative accuracy as a consequence of the transition from an earlier to a later theory.... Probably for the same reasons that make them particularly effective in creating scientific crises, the comparison of numerical predictions, where they have been available, has proved particularly successful in bringing scientific controversies to a close. (ET 213; emphasis removed)

This emphasis on quantitative accuracy, or perhaps more accurately *quantitative resolution* (understood as an ever-increasing number of decimal places required for reasonable agreement), is a remarkably robust and paradigm-neutral form of progress for Kuhn to have identified. And it is moreover a type of progress that has traditionally been neglected, along with a broader neglect of measurement and data in the philosophy of science until relatively recently. More generally, we hope that by situating Kuhn's views in SSR within both the broader philosophical context of his contemporaneous views on measurement and data, and the broader historical context of work being done by Kuhn's colleagues at Berkeley to determine the values of fundamental physical constants, we can gain a deeper appreciation of the extent to which Kuhn's philosophy remains relevant for the philosophy of metrology and philosophy of data today.

4   Normal Science
The Rise and Fall of Scientific Traditions

Pablo Melogno[†]

4.1   Introduction

Normal science is one of the most challenging and controversial notions introduced by Kuhn in *The Structure of Scientific Revolutions* (SSR), along with paradigms and incommensurability. It is one of the core concepts in Kuhn's philosophy, although it has not received the same exegetic attention as paradigms, incommensurability, and scientific revolutions.[1] In the 1960s, normal science was central in Kuhn's debates with Popper, Feyerabend, and other philosophers of science. But since the early 1980s, Kuhn's emphasis on semantic issues related to incommensurability left no room for the problems related to normal science. At the end of his career, the very notion had disappeared from Kuhn's language.

However, normal science and its implications have been the target of both critical approaches to Kuhn's work as well as friendly attempts of analytic elucidation. Both trends have shown how useful the concept is in helping us understand how science works beyond scientific revolutions and showing the challenges Kuhn raised in postulating scientific communities as a unit of analysis in the philosophy and history of science.

This chapter aims to clarify the role of normal science in Kuhn's philosophy. I will show that some basic features of normal science, such as problem-solving and the lack of criticism towards basic commitments, lead to a successful explanation of scientific progress. To do this, I examine normal science – emphasizing the main features of the concept, the role it plays in the notion of science defended by Kuhn, and how it allows us to articulate the social and cognitive dimensions of scientific practice. First, I review the characterization of normal science in SSR, emphasizing the role of conservative thought in developing a puzzle-solving tradition. Second, I show how Kuhn's notion of normal progress sheds light on the social nature of science, understood as a community-

---

[1] As Goodwin (2021) has highlighted, normal science has stood in the shadow of these more renowned Kuhnian concepts.

based activity, and on the role of consensus as a central feature of science. Third, I explore the connection between normal science and revolutionary progress, claiming that the conservative style of normal science is a condition for the rise of anomalies and scientific revolutions. Finally, I examine some of Kuhn's formulations on risk distribution and the tension between tradition and innovation.

I hope that this journey will make normal science more intelligible and show that it acts as a condition for accounting for progress and the cognitive success of science in the Kuhnian framework.

4.2   Normal Science in SSR

The concept of normal science has been the target of numerous analyses and reconstructions.[2] I will limit my exposition to the issues of interest for this chapter to avoid addressing topics that are well known to readers familiar with Kuhn's philosophy.

In Chapters II–IV of SSR, Kuhn introduces the notion of "normal science," understood as the research that a scientific community conducts based on achievements assumed as the foundation of the practice of the community (SSR-3 10). These achievements are accepted by the community without being exposed to critical review. They define the disciplinary field addressed by the members of the community, the lists of problems that they consider legitimate, the accepted procedures for dealing with these problems, the ontology of objects that make up the world, and the general laws or hypotheses that account for those problems. Kuhn loosely grouped these elements under the notion of "paradigm" and claimed that normal science is that which is practised by a scientific community under the aegis of a paradigm. In this way, the concepts of "paradigm," "normal science," and "scientific community" are inseparable.

The main activity of normal science is solving problems or puzzles within a problem-solving tradition to increase the articulation between nature and the theories accepted by the community. Kuhn compares solving scientific problems with piecing together a puzzle. Both cases entail solving a problem under the premise that this can be done following some rules taken for granted (SSR-3 48–50).[3] But as Paul

---

[2] Hoyningen-Huene (1989/1993, 2015a), Bird (2000b), Nickles (2003b, 2021), Worrall (2003), Marcum (2015), Goodwin (2021). On the origins of normal science in Kuhn's career, Mayoral (2011) is of interest.

[3] For a description of the problem-solving mechanisms derived from the collective division of labor, see D'Agostino (2010, chapter 3).

Hoyningen-Huene has pointed out (1989/1993, 172), the role of rules in normal research limits Kuhn's analogy between problem-solving and puzzle-solving. When playing a game or piecing together a puzzle, subjects follow rules that explicitly state which operations are acceptable and which are not. On the contrary, inclusion in a scientific community occurs not by assimilating explicit rules but by learning from successful cases accepted by the community, which provide a pattern for solving future problems by analogy.

The foundational successes that give rise to the normal scientific tradition – and which Kuhn later calls "exemplars" (SSR-3 186–187) – are powerful enough to unify the community and gain the trust of scientists and fertile enough to provide a series of unresolved problems that mark the horizon of the community's work.[4] The relationship between solved and unsolved problems comes as a consequence of the faithfulness that scientists place in the former: problems that have not been solved yet can be solved with the set of techniques and principles derived from successful exemplars that define the community's work.[5] Kuhn believes agreeing on the relevance of shared commitments is a requirement for normal science. Without consensus on fundamental questions, regular scientific research is impossible, and so is progress under an established tradition (SSR-3 11). This degree of consensus appears only in mature scientific disciplines and less so in social sciences or other areas. As Wray (2021c, 222) has pointed out, this "theoretical monism" makes it possible to explain the homogeneity of methodological and ontological commitments of mature disciplines and their effectiveness in solving problems.

Before scientists have begun to engage in a practice of normal science, they faced several fundamental disagreements. One of them is the lack of consensus regarding which facts are relevant to investigate. Without this agreement, problem-solving is highly ineffective since the range of potentially relevant facts is endless. In contrast, the consensus generated under normal science allows us to set a relatively stable domain of relevant facts on which scientists focus their energy. As I will emphasize below, the concentration of effort as a condition for progress is one of the fundamental components of normal science.

---

[4] A critical analysis of exemplars can be found in Nickles (2003b, 166–170).

[5] According to Hoyningen-Huene, "the expectation that a given research problem is soluble arises in part from its analogy to previously solved paradigmatic problems. This analogy initially only allows us to suppose that a solution to the problem is possible *in principle*" (1989/1993, 173). As Nickles explains (2021, 195), the scientist assumes that both sets of problems are of the same kind.

Under a normal scientific tradition, scientists can work effectively, taking the foundations of their discipline for granted. This moves the starting point forward, frees up debates and justifications about the framework adopted, and ensures a basis for convergence among scientists from the same community. In this way, scientists can spend most of their time solving the highly specialized – esoteric – problems that define their field (SSR-3 19–20). Their work would not reach this degree of specialization and they would not have as thorough an understanding of nature – or could do so but only at the cost of considerably more energy – if the group's fundamental commitments were not shielded from critical review.

Kuhn classifies the problems addressed by scientists in the normal scientific tradition as *factual* and *theoretical*. In both cases, their work focuses on determining significant facts, making adjustments to match theory with facts, and articulating the theory.[6] This problem-solving activity is esoteric to the extent that the community's degree of specialization makes its practice and language less and less intelligible to those who do not belong to it. Furthermore, normal research is esoteric due to the high degree of specialization and complexity of the work that the community's shared commitments allow.

The esoteric zeitgeist of normal science and theory articulation as the primary goal led Kuhn to a conservative and somewhat mechanical view of scientific work, which his critics repeatedly targeted:

Closely examined, whether historically or in the contemporary laboratory, that enterprise [i.e., normal science] seems an attempt to force nature into the preformed and relatively inflexible box that the paradigm supplies. No part of the aim of normal science is to call forth new sorts of phenomena; indeed those that will not fit the box are often not seen at all. Nor do scientists normally aim to invent new theories, and they are often intolerant of those invented by others. Instead, normal-scientific research is directed to the articulation of those phenomena and theories that the paradigm already supplies. (SSR-3 24)

In this context, several activities traditionally considered inherent to scientific practice are restricted to scientific revolutions, but not to the regular work of non-revolutionary periods. Exercises like the critical review of theories, crucial experiments, addressing problematic facts, creative innovation, and challenging inherited knowledge appear in the scientific agenda only when the specific historical conditions that cause a revolution arise. During a period of normal science, not only do these

---

[6] SSR-3 25–34. For a critical assessment, see Bird (2000b, 33–36, 49–54).

critical exercises not occur, but their occurrence would jeopardize the puzzle-solving goals pursued by the community.

This does not entail that normal science lacks a creative component. As conceived by Kuhn, problem-solving assumes the exclusion of fundamental innovations, that is, innovations that question the foundations accepted by the scientific community. But it does not exclude the exercise of creativity and innovation under the community's framework. Developing new instruments for measurement, introducing new mathematical tools, and expanding the available techniques to new domains are some of the operations that allow innovation and creativity under a normal scientific tradition. However, developing these creative elements is possible to the extent that it does not affect the methodological, ontological, and theoretical commitments that define the community.

In a negative sense, Kuhn's image of normal science breaks with a standard notion of science, according to which creativity and criticism of accepted knowledge are central traits of scientists, as their work seeks to produce unexpected novelties. In a positive sense, normal science allows us to confront these premises under a new vision, which highlights the importance of consensus for scientific progress, considers conservative thinking and resistance to novelties as functional features of progress, and accounts for the mechanisms that make up organized scientific work in a community. In sum, solving scientific problems cannot be reduced to a logical algorithm or to following explicit rules. Neither is it confined to the scientist's individual genius. On the contrary, the ability of scientists to solve problems is explained by the dynamics that define scientific communities.[7]

So far, I have restricted my attention to normal science characterized in SSR. In what follows, I will consider to what extent this characterization of normal science can account for scientific progress.

4.3   Normal Science and Normal Progress

At least two lines of criticism of normal science can be identified. Both were raised at the colloquium held in 1965 at Bedford College, mainly dedicated to discussing Kuhn's work. The first line of criticism questions the historical appropriateness of normal science and how it fits within the history of science. The second points out that normal science is either sterile or detrimental to accounting for scientific progress. The first trend

---

[7] As Thomas Nickles highlights, "in Kuhnian normal science, the problems and solutions are small enough that, given the constraints on them and the expertise of the investigators, discovery no longer seems like a gift from the gods or the powers of genius" (2021, 189).

has claimed that there are no episodes in the real history of science that coincide with what Kuhn calls "normal science" or that the distinction between normal science and revolutionary science is imprecise or not exhaustive.[8] The second trend points out that normal science renders scientific progress inexplicable and that if scientists acted as Kuhn describes, they would be endangering the very rationality of their work.[9] I will focus exclusively on this second critical trend because of space. Although it is not independent of the first, it is specific enough to warrant being treated separately.

Karl Popper was one of the main critics of Kuhn's notion of normal science. In Popperian critical rationalism, criticizing and revising established knowledge is a major driver of scientific progress (Popper 1963). Popper (1970) believed that Kuhn's normal scientists were fundamentalists who risked the progress and integrity of science. To the extent that they gave up the critical revision of their theories, they were giving up scientific rationality itself.

Kuhn's response to the Popperian offensive is well known: the features that Popper saw as intrinsic to science, such as critical thinking, breaking with prior knowledge, and conducting crucial experiments, belong to the revolutionary stage, not to the work scientists do under a consolidated tradition (Kuhn 1970a). Popper's mistake is taking a set of processes that only occur in revolutionary contexts as stable features of science.

Popper's attitude towards normal science faithfully reflects a relatively standard view that criticized Kuhn's philosophy. According to this view, criticism and continuous revision of the foundations are the basis of scientific progress. If scientists adopt a conservative attitude, they will never detect their errors, and therefore science will stagnate. Kuhnian normal science needs to be rejected because it does not account for scientific progress and, even more so, is at odds with it. If scientists systematically took the conservative attitude described by Kuhn, it would be the end of science.

The challenge posed by Kuhn's philosophy is to show that normal science is not only an obstacle to such progress but a condition for its realization. Recall Kuhn's distinction between normal and revolutionary progress. Normal progress results from developing a successful problem-solving tradition that aims to achieve better articulation between nature and the theories accepted by a scientific community. Revolutionary progress results from confronting rival paradigms, which

---

[8] Feyerabend (1970), McMullin (1993), Bird (2000b), Nickles (2021). For a discussion from a Kuhnian view, cf. Wray (2011b).

[9] Popper (1970), Lakatos (1970), Feyerabend (1970), Toulmin (1970).

occurs during a scientific revolution and results in a new paradigm replacing an old one.[10]

Both kinds of progress correspond to different stages of the historical dynamics of science. Normal progress implies an established problem-solving tradition developed around shared commitments. In contrast, revolutionary progress supposes a context of crisis and subsequent confrontation between paradigms, one of whose central features is breaking the fundamental commitments that allowed the work of normal science. Therefore, the problems of evaluating and comparing theories arise only in revolutionary contexts. In contrast, testing and assessing incompatible theoretical options is not part of the work of normal science.

Normal and revolutionary progress respond to different historical configurations. During a period of normal science, the main goal of the scientific community is puzzle-solving. In the context of a revolution, the main objective is to impose a given paradigm on a rival. Strictly speaking, there is a scientific community, but it is breaking down under the weight of anomalies. Given these differences, it is impossible to expect a single scientific attitude to be functional to both kinds of progress since they give rise to different aims and strategies.

In these terms, the methodological fertility of conservative and critical thinking hinges on the historical configuration of science. During a scientific revolution, the critical review of accepted theories and the exploration of alternatives incompatible with the fundamental commitments of the community are essential for progress. Similarly, conservative thinking and suppressing critical activity are fundamental for normal science to develop. This distinction between two types of progress allows us to say that there is no single "code of scientific ethics" but that the strategies that are functional to progress vary depending on whether it is a context of revolution or a normal context. This helps to clarify why, in a Kuhnian view, normal and revolutionary contexts do not require the same kind of attitude from scientists. Now it is necessary to establish why normal scientific progress requires a conservative mindset.

Normal science presupposes "social dynamics of collective inquiry" (D'Agostino 2010, 2). It is only possible to achieve the objectives of normal scientific progress if the fundamental commitments that define the community are shielded from critical review. This is the most

---

[10] Kuhn distinguishes both kinds of progress in SSR (Chapter XIII); in the Postscript he expands the notion of *revolutionary progress* (SSR-3 205–210). Revolutionary progress has received more attention and has generated more controversy than its normal counterpart, which is beyond the scope of this chapter. For the Kuhnian concept of scientific progress, see Hoyningen-Huene (1989/1993), Bird (2000b), Wray (2011b), and Marcum (2015).

controversial point in Kuhn's proposal. It could be accepted that normal science implies a different type of progress than that produced by revolutions; it could be assumed that such progress requires a specific social organization of scientific work. But we might challenge the fact that the consensus on the foundations and the absence of debate are the main features of this organization. Why does the specific working mode of normal science require suppressing criticism and excluding alternative views?

At least four reasons can be identified. The first refers to the trust that scientists place in community commitments. The deep exploration of nature and the successful resolution of the problems that the community considers relevant implies a high cost of work. Solving a scientific problem may require years of research, creating or improving instruments, and developing new mathematical techniques, among other things. In general, the generation that finally manages to solve the problem comes after the one that began to focus on solving it. Scientists can only carry out this sustained work in a context of unconditional trust in the fundamental commitments that define their discipline. Without confidence in these commitments, without suppressing criticism and alternative possibilities, it would be difficult to reach the depth required to solve the esoteric problems of mature science.

A second reason involves the concentration of effort. Once a normal scientific tradition is established, the scientific community can channel its working energy on a limited set of problems. This ensures the convergence and optimization of the energy of the community members. "By focusing attention upon a small range of relatively esoteric problems, the paradigm forces scientists to investigate some part of nature in a detail and depth that would otherwise be unimaginable" (SSR-3 24). The concentration of scientists' energy on a series of limited problems allows them to exploit, to their fullest, the potential of the theories accepted by the community. In contrast, calling into question the shared commitments would create endless ways of working that are incompatible with the concentration of effort needed for normal progress.[11]

John Worrall (2003, 77–78) points out the third reason in favour of the organization of normal scientific work in slightly Lakatosian language. A result contrary to the theory accepted by scientists may be due to the theory itself, auxiliary hypotheses, instruments, or other elements unknown to the experimenter. But there is no logical criterion for

[11] In the words of Alexander Bird, "normal science could not progress without the unquestioned acceptance of a theoretical foundation, just as civil society could not function without constitutional consensus" (2000b, 37).

knowing which of these variables is responsible for the lack of fit between theory and nature. Worrall claims that the organization of research provided by normal science guarantees the hierarchical preservation of the elements in play in these situations. The fundamental commitments of the group are protected from criticism, and research energies are distributed in the exploration of possibilities compatible with them: revising auxiliary hypotheses, improving instruments for measuring, and so on. In normal science, the administration of community energies and the distribution of risk are much more rational than if scientists went straight to attacking fundamental commitments when faced with negative data.

Finally, we must consider the relationship between individual inputs and collective outputs. The convergence of individual efforts on a series of firmly defined problems and based on principles that are not questioned offers results that none of the community members could achieve alone. In other words, the collective resolution of problems produces better results than the solutions provided by individuals, and it is impossible to create a context of research collectivization without a rigid consensus on the fundamental issues. In the words of D'Agostino, "once we understand that it is the community, rather than the individual, which makes and evaluates knowledge, we are in a position to understand how individuals working together in a community setting can be more efficient and effective makers and evaluators of knowledge than individuals conceived of as working alone and isolated could possibly be" (2010, 32).

This consideration of scientific work is a consequence of the change in the unit of analysis in history and philosophy of science initiated by Kuhn, one that takes scientific communities and not individual scientists as the basic elements of scientific knowledge production. Contrary to most of his predecessors, Kuhn believes that science is not an individual process but a collective enterprise. The trust in the commitments, the concentration of effort, the protection of the tradition, and the collectivization of work are community attributes. They are traits that are ascribed not to the individuals of the community but to a mode of organization that scientific communities adopt to guarantee the progress of science.

Worrall has also claimed that "Kuhn … should be seen not as advocating dogmatism, but rather as advertising the fact that 'commitment' to the sort of framework supplied by well-developed science brings enormous epistemic benefits; without such commitments, mature science would be incapable of making the progress it has in fact made" (2003, 81). Kuhn used the term "dogmatism" in "The Function of Dogma in Scientific Research," a lecture delivered in 1961 and published in 1963

(Kuhn 1963). After the emphatic criticism of Rupert Hall, Stephen Toulmin, and Hiram B. Glass (Reisch 2019, 266–270), he decided not to use the term again. In SSR-2, he chose to talk about group commitments, absence of debate, or attachment to tradition. Beyond this terminological change, Kuhn's proposal reveals that mature science requires a consensus regarding a set of basic commitments that are exempt from criticism and for which no alternatives are proposed. As Worrall rightly points out, such commitments are a necessary condition for progress in periods of normal science.

It is worth examining how scientific communities face anomalies and revolutions in this framework of unconditional attachment to shared commitments. This requires examining the relationship between normal science and revolutionary progress and clarifying the role of normal scientific work in triggering scientific revolutions.

4.4   Normal Science and Revolutionary Progress

Clarifying the connection between the organization of normal scientific work and the rise of anomalies that trigger scientific revolutions requires much more than a generic explanation. Indeed, it can be said generally that some anomalies are strong enough to undermine the trust that scientists place in community commitments. In this way, the conservative spirit of normal science begins to break down. The community consensus gives way to criticism of the established framework and, eventually, the development of alternatives. But this is insufficient since it only implies that normal science is not an obstacle to the appearance of anomalies and scientific revolutions. Kuhn's proposal goes beyond this general explanation: how normal science works is not an obstacle to revolutionary progress, but a positive condition for it to be achieved.

In "The Function of Dogma in Scientific Research," Kuhn claims that "though a quasi-dogmatic commitment is, on the one hand, a source of resistance and controversy, it is also instrumental in making the sciences the most consistently revolutionary of all human activities" (1963, 349). To what extent can a conservative and uncritical activity like normal science lead to a scientific revolution?

Regarding the relationship between normal science and progress, Kuhn believes that normal science works in such a way as to develop the maximum potential of scientific traditions. This means that normal scientific work increases the number and magnitude of problems that can be solved in a given context and creates the mechanisms to discover issues that cannot be solved in that context. Without the constraints of normal science, the problems that scientists seek to solve would be selected randomly. The successes that characterize normal scientific progress would be less likely, and problems that cannot be solved would be much harder to come by.

Normal science supports successful problem-solving and provides a model for exploring nature that is deep and detailed enough to bring to light unsolvable anomalies that will ultimately trigger a revolution. The conservative spirit of normal science is not only a condition for progress and the deep exploration of nature but also a pre-condition for the detection of anomalies and the emergence of scientific revolutions.

An anomaly does not impose itself against a scientific tradition due to intrinsic features. Kuhn points out several causes by which an anomaly can lead to a crisis: the impact on the foundations of the accepted paradigm, long-term resistance to attempts at resolution, and social pressure (SSR Chapter VII). In all three cases, the anomaly is only identified based on the relevance criteria accepted by the scientific community (ET 236). These criteria lead to the initial exploration of the problem and turn it into an anomaly once its resolution is deemed impossible. In this way, the community commitments adopted within normal science act, in the long term, as a detector of the anomalies that herald the crisis. The very emergence of anomalies would be unlikely without normal science criteria and the rigid problem-solving mechanics around these criteria.

Put more briefly, without the organization of scientific work required by normal science, the successful cases that drive a scientific tradition could not be achieved, but neither could the anomalies that cause its downfall be detected. To that extent, normal science supports the development of problem-solving traditions and also causes their crisis. Let us look at two historical cases that illustrate this point.

In the chemical revolution, the anomalies that led to the decline of phlogiston chemistry were linked to obtaining different types of air, increase in the weight of a substance after combustion, and oxide formation. These anomalies arose due to work done by chemists during the first decades of the eighteenth century, taking phlogiston as the principle of combustibility and as a defining element for many chemical phenomena.

The established consensus regarding phlogiston as a principle of combustibility and the lack of exploration of alternatives – until Lavoisier – allowed chemists to advance in several directions that led to the revolution brought about by Lavoisier's oxygen chemistry. Phlogiston chemists developed instruments that improved the precision of chemical measurements, succeeded in isolating various types of air, and tackled complicated questions such as the weight gain of substances after combustion or the formation of a new kind of air as a result of mercury oxidation. None of these problems would have been detected without a thorough exploration of the phlogiston theory. Without the framework provided by a normal scientific tradition, these problems would only be isolated chemical problems or would never have been detected. They only became anomalies and were the cause of a crisis because they were evaluated as such in light of the normal scientific tradition that made their emergence possible.

The Newtonian tradition provides an even more vivid illustration, linked to the anomaly of the perihelion of Mercury and the discovery of Neptune. Both episodes reveal the same methodological pattern in the behaviour of scientists, based on unconditional trust in Newton's mechanics and his explanation of planetary motion.

During the first half of the nineteenth century, it was known that both Mercury and Uranus described an orbit that did not coincide with Newtonian predictions, which presented astronomers with a typical case of a puzzle needing to be solved. In the case of Neptune, John Adams, Urbain Le Verrier, and James Challis, each on their own, assumed that a planet located beyond Uranus was interfering with its orbit, causing the deviations. The astronomical community focused on this hypothesis, calculations were made with the tools of Newtonian mechanics, and in 1846 Johann Galle discovered Neptune from Le Verrier's calculations. At no time did astronomers question either Newton's mechanics or the operational guidelines of the Newtonian tradition. They uncritically assumed that they could solve the problem without giving up the commitments accepted by the community. Thanks to this decidedly conservative attitude, they achieved one of the most striking discoveries in nineteenth-century astronomy.

The story of Mercury's perihelion has several points of contact with the discovery of Neptune, albeit with a slightly different outcome. As in the case of Uranus, Mercury's orbit had a deviation at perihelion that yielded values incompatible with Newtonian predictions. In 1843, three years before the first observation of Neptune, Le Verrier stated that the deviation responded to the presence of a planet whose trajectory disturbed Mercury's orbit, just as in the case of Uranus. This alleged planet was assumed to lie between Mercury and the Sun. Such was scientists' trust in Newton's framework that not only was the planet's position calculated, but before it was discovered, it was baptized with the name of Vulcan. Today we know that Vulcan does not exist, that the variation of Mercury's perihelion cannot be explained in a Newtonian framework, and that the problem could be solved only with the advent of the general theory of relativity.

The working strategy of normal Newtonian scientists in the successful case of Neptune and the failed case of Vulcan is exactly the same. In both cases, the astronomers did not question the commitments derived from the Newtonian framework and conjectured that there was a hidden variable that led to the deviation of the data from the expected result. In both cases, they conjectured that there was another planet interfering with the orbit of the observed planet and that Newtonian mechanics would account for the position of the new planet.

The discovery of Neptune was one of the greatest successes of nineteenth-century astronomy, and the perihelion of Mercury was a problem that could not be solved within the Newtonian tradition. Both were detected thanks to the dynamics of normal science and the intensive exploration of nature that depended upon the uncritical adoption of the Newtonian framework.

The moral of the story reinforces the idea that normal science provides the best way to exploit the potential of a theory in at least two ways. First, it focuses the energies of scientists, maximizes the chances of the theory being successful, and fits in with the facts of their domain. In that way, normal science contributes to normal scientific progress. Second, the concentration of energy also maximizes the chances of reaching the limits of the theory and bringing to light anomalies that cannot be resolved in the current framework; in this way, normal science contributes to revolutionary progress.

The material of this brief historical tour – this "fast and loose play with historical evidence," according to Thomas Nickles' happy expression (2003b, 157) – deserves a more detailed consideration, which we are not in a position to do here. This interpretation of anomalies is subject to comparison with the historical record as much as any other. Indeed, historical cases can be found that challenge the relationship between normal science and anomalies as I have characterized it.

But beyond nuances and potential counter-examples, if we ask ourselves about the explanatory capacity of normal science, the answer is that it is a helpful tool to account for both the rise of scientific traditions and their failure. Continuing the evolutionary metaphors that Kuhn liked so much, we can say that normal science creates the conditions for the development of scientific traditions and also for their demise. It creates the social and cognitive mechanisms necessary for traditions to develop until they exhaust their possibilities for growth and thus complete a life cycle.

4.5   Essential Tension and Risk Distribution

In the Postscript to SSR (SSR-2, SSR-3, SSR-4) and "The Essential Tension: Tradition and Innovation in Scientific Research" (ET

225–239), Kuhn sketches some considerations on the diversification of scientific work and the role of innovation in discoveries. Both texts include formulations closely tied to the picture of normal science that we have offered, so it is worth examining them briefly.

In the Postscript, Kuhn claims that

individual variability in the application of shared values may serve functions essential to science. The points at which values must be applied are invariably also those at which risks must be taken.... If all members of a community responded to each anomaly as a source of crisis or embraced each new theory advanced by a colleague, science would cease. If, on the other hand, no one reacted to anomalies or to brand-new theories in high-risk ways, there would be few or no revolutions. In matters like these the resort to shared values rather than to shared rules governing individual choice may be the community's way of distributing risk and assuring the long-term success of its enterprise. (SSR-3 186)

This passage gives the impression that, contrary to what I have said so far, normal science is an interplay of critical and conservative attitudes. Conservative attitudes keep the machinery of science running while critical attitudes fuel revolutions. In this way, scientific communities manage to spread the risk and strike a balance between tradition and innovation.[12] Generally speaking, this is effectively Kuhn's proposal. However, we must clarify some aspects to integrate this with what has been stated above.

The distribution of risk and the coexistence of critical and conservative strategies are constitutive features of scientific communities, but they do not operate equally in all historical contexts. When a normal scientific tradition is working well and racking up successes, conservative and trust-based strategies of shared commitments prevail. This occurs for two reasons.

First, these are theoretical contexts that lack an alternative to the established framework, so it is not within reach of scientists to try risky manoeuvres that challenge the foundations of tradition. In solving the problems of Mercury and Uranus, astronomers could adopt more or less novel strategies. Still, it was not possible for them to think about the problem in a non-Newtonian way.

Second, conservative strategies give rise to alternation with critical strategies only when the community has recognized an anomaly significant enough to disperse the efforts of scientists on different paths of exploration. In a normal scientific context, in which the community

---

[12] D'Agostino (2010, 83–89) has put forward a pluralist analysis of risk distribution in scientific communities, focusing on how collective work increases the efficiency of scientific research.

manages to solve the problems in its field successfully, it would not make sense for scientists to dive into developing risky strategies without a reason.

This is the stance of Kuhn also in "The Essential Tension: Tradition and Innovation in Scientific Research," when he claims that "except under quite special conditions, the practitioner of a mature science does not pause to examine divergent modes of explanation or experimentation" (ET 232). Under these special conditions, alternatives to community consensus arise in response to an anomaly. When some of these alternatives produce a finding incompatible with the fundamental commitments of the community, we are at the dawn of a crisis.[13]

For this reason, Kuhn describes the "essential tension" as a relationship between successive states of convergent and divergent thought, but not as a single state where both forms coexist: "revolutionary shifts of a scientific tradition are relatively rare, and extended periods of convergent research are the necessary preliminary to them" (ET 227). In other words, the essential tension between tradition and innovation must be understood diachronically and not synchronously.

In these terms, risk distribution and developing critical attitudes are the tools of normal science activated when anomalies appear. Still, they are not permanent mechanisms for organizing scientific research. The concentration of energy which is typical of normal science is the flip side of risk distribution. The former sustains the regular operation of successful problem-solving traditions, and the latter is activated when an anomaly challenges problem-solving abilities. Kuhn also highlights that the anomalies that trigger the revolution arise only with the work done within the normal scientific tradition: "Again and again the continuing attempt to elucidate a currently received tradition has at last produced one of those shifts in fundamental theory, in problem field, and in scientific standards to which I previously referred as scientific revolutions" (ET 234).

As I have already pointed out, all this does not mean that working under a normal scientific tradition is incompatible with certain levels of innovation and the production of novelty. D'Agostino (2010, 130), for example, has pointed out that to solve a normal scientific research problem, scientists can break the problem into several derived sub-problems and distribute the efforts of those working on the problem across the sub-problems. This implies a division of labour and a diversification of effort. However, the division and diversification are deployed in the framework of the same base problem, and the resolution strategies of the sub-problems respond to the commitments agreed upon by the

---

[13] As Nickles claims, "normal science does depend heavily on its peculiar form of social organization and social cognition, we cannot create it at will" (2003b, 145).

community. Under these conditions, a certain degree of risk-sharing and diversification of effort is not incompatible with consensus on the community's basic commitments.

In short, both risk distribution and the tension between tradition and innovation must be understood dynamically as processes that account for scientific revolutions but not as stable characteristics of normal science.

4.6   Concluding Remarks

The notion of normal science introduced by Kuhn breaks with two widely held ideas: (1) that the production of scientific knowledge results from logical operations and (2) that individual genius and creativity are the engines of science. It is equally opposed to the image of the critical scientist, who constantly generates innovations and promotes divergent thinking. Kuhn's proposal shows us that a critical spirit and risky strategies are not invariant features of science but surface as part of the historical processes typical of scientific revolutions. As an alternative, Kuhn understood that the unit of analysis of scientific work is not individual scientists but communities. Science is not an individual game but a community endeavour. It is the communities that adopt specific commitments that define their identity and allow them to establish a problem-solving tradition.

Kuhn's great discovery consists in having shown that normal science implies a specific form of collective organization of work without which scientific progress is not possible. Relying on community commitments, concentrating efforts on problem-solving, suppressing criticism, and making research collective are the main features that scientific communities adopt when they organize themselves into normal scientific traditions.

This mode of organization contributes to scientific progress in the two meanings that Kuhn distinguished. As for normal progress, it allows us to accumulate new successes and to optimize the adjustment between (1) the theories accepted by the community and (2) nature. This is achieved by focusing our efforts on the problems considered relevant for adjusting theories and on the reluctance to explore alternatives to the accepted framework.

The organization of normal science is also a condition for revolutionary progress. Troubleshooting done by normal scientists allows for investigations deep and detailed enough to bring to light anomalies that community-accepted strategies cannot resolve. In these terms, the conservative style of normal science is a condition both for developing normal problem-solving traditions and for the emergence of scientific revolutions.

5   "Textbook Science" before and after *Structure*

Alexandra Bradner

> One of Kuhn's most significant, but often neglected points is that scientists' descriptions of scientific episodes are a source of considerable obfuscation to sociological and historical understanding of science.
>
> —Steve Woolgar (1981, 256)

> It is very difficult, if not impossible, to give an accurate historical account of a scientific discipline.... It is as if we wanted to record in writing the natural course of an excited conversation among several persons all speaking simultaneously among themselves and each clamoring to make himself heard, yet which nevertheless permitted a consensus to crystalize.
>
> —Ludwik Fleck (1935/1979, 14–15)

5.1   Introduction

"Textbook science" is the caricature of science with which one is left after reading about discoveries, experiments, and theories in books written by scientists for teaching purposes. From the very first paragraph of Thomas Kuhn's *The Structure of Scientific Revolutions* (SSR), we are introduced to textbook science as a misleading "tourist brochure," one that aims to persuade students that good scientific laws and theories are justified responsibly through observation and controlled testing, that scientific development is a process of accretion, and that the methods through which we connect observed facts to broader explanations and predictions are securely rational (SSR-2 3). Though Kuhn was trained as a theoretical physicist, he began to doubt the message of this brochure as a graduate student, after assisting in just one course in the history of science taught by Harvard president James B. Conant. Kuhn writes in the Preface to SSR that Conant "initiated the transformation in my conception of the nature of scientific advance" (SSR-2 xi). Conant was inspired as a graduate student himself to think beyond his narrow training in chemistry by Harvard historian of science George Sarton

95

(Hershberg 1993, 407–408).[1] Sarton later lobbied President Conant for departmental resources in the early 1940s on the dual ground that the study of the history of science would improve the public's understanding of science in an age when "science is now more than ever under fire," and that it would bridge "the widening abyss between science and the humanities" (Hershberg 1993, 408). With funds from the Carnegie Corporation in New York, Conant developed a new undergraduate course for Harvard, Natural Sciences 4: "On Understanding Science," which he taught for three years with a collection of young scientists that included Kuhn. The course used an edited collection of original papers from the history of science to show students "how scientific research is really done" (Hershberg 1993, 409). Conant wanted his future civic leaders not to idolize science as an impartial adjudicator with direct access to reality, but to see the actual practice of science as a model for how we might solve complicated social problems using certain habits of mind (Conant 1947/1952, 25–27). Understanding the "Tactics and Strategy of Science" is something different from being informed about science (ibid., 30–31). Case histories of science, or "examples of science in the making" (p. 28), reveal that scientists solve problems by stumbling through "thickets of erroneous observations, misleading generalizations, inadequate formulations, and unconscious prejudice" (p. 30). Conant writes that he wants to convey "how new concepts evolve from experiments, how one conceptual scheme for a time is adequate and then is modified or displaced by another. I should want also to illustrate the interconnection between science and society about which so much has been said in recent years by our Marxist friends" (p. 32). Conant explains that there are two approaches to the teaching of any human activity, the historical and the philosophical: "one is to retrace the steps by which certain end results have been produced, the other is to dissect the result with the hope of revealing its structural patterns and exposing the logical relations of the component parts" (p. 27). At the time, the latter "philosopher's interpretation" of science (p. 28) was represented by logical empiricism, known to Conant through a group of local Harvard philosophers, friends, and refugee scholars who had fled Hitler (Hershberg 1993, 410–411). For Conant, and then for his young

---

[1] Hershberg explains in a footnote that this early course was Natural Sciences 11, a precursor to Natural Sciences 4, which Conant, Kuhn, and others would later develop together (Hershberg 1993, 860). During a break from Kuhn's dissertation research in physics, Conant asked Kuhn to prepare case studies for Nat Sci 11 in the history of mechanics, which was outside Conant's expertise. The process of preparing those teaching materials was the project that transformed Kuhn from a physicist to a historian.

colleague Kuhn, the idealized version of theory confirmation divined a priori by the logical empiricists does not capture how the successful practice of scientific theorizing in fact proceeds (Wray 2021a, 33–34). Kuhn's SSR, published in 1962, a decade after his paradigm-shifting teaching experience with Conant, was written as a correction to the logical empiricists' picture of science. "My most fundamental objective," Kuhn writes, "is to urge a change in the perception and evaluation of familiar data." This book will be "a quite different account of science that can emerge from the historical record of the research activity itself" (SSR-2 1).[2]

Much has been made among philosophers and historians of science about the distinction between the whiggish, internalist history of science that one might learn from a STEM textbook steeped in the principles of logical empiricism, on the one hand, and the externalist history of science that one might learn in a history or sociology classroom using a collection of individual case studies, on the other. When reading a case study of a single experimental episode, students consider the integrity of the science "in its own time" (SSR-2 3). They hone in on the site-specific standards at play within the local context. In contrast, when learning about the scientific method from a textbook, the details of any particular case are irrelevant. Students can access only the forms of scientific reasoning that the cases share in common. The internalism/externalism distinction finds its roots in Auguste Comte's positive philosophy. Comte writes in 1830, long before Conant, that every science can be presented according to two distinct methods. The ahistorical, theoretical method attempts to rationally reconstruct science as a system of ideas "as it might be conceived of today by a single mind ... placed at the right point of view and furnished with sufficient knowledge" (Comte 1988, 46). Through this method, scientific theories are presented to students as the conclusions of arguments within a system of rationally connected claims. The claims of science are assembled into a collection of proofs. The historical method, in contrast, presents knowledge "in the same order as that in which the human mind actually obtained it, following as far as possible the actual track pursued" (Comte 1988, 46). Through this method, scientific theories are presented to students as the contextually conditioned effects of human decisions and actions. The events of science are assembled into a collection of causal chains. Comte notices, like Kuhn and Conant, that science is taught almost exclusively through the theoretical method. He writes that "a modern geometer has usually finished

---

[2] Kuhn and Conant shared a preference for the historical method of presenting science. For differences between Kuhn and Conant on other concepts and themes, see Wray (2021a).

his education without having read a single original work" (Comte 1988, 47). Unlike Kuhn and Conant, however, Comte finds the theoretical method natural to the human mind and practically efficacious.

In SSR, Kuhn argues famously against internalism that it is historically inaccurate and logically underarticulated, for the methodological direct-ives of science alone underdetermine the answers to many kinds of scientific questions, in particular, questions of theory choice (SSR-2 3). Once a scientist is faced with a collection of incompatible candidate answers to the same question,

> the particular conclusions he does arrive at are probably determined by his prior experience in other fields, by the accidents of his investigation, and by his own individual makeup … An apparently arbitrary element, compounded of personal and historical accident, is always a formative ingredient of the beliefs espoused by a given scientific community at a given time. (SSR-2 4)

Textbook science misses one of the central drivers of theory choice by ignoring these accidents of history, culture, and psychology. Instead, textbooks written according to Conant's philosopher's interpretation of science or Comte's theoretical method record only the work of past scientists that relates to the text's paradigmatic problems. As a result, the history of science looks linear or cumulative (SSR-2 136–40). Kuhn emphasizes later in SSR:

> From the beginning of the scientific enterprise, a textbook presentation implies, scientists have striven for the particular objectives that are embodied in today's paradigms. One by one, in a process often compared to the addition of bricks to a building, scientists have added another fact, concept, law, or theory to the body of information supplied in the contemporary science text. But that is not the way science develops. (SSR-2 140)

5.2   What Textbook Science *Can* Do

Few STEM textbooks are written to reflect the kind of historiography Kuhn chooses for SSR. They do not capture the extra-logical drivers of scientific revolutions. Yet, ironically, textbook science plays an essential role in Kuhn's philosophy. Traditional textbooks transmit accepted theory by identifying the entities that compose the universe, explaining how they interact with one another, listing the techniques and methods we can use to study them, and, most importantly, establishing the kinds of questions we can ask about them (SSR-2 5, 10). In this way, textbooks are the repository of normal science. They assist in the constitution of paradigms by offering training materials for new community members; reaffirming the beliefs of old members; stabilizing discourse after a revolution, by hiding that there ever was a revolution; and providing accessible storage for the community norms and exemplars that are used to legitimize the conclusions of scientists working within the paradigm.

According to Kuhn himself, his account of textbook science and its constitutive role in normal science owes much to Polish microbiologist Ludwik Fleck. In explaining to readers of Fleck's 1935 *Genesis and Development of a Scientific Fact* which of Fleck's ideas were adopted in SSR, Kuhn writes, "I am, for example, much impressed by Fleck's discussion (Chap. 4, Sec. 4) of the relation between journal science and vademecum science. The latter may conceivably be the point of origin for my own remarks about textbook science" (Fleck 1935/1979, ix). In the section Kuhn cites, Fleck discusses the special structure of the thought collective of modern science and introduces science education as the process through which an individual moves from the exoteric circle of laymen to the esoteric circle of experts. There are four types of science, each one with its own literature: (1) journal science for specialized experts, (2) vademecum (handbook) science for general experts, (3) popular science for the exoteric circle, and (4) textbook science to initiate students into the esoteric circle. Fleck finds journal science to be provisional, disjointed, and personal (p. 118): "The fragmentary nature of the problems, the contingency of the material ..., the technical details, in short, the uniqueness and novelty of the working material tend to associate it inseparably with the author" (p. 119). The vademecum, in contrast, "governs the decision on what counts as a basic concept, what methods should be accepted, which research directions appear most promising, which scientists should be selected from prominent positions and which should be simply consigned to oblivion" (p. 120). The vademecum is common property and "becomes an axiom, a guideline for thinking" (p. 121). Fleck writes that mastering this form of scientific communication results in the "formation of a special readiness for directed perception and specialized assimilation of what has been perceived" (p. 120), which he calls a thought style. Scientists are socially constrained by the versions of concepts presented in the vademecum and cannot think any other way (p. 122). That is the genesis of a scientific fact. Fleck has little to say about textbooks, but one can assume that he takes this form of scientific communication to be like a vademecum for students, instead of general experts. Last, popular science for nonexperts is "vivid, simplified, and apodictic" (p. 113). Where vademecum science is a simplification of journal science, popular science is *artificially* simplified.

Fleck argues that ideas and experiences are shared outward from the esoteric circle of experts to the exoteric circle of laypeople and back again, and it is through this process of social consolidation that scientific facts emerge: "once a statement is published it constitutes part of the social forces which form concepts and create habits of thought. Together with all other statements it determines 'what cannot be thought in any other way.'... There emerges a closed, harmonious system within which the logical origin of individual elements can no longer be traced" (p. 37). Facts are proven not by isolated experiments but by the broadly based experience of the members of a thought collective who all share the same thought style. Thought styles are built up from many successful and unsuccessful experiments, extensive practice and training, and, most importantly, the adaptations, revisions, transformations, and selective reductions of concepts, as those concepts are passed within and among the esoteric and exoteric circles in journals, vademecums, textbooks, and popular publications. Fleck is especially interested in the ways in which those changes in the concepts of the thought collective are conditioned by the varying power relationships that can exist between the esoteric and exoteric circles. Writing in 1935, he finds that the esoteric circle dominates in religious thought collectives, while the exoteric circle dominates in scientific thought collectives (pp. 105–106). He is also struck by the way in which the requirements of instructional contexts influence the modification and consolidation of concepts. The only way to responsibly discuss the concept of syphilis, according to Fleck, is to consider the history of the connections between the concept and the training of the individuals who have to employ it:

Without this experience the concepts of syphilis and that of serum reaction could not have been established and research workers could not have been *trained* to practice accordingly.... The perfection of the Wasserman reaction can be seen from this point of view as the solution to the following problem: how does one define *syphilis* and set up a *blood test*, so that after *some experience* almost any research worker will be able to demonstrate a *relation between them* to a degree that is adequate in practice? (p. 98)

Facts represent a "stylized signal of resistance in thinking" for Fleck, because they arise from entrenched thought styles to stabilize the chaos of journal science.

The justification of a scientific conclusion within a thought style is not the linear process of deducing a later discovery from an earlier one using rules. Instead, early experiences become condensed into authoritative episodes as these experiences are exchanged within cycles of communication between experts and novices. Here we can see the sources of Kuhn's paradigm-constituting exemplars. By reflecting on Fleck's four vehicles of scientific communication and how they are used in teaching, we can fill in Kuhn's somewhat mysterious notion of the exemplar.

Consider what is involved in transferring information from a scholarly journal to a vademecum or handbook for other experts. Philosophers might think about how one produces a *Stanford Encyclopedia* entry, a *Cambridge Companion* chapter, or *Philosophy Compass* paper from a collection of scholarly journal articles for an audience of other professional philosophers who are not trained in that particular subfield. The goal is to include enough information in the handbook to help the other philosopher responsibly teach the material or begin a literature review, but there is no call to include the level of detail in the handbook that a professional philosopher would need to write a scholarly paper. Next, consider what is involved in transferring that information from the handbook to either a textbook for introductory students or a ten-minute YouTube video for a popular audience. Finally, consider how that translation process and the feedback from that process might influence the next scholarly journal article.

Within Mill scholarship, for example, there is a long-standing scholarly debate among experts as to whether or not Mill is to be interpreted as a rule-utilitarian (see, e.g., Urmson 1953; Smart 1956; Mandelbaum 1969). In academic journals, this debate is waged in articles of ten or more pages, each one arguing for a single interpretation. *The Stanford Encyclopedia of Philosophy* bibliography on this topic lists more than 150 of these scholarly sources (see Hooker 2015). Along with the *Stanford Encyclopedia*, the *Cambridge Companion* series is one of the go-to handbooks to which professional philosophers refer when they are assigned to teach courses outside their subfields. The single chapter on Mill's utilitarianism in the *Cambridge Companion to Mill* distills this seventy-year debate down to just four pages (see Donner 1998, 278–282). When initiating students into the esoteric circle, an instructor might use a wonderful textbook with both summaries written by the editor and selected original readings, like Elliott Sober's *Core Questions in Philosophy* (2009). Sober devotes one page to rule-utilitarianism, where he presents it as a standard reply to the lonesome stranger example, in which a sheriff frames a stranger with no connections for a crime that will soon produce a riot in the town, if no one is arrested. The *rule* "never punish the innocent" has better consequences than the rule "punish the innocent when it is convenient," so interpreting Mill as a rule-utilitarian in this way saves his philosophy from generating unintuitive verdicts on cases like the lonesome stranger. The textbook *Fundamentals of Philosophy* summarizes the difference between act- and rule-utilitarianism in half a page, presenting it again as a solution to a problem: rule-utilitarians can find ways to conclude that lies that generate lots of pleasure are wrong, where act-utilitarians cannot (see Stewart et al. 2013, 292–293). Moving to the final form of scientific communication in Fleck's account, popular versions of Mill's philosophy written for laypeople usually center exclusively on the tagline "the greatest good for the greatest number" or invoke Mill to support some call to redistribute wealth (see, e.g., Fiske 1987; Cowan 2013; and Williams 2020). For a Mill scholar, happening upon a public sphere reference to the object of your life's purpose in the *New York Times* must feel affirming and fun. "That is where the expert obtains his faith," according to Fleck (1935/1979, 115).

There are a few interesting takeaways here. Notice how extensive the scholarly literature is, and how deeply isolated and recessed the complications and disagreements of the scholarly conversation are, even from other professional philosophers in different subfields. Now consider the role that the vademecum plays in occluding all of that, but, at the very same time, how important these texts are in creating fellow feeling within the discipline of philosophy writ large. The vademecum "governs the decision on what counts as a basic concept, what methods should be accepted, which research directions appear most promising, which scientists should be selected from prominent positions and which should be simply consigned to oblivion" (Fleck 1935/1979, 120). Next, think about the textbooks. Notice how they aim to teach students by presenting a model problem and solution. At the end of the term, the philosophy professor will ask each student to produce a paper, in which each student must rehearse the same lonesome-stranger/rule-utilitarian problem solution for a grade. But better grades will be earned for papers that are not straight summaries. The best students will be able to apply the lonesome-stranger/rule-utilitarian exemplar to a new situation. There is no algorithm to help students deduce the kind of similarity they need to locate between the exemplar and the new telling, in order to earn the professor's approval. The task requires judgment from both parties, judgment conditioned by the local details of the semester. In grading the papers, the professor will be thinking about how helpful their lectures were, what topics had to be omitted for time, and whether there had been any big distractions on campus that term. In trying to meet the professor's expectations, the students will be thinking about which points were emphasized in class, how the issue is presented on YouTube and Wikipedia, what the professor seems to think is reasonable, and the grading rubric. Whatever notions of "proper" for the similarity relations that might appear are constructed on the spot, in context, in response to the papers and the professor's experience of reading this particular set of student papers. Once the student has produced something that satisfies the professor, a standard for what constitutes an acceptable similarity is locked into place. As more students produce more satisfactory work over many more courses and many more years, the standard becomes clearer and more entrenched, and an academic subfield grows up around that consolidated set of acceptable problem solutions and the community that has produced the set.

Fleck imagines that those mopped-up standards then influence and structure the messier scholarly debate, perhaps as the professor moves from their classroom back into their scholarly work. Finally, there is a story to tell about how the production of the popular newspaper articles and YouTube videos in the exoteric sphere affect the layers of the esoteric sphere. Such influences enter in through students, who tend to be more aware of popular culture than their professors. But professors committed to public philosophy write popular treatments of philosophical topics, and professors looking for engaging classroom materials often spend time searching through the popular press. When scholars come across something written on the object of their life's work in the public sphere, it is a thrill. Fleck argues that the simple, direct, self-evident, and emotive quality of communications in the exoteric sphere inspires and motivates the experts of the esoteric sphere: "If we move still further away from the esoteric center toward the exoteric periphery, thinking appears to be even more strongly dominated by an emotive vividness that imparts to knowledge the subjective certainty of something holy or self-evident. No more thought-constraining proofs are demanded, for the 'word' has already become 'flesh'" (Fleck 1935/1979, 117). "Truth is thus made into an objectively existing quality" (p. 116).

When Kuhn says that his notion of textbook science was inspired by Fleck's four types of science and the associated four forms of scientific communication, I believe the account just presented details the way in which we should understand Kuhn's statement. For Fleck, scientific facts emerge from focusing, pitching, clarifying, honing work, and almost all of this work is *pedagogical*, in some form. The practice of teaching, and the communication tools employed in that practice, become the drivers of what used to be called justificatory work in science and is now better referred to as consolidation or stabilizing work. Textbook science consists of the stable, consolidated exemplars that emerge when experts mop up their journal articles for vademecum, and nonspecialist professionals translate those handbooks into teaching tools. Normative standards for the collective emerge when students have to use the exemplars under the direction of their teachers in assessment contexts. Through the process of writing new papers on the lonesome-stranger/rule-utilitarian problem solution exemplar, students come to understand how to practice philosophy. As Kuhn writes in the Postscript to SSR,

The student discovers with or without the assistance of his instructor, a way to see his problem as *like* a problem he has already encountered.... After he has completed a certain number, ... he views the situations that confront him as a scientist in the same gestalt as other members of his specialist's group.... He has meanwhile assimilated a time-tested and group-*licensed* way of seeing. (SSR-2 189; second emphasis added)

Kuhn deemphasizes learning through the learning of rules throughout Chapter V in SSR, "The Priority of Paradigms." In his Postscript, following Michael Polyani, he writes that what results from learning to apply exemplars "is tacit knowledge which is learned by doing science rather than by acquiring rules for doing it" (SSR-2 191). Students learn what philosophers do, not by memorizing the steps of a method but by running through an original-to-them and satisfactory-to-the-professor version of the exemplar: "If, for example, the student of Newtonian dynamics ever discovers the meaning of terms like 'force,' 'mass,' 'space,' and 'time,' he does so less from the incomplete though sometimes helpful definitions in his text than by observing and participating in the applications of these concepts to problem-solution" (SSR-2 47). Importantly, the students could never learn how to practice philosophy by wading through the 150 journal articles of Fleck's esoteric circle. It would be impossible for them to see the forest through the trees, and, without the vademecum and textbook, there would *be* no forest. Disciplinary standards, thought styles, forms of explanation, and so on do not arise from intricate conversations among experts. Disciplinary standards arise when those experts are forced to distill their conversations into vademecums for their out-of-field, expert colleagues, and when those colleagues then place that content in front of students in classroom contexts and situate that content as something valuable or good, or that the students could get wrong. It is in the classroom context that the normativity arises: because a student could get *wrong* a case in which rule-utilitarianism is proposed as a solution to a problem, there is a right way to practice this subfield of philosophy.

So, that is what textbook science can do, and how it plays a constitutive role in normal science. We can find textual support for an exemplar reading that centers pedagogy in Kuhn's writings. In a 1961 letter responding to Conant's comments on a draft of SSR, Kuhn characterizes a paradigm as "the vehicle of scientific education and the locus of scientific commitment" (Kuhn 1961a, 3). He makes it very clear that a paradigm does not consist in the theoretical, factual, conceptual, empirical, or instrumental components on which the scientific community might agree, for many of these components "are often inseparably mixed together" in the problem solutions (ibid.). *Philosophers* might try to analyze these components separately after the fact, in a number of different ways. But, as Kuhn explains: "Scientists do not have to make such analytic divisions themselves or to agree upon how they are to be made except to the extent as such divisions facilitate science pedagogy.... What they agree about is that such-and-so *is* a problem solution.... The problem-solution, the achievement, comes first" (ibid.).

Significantly, Kuhn emphasizes that he was not able to write SSR until he had grasped this "fundamental" point (ibid.). Later, in the final version of SSR, Kuhn writes:

Close historical investigation of a given specialty at a given time discloses a set of recurrent and quasi-standard illustrations of various theories in their conceptual, observational, and instrumental applications. These are the community's paradigms, revealed in its textbooks, lectures, and laboratory exercises. By studying them and by practicing with them, the members of the corresponding community learn their trade. (SSR-2 43).

The intimate relationship between textbook and paradigm is clear: a paradigm is a set of commitments about how to teach something. Notice, however, that textbook science cannot deliver any collective into a new paradigm. We can train with a typical undergraduate chemistry textbook under a chemistry professor and learn how to work professionally in a pharmaceutical lab. But that chemistry textbook is not articulated enough to deliver us into the future of chemical scholarship. That can happen only deep within the esoteric sphere. Exemplars are too watered down to determine next steps. Instead, we just have to wait for a young, insightful expert to have a new idea, an idea that somehow gets published in that cacophonous landscape of scholarly journals, and then we will be off and running.

5.3   The Study of "Textbook Science" in Science Education

Over the last twenty years, a collection of researchers from several disciplines – historians of science education, history of the book scholars, scholars of scientific rhetoric, and a few scientists themselves – have foregrounded the science textbook as a subject of scholarly interest. Given the important role that textbooks play for Kuhn, Fleck, and Conant, it is curious that philosophers have not paid more attention to this work. If it is reasonable for Kuhn to say that textbook science makes normal science possible, and if it is reasonable for Fleck to say that the exchange and consolidation of scientific experiences through some expert-to-novice communication cycle creates scientific facts, it would be important to know what is happening within those textbooks and communications. Historian Kathryn Olesko affirms that "the historiography of science pedagogy to date has highlighted the ways in which educational settings sustain clusters of values, mental habits, and material practices that make possible the epistemological dimensions of scientific activity" (Olesko 2006, 864). Conceptual studies in this field ask whether STEM textbooks should continue to present textbook science or evolve to reflect the "nature of science," the way science is actually practiced. Empirical studies in this field review actual textbooks to see if they have taken up Kuhn's recommendations.[3] Studies of this kind have been published on textbooks in epidemiology (see Bhopal 1999), astronomy (Shipman 2000), psychology, physics, chemistry, and microbiology, among many other disciplines.

In one recent pedagogical study in psychology, scientists William O'Donohue and Brendan Willis review thirty psychology textbooks to see whether any of them address the "second demarcation problem" of whether psychological science is appreciably different from the other sciences. They find that hardly any of the textbooks engage topics in the philosophy of science: three textbooks cover Kuhnian paradigms; three discuss whether psychology is a preparadigmatic science or a mature science; six address the role of competing theories and evaluations in science; three introduce social constructionism; and only four note that the definition of science can be a controversial topic (O'Donohue and Willis 2018, 58). The authors conclude that "despite the variance observed across texts, these data ... generally suggest that students are being presented a simplistic notion of science as having a relatively straightforward and settled characterization. Few texts mention that science is difficult to define, or that there are multiple proposed accounts of science" (ibid.). The authors worry that most psychology textbooks adopt a logical empiricist account of science, "insinuating that the scientific method for physics is (or ought to be) the same scientific method that is used in psychology. Seeing science in an overly positivistic and simplistic light is known as the problem of scientism and one wonders if there is a problem in these texts along these lines" (ibid.).

In the recent anthology *Critical Analysis of Science Textbooks: Evaluating Instructional Effectiveness*, only three of the fifteen essays engage Kuhn's concerns about textbook science, and one of those, only briefly. Kostas Dimopoulos and Christina Karamanidou's "Towards a More Epistemologically Valid Image of School Science" begins with a comparison chart (Table 5.1) that distinguishes textbook science, or what the

---

[3] For the early history of the scientific community's uptake of SSR, see Loving and Cobern (2000), Bell et al. (2001), and Matthews (2003).

Table 5.1 *Epistemological differences between textbook science and scientific practice in Dimopoulos and Karamanidou*

School science (textbook science)	Science in the public domain (scientific practice)
Static, final	Dynamic, in the making
Ahistorical	Evolutionary, innovative
Beyond doubt, above controversy	Under negotiation, controversial
Free from conflicts of interest, insulated from motives, a product of cognitive effort exclusively	Under the influence of interests
Universally applied	Locally applied, depending on the context
Linear, deterministic	Nonlinear, reflective
Generated by individual scientists	Generated by multidisciplinary teams of collaborating experts
Linear path to a successful conclusion	Regressions between successes and failures

authors call "school science," from "science in the public domain," which captures authentic scientific practice (Dimopoulos and Karamanidou 2013, 62). But the paper moves away quickly from these issues to discuss various forms of textbook rhetoric.

Ian Binns' "A Qualitative Method to Determine How Textbooks Portray Scientific Methodology" reports the review of two popular STEM textbooks, one physics text and one biology text, to see whether they employ a "traditional" or "broad" view of the scientific method:

The traditional view emphasized that scientists follow a stepwise procedure referred to as "the scientific method," scientists must test hypotheses, or scientists only conduct experimental research. The broad view emphasized that scientists use a variety of methods to conduct their work and that there are multiple types of research, including descriptive research, correlational research, epidemiological research, and experimental research. (Binns 2013, 243)

Table 5.2, reproduced from the paper, details the criteria Binns uses to determine whether a text counts as traditional or broad.

Binns finds that neither of the two textbooks studied places much emphasis on the scientific method and neither book describes how scientists actually reason in practice (p. 254). The student version of the physics textbook emphasizes the traditional view, while the teacher version has a mixed message, which makes it difficult for the teacher to grasp and transmit any lesson about methodology. The student version of the biology textbook emphasizes the broad view, while the teacher version of

Table 5.2 *Binns' traditional versus broad views of scientific methodology*

Traditional view of scientific methodology (textbook science)	The broad view of scientific methodology (scientific practice)
The "scientific method"	Multiple methods, no single scientific method
Scientific method proceeds step by step	Serendipitous discoveries, not a set of sequential steps, not rigid
Experimental research only, laboratory work. Components of experiments: hypothesis-testing, controlling variables, etc.	Various types of scientific research: observational/descriptive research, field studies, correlational research, epidemiological research, experimental research

Source: From Binns 2013.

the biology text strongly emphasizes the traditional view (pp. 247 and 253). Binns concludes that STEM textbooks are starting to show improvement, in that they are beginning to recognize the existing and historical plurality of scientific practice in the broad view (p. 254). But this was a study of only two texts, and only one of the four versions had made such progress.

Science education authority Mansoor Niaz has written a number of papers and a monograph exploring the image of science depicted in chemistry textbooks. His *Chemistry Education and Contributions from History and Philosophy of Science* offers a comprehensive account of the extent to which STEM educators have incorporated history and philosophy of science (HPS) material into their chemistry courses. In Niaz's comprehensive review of fifty-two studies published from 1996 to 2010 in major science education journals (*International Journal of Science Education, Journal of Research in Science Teaching, Science Education,* and *Science & Education*), he found that "most biology, chemistry, physics, and school science textbooks lack an HPS perspective" (Niaz 2016, 32). In one particularly powerful study published in 2011, Niaz and Arelys Maza review the introductory chapter of every university-level general chemistry textbook published in the United States (see Niaz and Maza 2011). They rate the seventy-four textbooks according to whether each one offers a satisfactory treatment, a mere mention, or no mention of each of the following nine post-Kuhnian criteria:

1. Scientific theories are tentative in nature.
2. Laws and theories serve different roles in science (theories do not become laws even with additional evidence).

3. There is no universal step-by-step scientific method.
4. Observations are theory-laden.
5. Scientific knowledge relies heavily, but not entirely, on observation, experimental evidence, rational arguments, creativity, and skepticism.
6. Scientific progress is characterized by competition between rival theories.
7. Scientists can interpret the same experimental data differently.
8. Development of scientific theories at times is based on inconsistent foundations.
9. Scientific ideas are affected by their social and historical milieu. (Niaz and Maza 2011, 4–8)

Eight of the seventy-four textbooks are rated satisfactory on two of the criteria; two textbooks are rated satisfactory on three of the criteria; and one 1978 textbook, Ernest Toon and George Ellis' *Foundations of Chemistry*, is rated satisfactory on five of the criteria (Toon and Ellis 1978). But that was the extent to which the chemistry textbooks satisfactorily incorporated Kuhnian themes from the history and philosophy of science. The most frequently mentioned criteria were 1, 3, and 5, with only approximately 30 percent of the textbooks mentioning those issues (Niaz and Maza 2011, 13). Additional data, about the extent to which the nine criteria were merely mentioned (versus treated satisfactorily), is available in the paper. Niaz and Maza conclude, "Most textbooks in this study provided little insight into the nine criteria used for evaluating the presentation of nature of science" (p. 26).

In another example from the *Critical Analysis of Science Textbooks* volume, Niaz and Bayram Coştu study the textbook presentations of four classic experiments to see if they mention certain well-known Kuhnian complications, that is, certain nonepistemic, personal, and historical accidents that drive the downstream decision-making of the scientists involved (Niaz and Coştu 2013). Almost none of the textbooks mention Kuhnian complications. The authors identify six criteria that a textbook would have to fulfill in order to count as presenting a Kuhn-approved presentation of Millikan's oil-drop experiment. They find that of twenty-seven chemistry textbooks, "none of the textbooks mentioned … one of the most important feature of Millikan's methodology, … that is, in the face of anomalous data, a scientist perseveres with his guiding assumption, holding its falsification in abeyance – in other words, suspension of disbelief" (p. 206). Niaz reports similar results in several other studies (Niaz 2001).

Thus, though SSR has been with us for sixty years, the authors of STEM textbooks have largely ignored Kuhn's call to present a different account of science. Textbook authors are ignoring similar calls from state governments and prominent professional organizations. The 2013 Next Generation Science Standards state that "the only consistent characteristic of scientific knowledge across the disciplines is that scientific knowledge itself is open to revision in light of new evidence" (NGSS 2013, 2). Many position statements from the National Science Teachers Association, the 1990 "Science for All Americans" statement and the 2009 "Benchmarks for Science Literacy" statement, both from the American Association for the Advancement of Science, and the 1996 National Science Educational Standards also argue for the inclusion of history and philosophy of science content in STEM courses. We are not bridging the gap between the sciences and the humanities, as Conant and Sarton had hoped, by incorporating the history and philosophy of science into STEM courses, which has left both STEM and non-STEM majors with little understanding of how science really works. The social stakes of mass ignorance are becoming increasingly high, as the technical expertise required to fully understand the STEM disciplines is pulling away from what it is possible to transmit through a typical K-12 public education. Non-STEM students do not understand science well enough to be responsible voters and policy makers, and STEM students cannot think beyond their lab benches. Something should be done. But scientists are still producing the tourist brochure, and philosophers want to write, primarily, about issues of justification and normativity for one another.

5.4   Moving Forward through a Historical Exemplar

Conant proposed a solution to this problem in the first half of the twentieth century. He developed the Harvard courses Nat Sci 11 and then Nat Sci 4, based on a collection of historical case studies edited by Conant and his colleagues, which were later published for a wider audience (see Conant and Nash 1948/1957). Instead of incorporating humanities and social science content into STEM courses by adding "nature of science" content into STEM textbooks, Conant's solution was to create a separate course for non-STEM undergraduates. He was committed to the belief that scientists solve problems from a "special point of view" and that society would benefit if its future leaders could develop an understanding of the scientist's tactics and strategies (ibid., vii). The average layperson, without the requisite technical and mathematical background, cannot come to an understanding of science by visiting a contemporary lab or talking to a scientist (p. viii). But by reading a historical case study about a series of experiments in the simple, early stages of a particular science, the average layperson can follow the rationale for each decision. The cases cover instances of new conceptual schemes replacing old ones and conflicts between rival conceptual schemes. Like Niaz and other leaders of the present-day "nature of science" movement in science education, Conant hopes that readers will come to appreciate the messy, sometimes contingent, nonlinearity of scientific investigation and change. He writes that

a study of these cases makes it clear that there is no such thing as *the* scientific method, that there is no single type of conceptual scheme and no set of rules specifying how the next advance will be made through the jungle of facts that are presented by the practical arts on the one hand and by the experimentation and observation of scientists on the other. (p. xi)

Many initial scientific hypotheses are the result of inspired guesses, intuitive hunches, and brilliant flashes of imagination (p. xi). And there are times when we cannot test a theory due to some contingent lack of instrumentation or material (p. xiv). Still, the story Conant aims to tell is one in which the scientists of each episode slowly transform speculative concepts into empirically responsible ones by testing the predicted consequences of their working hypotheses. He even ends up offering a definition of science, though it is a spare one. "Almost by definition, I would say, science moves ahead" (p. 37). What distinguishes science from revealed religion is that science is fruitful. Good theories save the phenomena, but they also drive the practice forward. A scientific theory tells its adherents that they are on the right track by leading them toward further questions, by giving them new things to do and find. Dead ends are the sign that you have not hooked onto reality or, at least, that you have not hooked onto something humanly tractable. Conant points out that this feature of science is an argument for the historical, versus theoretical, method of teaching science. Only the historical method can follow the dynamics of scientific change. The theoretical method only mines for reasons.

Conant himself produced half of the eight cases in the two-volume set. The development of each case involved much more than the selection of the original texts. A ten-page introduction to the case teaches readers all they need to know about the science and offers a helpful timeline of related scientific events leading up to the case. The introductions also do a nice job of articulating the puzzle each scientist hopes to solve, drawing readers into a mystery. In one of Conant's own cases, Antoine Lavoisier discovers that sulfur and phosphorus gain weight, rather than lose weight, in burning. He speculates that this might occur in all substances that gain weight in combustion and calcination, including metallic calxes: "*something* was taken up from the atmosphere in combustion and calcination. This was exactly opposite, be it noted, to the phlogiston doctrine. But what was the something?" (p. 73). The prose is simple and clear, written in the style of a newspaper article, rather than in the style of an academic journal article. The editor of the case remains a presence throughout, even after the introduction, as the original texts are interspersed with running commentary. The editor provides the publication context for each of the original texts, delivered to the French Academy of Science or elsewhere. The whole thing reads like an episode of a mid-century radio serial or the television show *Law & Order*. At the end of the case, a timeline puts the episode into the larger sociopolitical context of the time. An annotated bibliography of ten or so sources points ambitious readers toward more material without intimidating them.

The eight cases are works of art. They stand up. I would look forward to assigning them in a 200-level philosophy of science course, and I believe they would appeal, as readings, to contemporary students. The problem is, first, that Conant has left us with only eight, and, second, though I believe the cases would be interesting for anyone to read, I do not believe they would precipitate deep in-class discussion among a typical population of general education students. Conant writes that he was inspired by the way in which military scientists teach the tactics of war through cases and by the case method of Harvard's law and business schools, all of which are well known for keeping students engaged in class (p. 31). But those classes succeed, first, because they address social, ethical, and political subject matter and, second, because engaging with those cases requires, on the student's part, only an easy reaction, rather than a principled objection. In preparing their materials for Nat Sci 4, Conant and his collaborators were thinking like scholars, imagining the kinds of texts that would respond to the problem of science illiteracy. But they might have done more to work out the in-class experience, that is, what might actually succeed in class.

Turning to case studies as a way to expose STEM students to the history and philosophy of science raises several concerns, some articulated persuasively by Comte and others by contemporary classroom educators:

- When instructors use case studies to teach science, they are employing Comte's historical method. Science is presented as a series of events, one following upon another in time, rather than one deriving from another conceptually or in virtue of a rule. But the historical method, with all of its twists, turns, and details – that is, with all of its history – can be overwhelming for typical students. Comte writes: "The general problem of intellectual education consists in enabling an individual of usually but average ability to reach in a few years the same stage of development that has been attained during a long series of ages by the efforts of a large number of superior thinkers, who have throughout their lives concentrated their attention upon the same subject" (Comte 1988, 47). Typical students are in no position to absorb the long sequence of events that has led up to any particular scientific subfield, as it is currently practiced. Comte explains that "it would be quite impossible to attain the desired end if we tried to compel each individual mind to pass successively through the same intermediate stages that the collective genius of mankind has necessarily had to traverse" (p. 48).

- Related, when one teaches through the historical method, one realizes that the sciences that we separate from one another in the theoretical method are often "developed simultaneously and under the mutual influence of one another" in the historical method (Comte 1988, 48). Moreover, we find that sciences often develop alongside an art or in concert with some particular shift in social organization. To embrace the historical method is to require students, unreasonably, to make "a direct study of the general history of humanity" (ibid., p. 49). As an instructor, one could make some strategic decisions about which extra-scientific historical elements to include and which to exclude, but to do this would be to betray the historical method and, once again, present an isolated, purely hypothetical, abstract, and false image to students (ibid.).

- In contemporary classrooms that emphasize active, learner-centered pedagogies, instructors often find that case studies fail to generate sustained in-class discussion. When one teaches the Tuskegee syphilis study in a bioethics course, for example, one can generate a ten- to twenty-minute discussion about the exact nature of the ethical violation. But, after that, the discussion fizzles into the general consensus that the study participants were horribly exploited. Case studies can also shut down discussion through their particularity. Students are hesitant to say anything about eras, geographical regions, or subfields about which they do not know. A targeted case study might engage one or two students who are familiar with the context, but the rest of the class will sit back and let those students take the lead in discussion.

- When case studies do generate in-class discussion, the conversation often neglects the specific philosophical character of the underlying epistemic or ethical violations. Consider the difference between a bioethics course taught in a medical school and a bioethics course taught in an undergraduate philosophy department. In the philosophy department course, students are expected to master the philosophical issues and approaches that cases invoke: whether there is value to be had beyond market value, whether we should redistribute legitimately acquired property to benefit a collective, whether governments should constrain autonomous individuals only to prevent harms or to promote some global notion of the good, and what Anderson, Nussbaum, Held, Rawls, Nozick, Mill, Kant, and Aristotle have to say about all of these things. In the medical school course, students share their intuitive reactions to each case and see if they can come to some policy consensus as a class regarding the appropriate response for a physician. Taught out of its disciplinary context, the medical school course cannot responsibly convey the philosophy that lies beneath.

- The art of case study teaching is in the selection of the cases. To count as instances of the historical method, the cases must reflect the fullness of the historical record, revealing all of the accidents of history, culture, and psychology that are omitted by the theoretical method. At the same time, the historical episodes must be carefully vetted so that each one conveys an engaging, relevant, and unique lesson in the history and philosophy of science: how and why anomalies are ignored; what has been done in the past when two candidate theories, one accurate and one with broad scope, conflict; what factors make it possible for a new paradigm to emerge in a revolutionary crisis; and so on. To meet contemporary standards of inclusive teaching, cases must feature scientists with a range of identities. Finally, cases should be organized chronologically into a narrative that culminates in the grand learning objective of the course. Locating, sourcing, editing, and testing such a collection of cases requires years, possibly even decades, of research on the part of a philosophy PhD with historical and scientific credentials, a history PhD with philosophy and scientific credentials, or a STEM PhD with philosophy and history credentials. Conant and Kuhn had the third magical combination. But such expertise is extremely rare. Conant himself writes that "the greatest hindrance to the widespread use of case histories in teaching science is the lack of suitable case material. Limitations in this regard would virtually require the instructor to choose his topics quite as much because of the availability of the printed material as because of their intrinsic pedagogical value" (Conant 1947/1952, xiv). It is difficult enough to take one or two history of science graduate seminars as a doctoral student in philosophy, and nearly impossible to simultaneously pursue a STEM credential. Even if this triple threat were manageable, one's historical and scientific expertise would still be too narrow to put together an entire course full of cases worth their class time. Conant's two-volume set is a gift to science education. But it would be challenging to produce additional textbooks of the same quality.

- Case studies can be difficult to motivate. The best way to generate class discussion of a case is to ask students to imagine what *should* have transpired. But that runs the risk of invoking some missing, more rational, algorithmic methodology for science that Conant is trying to replace with his looser notion of scientific practice. In other words, if cases are to be presented as genuine instances of successful scientific practice, they should be introduced as calcified facts of history, without any counterfactual speculation as to what should have transpired or how the contingencies of the case will affect future experiments and episodes. History does not progress according to rules or guiderails that enable us to justify beliefs about the future. Kuhnian history, particularly in moments of crisis, can be one damn thing after another. But without the normative layer, it can be difficult to convey to students why they should care. What is the point of mastering some esoteric, stand-alone case? Only to note that this is sometimes how science proceeds? Your students might have accepted the case as a simple footnote to any single day's lecture in a more typical STEM course.

Both Conant and Kuhn champion Comte's historical method of presenting science by producing texts that trace the dynamics of scientific change. Yet both former scientists allow for the scientists themselves to keep the STEM textbooks that follow Comte's ahistorical, theoretical method. Their Nat Sci 4 course was created for *non*scientists. I agree with Conant and Kuhn, to the extent that different disciplines, different courses, and different student populations have different learning objectives. As Comte writes, the "historical study of the sciences should be looked upon as entirely separate from its proper and theoretical study" (Comte 1988, 49). STEM instructors want their students to learn how to justify scientific claims, so the theoretical method presents scientific theories as the conclusions of arguments, rather than as effects of causes. STEM instructors are presenting an ideal version of science to which their students might aspire. As Vasso Kindi argues, textbook science "is conducive to forming the scientists' course of action" (2005a, 730). Travel brochures are useful for potential tourists hoping to plan an efficient week-long trip through the highlights. Such brochures can even make the trip possible for new visitors. History teachers, on the other hand, want their students to learn how to craft original narratives, while developing an opposing appreciation for the openness of contingency, so scientific theories are presented as effects, rather than conclusions. They want to model for their students the way in which an accomplished historian sifts through artifacts for relevance and a storyline. It is perfectly reasonable for an educational institution to offer more than one form of training. We need not abandon the internalist or the externalist history of science, the theoretical or the historical method, for both make possible a valuable form of teaching.[4] The theoretical method produces an image of science that is no less true than the historical one. Both methods are practiced by human beings who cannot echolocate or see very small things, so neither method delivers the textbook author to some metaphysical truth beyond appearances. Each method lives within an entrenched discipline and, thus, recognizes some constraints. But I am not sure that the best way to bridge the gap between the STEM disciplines, on the one hand, and the humanities and social sciences, on the other, is to drop nine post-Kuhinan information boxes into the pages of every STEM textbook (although I would welcome that baby step). Instead, we should ask both STEM and non-STEM students to take courses on both sides of the gap, leaving the instruction of each presentation mode up to the appropriate expert. STEM students should study the history and philosophy of science, but not as those humanities would be taught as a sidebar by an immunology professor. They should take required history and philosophy of science courses with doctoral-level scholars in the history and philosophy of science. Meanwhile, non-STEM students should take courses like Nat Sci 4 with materials produced by professional scientists, in order to become informed citizens in this technological age. In short, changes in science education should be made at the curricular or degree program level, not at the course level.

5.5   A New Paradigm

Conant envisions the teaching of science as succeeding through some central *book* written for students. But there is a movement afoot to abandon the textbook altogether as a teaching tool, or at least radically modify it. In 2006, fifty-four award-winning STEM educators, technology professionals, corporate publishers, program officers, and professional organization administrators gathered for a three-day workshop, "Reconsidering the Textbook," at the National Academy of Sciences in

---

[4] This is not Steve Fuller's doctrine of "historiographical segregationism," the view that there should be "a double truth ..., one for the elites and one for the rabble" – one responsible history for historians and philosophers, and one tidy history for scientists, who will need clarity and trust in some authority to conduct normal science (see Fuller 2000b, 27).

Washington, DC, funded by the National Science Foundation's Division of Undergraduate Education (award number 0549185). The group convened to discuss whether 1,200-page introductory STEM textbooks and their "encyclopedic style of presentation" were appropriate teaching tools, given the latest educational research (Bierman 2006, 1). Textbooks are peer-reviewed and reflect their field's consensus, so they have a great deal of authority, perhaps the most authority, outside a peer-reviewed research paper in a prestigious journal. But they are not easily revised to reflect the latest science; they are expensive; they are pitched to only one level of understanding (introductory, advanced, etc.); and it is not clear that they can best support the active-learning, inquiry-based, or just-in-time pedagogies that support learner-centered teaching.[5] The group recommends that future classroom materials be more clearly organized for the introductory student, searchable, flexible, and, thus, digital:

The textbook of the future will be more than a static printed volume. It will function as a guide, interweaving and coordinating a variety of different learning resources including animations, simulations, and interactive exercises. Such a package of resources will be easily searchable and thus learner-accessible with a flexible electronic interface. The textbook, whether printed or electronic, will be the organizing hub of an integrated learning environment where the student experience is key. The goal here is to retain the core stability and authority that make the textbook so valuable while at the same time providing the flexibility, timeliness, and inquiry-focused approach that the web and other electronic resources provide. (Bierman 2006, 2)

The group imagines an "adaptable, customizable, and responsive" resource that can be modified by the instructor to fit the goals of each particular course, but easily expanded on the spot to direct students to deeper levels of content (Bierman 2006, 3). Intelligent software should respond interactively to feed developmental exercises to students who are struggling and advanced content to students who have mastered the required material. Course materials should be offered in a number of different formats, in order to satisfy the needs of universal design for learning. The digital format will support the use of animations, to engage students in the study of complex, time-dependent phenomena; games, which students can play against one another; linkages to original documents, internship opportunities, and career options; place-based learning environments for on-location fields such as geology, biology, and

---

[5] This last objection to textbook-centered pedagogies can be traced back to John Dewey's call for interactive, experiential forms of learning in *Democracy and Education: An Introduction to the Philosophy of Education* (New York: Macmillan, 1916).

engineering; and ongoing student feedback, to inform instructors and authors in real time of students' needs (Bierman 2006, 4–5). Unlike traditional textbooks, which are typically written by one to three authors, each one of whom can only cover one or two subfields of content, a collection of narrower digital resources could capitalize on expertise, by asking singular authors to create smaller, targeted resources for their individual subfields and by making it possible for these authors to comment on and critique one another's materials. Open-source course materials will be less expensive for students, and cloud-based course materials will never go out of print.

Questions certainly remain about this model. Will professors create all of this targeted material without compensation just to benefit their field and its students? Through what central portal will instructors access and navigate all of these resources? Who will curate, maintain, and update these materials? Or will students and teachers have to rummage through thousands of dead webpages, looking for relevant, up-to-date resources? More interestingly, if the character of science education influences the practice of science, how will that practice change with new educational materials? It could be that we are unwilling to trade the authoritative centrality of the traditional textbook publishing system for the engaging flexibility of a more diverse, modular, digital system. But there might be advantages in returning to a state in which doctoral-level scholars vet their own teaching materials, instead of assigning whatever some small handful of corporate publishers has on offer or whatever their institution requires. The doctoral credential is a meaningful one that connotes mastery of a discipline. Professors know their subfields; they know what they want their courses to accomplish; and they know their local student populations. It would take more time, effort, and creativity to select from among a sea of digital course materials, instead of simply ordering the 1,200-page introductory textbook suggested by a department, dissertation advisor, or previous instructor. But faculty would be realizing the point of their terminal degrees: they would be using their hard-earned judgment to determine what the next generation needs to know, given the current state and future trajectory of science and society.

Despite the fact that most STEM faculty still teach with the books of "textbook science," albeit now in digital versions, there are pockets of science educators working to usher in a new era. In the fall of 2020, the National Science Foundation funded a project through its Division of Undergraduate Education to provide STEM instructors with some inspiration. The "Workshop on the Substance of STEM Education" (Stem Futures) – or STEM Futures project – brought together 105 science educators from fifty-three institutions and twenty-nine states and assembled them into twenty-five teams. Each team was asked to watch three preparatory webinars and develop a curricular innovation that met three associated desiderata. The curricular projects had to: (1) teach learners to view their STEM work within a broader human context, that is, through a social, historical, or ethical lens; (2) incorporate the latest pedagogical research in the meta-cognition of learning, that is, emphasize active-, student-centered learning, collaboration, and creativity; and (3) retain the traditional goal of teaching students the foundational principles of the STEM disciplines. Requirement number 1 asks STEM educators to look beyond textbook science to incorporate the history and philosophy of science into their teaching. Teams had four days to prepare a presentation that included a rationale for their curricular innovation and two to three sample assignments. Interested readers can find descriptions of the many curricular products at https://stemfutures.education.asu.edu/curricular-products/. The hope is that by creating a permanent web resource of curricular innovations, any interested STEM educator can browse the website for free and locate a lesson, activity, syllabus module, course, or degree credential that might fit the local needs of their student population, department, or institution. Because there are so many innovations posted to the site, the hope is that viewers will be inspired by the variety to come up with their own ideas as well.

I believe Conant and Kuhn would be the first to see that we are living through a paradigm shift in undergraduate education and in STEM education, more specifically. In the old paradigm, professors would walk into class, lecture on their scholarly interests, and test their students to identify high performers for graduate schools and corporations. The goal of this form of education is credentialing, and professors teach, in the end, to service graduate schools and employers. The questions professors ask themselves are: "How can I clearly project and disseminate my expertise? How can I fit my teaching responsibilities into my research interests? How can I identify and champion the very best students?" One answer to the first question is the 1,200-page introductory textbook.

In the new paradigm, professors must begin by learning about their student population, what the students can and cannot already do, and what they need. We meet our students where they are, with the aim of moving every individual one step forward, whatever that means for each individual student. We look for course materials that will engage them, and we try not to leave anyone behind, by responding, where we can, to existing inequities and injustices, both interpersonal and structural. The goal of this form of education is student learning. The questions we ask ourselves are: "How are students with full-time jobs, from historically underrepresented populations, with disabilities, without quantitative skills, or other issues going to succeed in my course? How can I marshal my expertise to respond to this local diversity of needs? How can I keep everyone engaged? How can I use my skills and privilege to prepare my students to contribute to a functional society?" These are new questions, and a 1,200-page introductory textbook that presents science as a series of conclusions, instead of a series of historical effects, cannot answer any of them. We recognize that we are in a prerevolutionary crisis, because professors are now being fired for excelling according to the old standards – for following the old rules – and colleagues of theirs from the same generation cannot understand this (see Saul 2022, 1). The older generation sees injustice, where younger teachers see someone who does not understand the point of the game.

In the coming years, we are all going to have to reconstruct our teaching practices and course materials from new fundamentals (SSR-2 85). At present, during this time of crisis, there is no normal science – no vademecum – to guide us toward alternative answers. But STEM educators are starting to experiment with more engaging course materials and more inclusive pedagogies, such as problem-based learning, group problem-solving in lecture, interactive computer learning, ungrading, mastery learning, group tests, scaffolded assessments, and inquiry-based labs, among many others (see Handelsman et al. 2004). As Kuhn writes in SSR, "an occasion for retooling has arrived" (SSR-2 76).

6   Thomas Kuhn, Normal Science, and Education

Hasok Chang

6.1   Introduction

Understanding how scientists were trained was a central part of Thomas Kuhn's study of the nature of science, and Kuhn's views have been very influential in the field of science education, as documented by Michael Matthews (2000, 2022). Yet the nature of science education does not seem to command sufficient attention from the philosophers of science who engage with Kuhn's ideas, with a few notable exceptions such as Hanne Andersen (2000, 2001b). An anecdotal indication of this neglect is the fact that there is no index entry on "education" in a number of well-known monographs and edited collections on Kuhn.[1] This is, of course, not to say that these texts do not discuss education at all, but it is a clear indication that education has not been a standard item of concern among philosophers who engage with Kuhn's views of science. This is probably also the case among historians and sociologists of science who have used or criticized Kuhn's ideas, though there are notable exceptions such as Andrew Warwick (2003). Even Kuhn himself did not expand greatly on the nature of science education in his later works, in which concerns about education took a back seat as he focused on producing detailed historical accounts or refining and developing the philosophical ideas that had only been presented in suggestive forms in *The Structure of Scientific Revolutions* (SSR).

This was not always so. Even though the general furor surrounding Kuhn's views on science focused heavily on his view of revolutions and incommensurability and the implied threats of relativism, in the early stages much of the critical concern was on normal science, and the education of scientists was at the core of this concern, as recounted in the insightful account by John Worrall (2003). The critics who focused

---

[1] These include Hoyningen-Huene (1989/1993), Horwich (1993), Bird (2000a), Fuller (2000b), Nickles (2003a), Kindi and Arabatzis (2012), Richards and Daston (2016), and Wray (2021a, 2021b).

on normal science worried that Kuhn at least implicitly endorsed the dogmatic state of science education that he described. This was apparent in the famous 1965 conference in London, the papers arising from which were collected into the volume *Criticism and the Growth of Knowledge*, edited by Imre Lakatos and Alan Musgrave (1970). John Watkins, who was given the role of chief respondent to the opening paper that Kuhn gave at this conference, emphasized his objection to Kuhn's "view of the scientific community as an essentially closed society, intermittently shaken by collective nervous breakdowns followed by restored mental unison" (Watkins 1970, 26). He also gave a structured argument as to why he believed such a closed society would not be able to produce substantially new thoughts that would lead to the emergence of a new paradigm (ibid., 34–37). In a more descriptive vein than Watkins, Paul Feyerabend (1970, 207–208) opined that Kuhn had exaggerated the extent to which a paradigm typically achieves the kind of total dominance of a field, and the length of time for which any such dominance is maintained. In contrast to Watkins' and Feyerabend's view that science could not actually be as Kuhn describes it, Popper worried that something like Kuhnian normal science would actually become normalized. And it was in Popper's critique that the role of education was brought out explicitly:

"Normal" science, in Kuhn's sense, exists. It is the activity of the non-revolutionary, or more precisely, the not-too-critical professional: of the science student who accepts the ruling dogma of the day.... In my view the "normal" scientist, as Kuhn describes him, is a person one ought to be sorry for.... The "normal" scientist, in my view, has been taught badly.... He has been taught in a dogmatic spirit: he is a victim of indoctrination.... I admit that this kind of attitude exists.... I can only say that I see a very great danger in it and in the possibility of its becoming normal ... : a danger to science and, indeed, to our civilization. And this shows why I regard Kuhn's emphasis on the existence of this kind of science as so important. (Popper 1970, 52–53)

In this chapter I wish to address the two concerns expressed by Popper, and consider what can be done in science education in order to address them. In the course of this discussion I will take in and build on Kuhn's thoughts in relations to these issues. As Steve Fuller (2003) has emphasized, Popper's first concern (also expressed by Watkins) was that the closed nature of normal science would prevent the kind of innovative thinking that was needed in the creation of a new paradigm. This was a worry fully anticipated by Kuhn, though his answer may not have been entirely satisfactory. Popper's second concern, shared by Feyerabend, was that society needs to maintain the freedom of critical thinking, and science ought to lead the way in this regard. Kuhn really did not have much to say about this concern, and we need to go beyond him, while paying full respect to his hard-earned insights about the requirements of scientific training. I want to search for ways of ameliorating dogmatism in science education while respecting the necessities of professional training. This discussion will also have relevance to the place of all expert communities in democratic societies.

6.2   What Did Kuhn Say about Science Education?

Before trying to develop my own ideas, I will start by briefly reviewing what Kuhn did say about normal science and science education. It does seem obvious that the picture of normal science that Kuhn portrayed was a highly regimented and closed one. In his opening paper for the 1965 conference in London, Kuhn highlighted the dogmatism of normal science in conscious opposition to Popper's views:

> it is normal science, in which Sir Karl's sort of testing does not occur, rather than extraordinary science which most nearly distinguishes science from other enterprises. If a demarcation criterion exists (we must not, I think, seek a sharp or decisive one), it may lie just in that part of science which Sir Karl ignores.... In a sense, to turn Sir Karl's view on its head, it is precisely the abandonment of critical discourse that marks the transition to a science. (Kuhn 1970a, 6)

This must have only served to amplify the worries of his critics arising from their reading of the first edition of SSR (1962).

What was Kuhn's view on the role of science *education* in the maintenance of normal science? It is fair enough to say that Kuhn regarded science education as a process of indoctrination, even though he did not use that word. At the heart of science as portrayed by Kuhn, there is a regimented process of training that equips scientists for the rigorous and disciplined practice of normal science. This training process not only teaches the future scientists a set of specific beliefs and skills belonging to the ruling paradigm in the field in question but imparts an uncritical attitude. He did make this view explicit in SSR, even though he did not give an extensive discourse on education there. For example, he emphasized the process of "learning by finger exercise," the puzzle-solving training in handling paradigmatic problems, noting that such training continued "throughout the process of professional initiation" from freshman courses to doctoral research (SSR-2 47).

To my knowledge, the clearest statement Kuhn made about the nature of scientific training was in his paper "The Essential Tension," originally published in 1959 and reprinted in 1977 (ET 225–239) in the well-known collection whose title he took from this paper (ET). The paper was originally presented at a conference "on the identification of scientific talent" at the University of Utah, and Kuhn began by cautioning against an overemphasis on "divergent thinking" in basic science. Kuhn wondered "whether flexibility and open-mindedness have not been too exclusively emphasized as the characteristics requisite for basic research" (ET 226).

In that paper, given three years before the publication of SSR, the concept of normal science is already clearly present: "normal research, even the best of it, is a highly convergent activity based firmly upon a settled consensus acquired from scientific education and reinforced by subsequent life in the profession" (ET 227). Kuhn continues by noting that the "single most striking feature" of education in the natural sciences is that "it is conducted entirely through textbooks" and notes that this sets the natural sciences apart from other creative fields (ET 228). How he describes the nature of this education is worth quoting at length, which I will do step by step:

Typically, undergraduate and graduate students of chemistry, physics, astronomy, geology, or biology acquire the substance of their fields from books written especially for students. Until they are ready, or very nearly ready, to commence work on their own dissertations, they are neither asked to attempt trial research projects nor exposed to the immediate products of research done by others, that is, to the professional communications that scientists write for each other. (ET 228)

This education is not learning by doing, but learning through "paradigms" in the sense of exemplars: "science textbooks do not describe the sorts of problems that the professional may be asked to solve and the variety of techniques available for their solution. Rather, these books exhibit concrete problem solutions that the profession has come to accept as paradigms" (ET 229). The situation may be shifting somewhat with undergraduate research now in vogue, but it would have been a largely accurate description of the state of affairs in the mid-twentieth century. In a paper consciously inspired by Popper's critique of Kuhnian normal science, Berry van Berkel et al. (2000) criticize standard chemistry education in schools as normal science education, which is out of touch with chemical research; however, in Kuhn's view this disconnection from current research was a paradoxically essential feature of normal science education.

Kuhn also emphasizes the monism that he regarded as inherent in science education, and interestingly introduces this point by pointing to the exclusion of history:

There are no collections of "readings" in the natural sciences. Nor are science students encouraged to read the historical classics of their fields – works in which they might discover other ways of regarding the problems discussed in their textbooks, but in which they would also meet problems, concepts, and standards of solution that their future professions have long since discarded and replaced. (ET 229)

So the history of science, except mythical tales that bear little connection to how actual developments unfolded, is carefully kept out of textbooks, lest it should distract the students with alternative perspectives and problems. Most of all, the impression must be avoided that there are multiple competing approaches that are all legitimate within a given scientific field. When it is undeniable that there are multiple approaches, they are presented as treatments of "different subject matters, rather than … exemplifying different approaches to a single problem field" (ET 229). This is consonant with Kuhn's later observation that science often develops by a process akin to biological speciation, through which fields split into subfields that do not share the same paradigm, as in the split between physical chemistry and organic chemistry, or even the physics of atoms and the chemistry of atoms.

Such a process of education is bound to produce a particular type of intellect. Kuhn admits that the education he describes is "a dogmatic initiation in a pre-established tradition that the student is not equipped to evaluate" and anticipates that "even the most faintly liberal educational theory must view this pedagogic technique as anathema" (ET 229). So when Popper declared that the normal scientist "has been taught in a dogmatic spirit," he was not telling Kuhn anything he had not realized. And Kuhn said it himself, in print, at least six years before Popper uttered his condemnation. But Kuhn admitted the dogmatism of normal science education so freely only because he thought that, somehow, it did not retard scientific innovation. I will consider Kuhn's arguments concerning this point in more detail later.

6.3   Why Separatism Would Not Work

As I enter into my own thoughts about science education in the light of what Kuhn has taught all of us, one thing to recognize at the outset is that for Kuhn science education seems to have been synonymous with the training of scientists. But in nearly all modern societies a diverse array of people receive science education. In fact, only a tiny proportion of students who are taught science at school go on to become professional scientists, and this is the case even for education at the university level. So science education is plainly not synonymous with the training of professional scientists. There are different purposes that science education can and should serve, and they need to be considered separately. It is understandable that Kuhn focused on the training of professional scientists, since his main concern as a historian and philosopher of science was to understand what made scientists how they were. However, his silence on other aspects of science education is frustrating.

This situation is also very ironic, given that Kuhn's initial work in the history of science was fostered in the context of the teaching of science through history in James B. Conant's program of general education at Harvard. Kuhn's relationship to Conant, his work in Conant's teaching program, and the sociopolitical contexts and drivers of Conant's work have been extensively studied and debated (see Reisch 2019, 2021a; Fuller 2000a, 2000b; Wray 2021a, chapter 2). The immediately pertinent point here is that Conant's program was expressly designed for students who were not destined to become professional scientists but "graduates who are to be lawyers, writers, teachers, politicians, public servants, and businessmen" (Conant quoted in Wray 2021a, 32). Conant's view concerning the teaching of science as part of general education was twofold: what we need to teach the public about science is not so much its specific content, but the nature of scientific knowledge and the scientific process of investigation; and the best way to impart this lesson is teaching the historical development of science. Both aspects of Conant's views were shared by some other pioneers of history of science such as George Sarton, but they were quite contrary to Kuhn's view of normal science education. Wray (2021a, 33–43) has given a convincing argument about just how much of Conant's view of the nature of science Kuhn had absorbed, which makes it even more striking that on the subject of science education Kuhn does not seem to have taken much from Conant at all.

Of course, it could be that Kuhn simply considered Conant-style general education of science to be an entirely different enterprise from the training of scientists. In that case, what we need to recognize is that it would be unwise, even illegitimate, to apply Kuhnian insights about the training of normal scientists to the question of what students who are not scientific apprentices should learn about science, and how. The training of scientists rightly focuses on the teaching of the fundamental ways of thinking and methods of problem-solving that are central to the current ruling paradigm in their intended field of specialization. In contrast, science education for the nonscientists needs to go beyond the imparting of factual and theoretical knowledge and problem-solving skills, to pay attention to the nature of scientific knowledge and practice and to the cultural value and valency of science. These are points so well rehearsed in general debates in science education that I would not presume to add anything new to them.

However, the picture is not so simple, for there is a niggling question: Are professional scientific training and general science education really so separate? Even among the university students who get degrees in the natural sciences, probably the majority do not become professional research scientists. So the fact is that a large number of students who do not become scientists are still receiving the kind of science education designed for those who will become scientists. Such education often tends to generate a fear and dislike of science, much more than it imparts any lasting scientific knowledge or the kind of appreciation of the nature of science that Conant and others had hoped for. So there might be an argument for an educational reform designed to separate scientific and nonscientific students earlier on than we currently do, and giving each group an appropriate kind of science education. Or perhaps those on the professional track should receive a double dose of education, since one would hope that the professional scientists also receive the broad cultural education about science that Conant-style general education seeks to impart.

There are, however, two difficulties with such a separatist proposal. The first is what I will call the problem of "Sagan undergraduates," in honor of the great American science writer and documentarian of the late twentieth century, Carl Sagan. I confess to having been a Sagan undergraduate myself, by which I mean a young student of science who is inspired to enter into science by the kind of noble image of science that the likes of Sagan promoted: an exhilarating enterprise of learning and research, which is extremely useful for both its practical applications and the deep sense of understanding that it generates; this scientific enterprise is also imbued with humanist values of progress, objectivity, curiosity, open-mindedness, and critical thinking. It is a common phenomenon that Sagan undergraduates become discouraged and disillusioned by the reality of normal science education that confronts them by the time they reach upper-level undergraduate courses. The intellectual values that drew them into science are not nurtured in "Kuhnian" scientific training; unless they *also* happen to be good at normal-scientific puzzle-solving *and* relish it as their daily life, they are likely to leave science. The problem is that there is a mismatch between what should motivate people to want to learn science, and what enables them to succeed at it.

From a ruthless sort of social-engineering perspective, one could consider Sagan undergraduates as misfits that we should avoid producing. We could fix the mismatch between motivation and ability by determining who goes into science strictly by normal-scientific criteria. That is, we could turn early science education into a baby version of Kuhnian science education and only allow those who perform well in it to go into science at higher levels. This way, those who are inspired by the ideals of science but not very good at normal science will not be tempted to go into science professionally, but instead be given access to general science education and learn something of cultural value about science. But now the second difficulty with the separatist proposal rears its head: Will a scientific community consisting only of those who do well in normal science education be the most effective one? In other words, the separatist proposal makes sense only if normal science education as Kuhn described it is indeed the best way to train scientists. And that was precisely the assumption that some of Kuhn's critics were questioning, as mentioned above.

6.4   Normal Science, Innovation, and Pluralism

So now I return to the question concerning the training of scientists designed to equip them for their narrowly conceived scientific work. Here we face the question of innovation, which Popper and Watkins pointedly raised: How would dogmatically educated people be capable of coming up with truly new ideas? As briefly mentioned already, this was a worry that Kuhn had fully anticipated in "The Essential Tension," even before SSR was published. If Popper and others thought that Kuhn was blind to this concern, that is first of all an indication that they were not very familiar with the whole body of his work. ("The Essential Tension" was available only in a little-known conference proceeding volume until it was reprinted in the 1977 collection *The Essential Tension: Selected Studies in Scientific Tradition and Change* [ET].) On the other hand, it is also an indication that in SSR Kuhn had not said enough to address the concern of which he was quite well aware.

Kuhn's response to the worry begins with the observation that normal science education is actually quite compatible with scientific innovation: "this technique of exclusive exposure to a rigid tradition has been immensely productive of the most consequential sort of innovations" (ET 229–230). More than that, he implies that scientific innovation is *caused* by normal scientific education: "the various fields of natural science have not always been characterized by rigid education in exclusive paradigms, but ... each of them acquired something like that technique at precisely the point when the field began to make rapid and systematic progress" (ET 230). And he is talking about not just the kind of orderly cumulative progress that is to be expected in normal science but even the kind of innovations involved in scientific revolutions. As he put it evocatively in SSR: "research under a paradigm must be a particularly effective way of inducing paradigm change" (SSR-2 52). And here is the basic explanation of this curious historical phenomenon that he offered:

At least for the scientific community as a whole, work within a well-defined and deeply ingrained tradition seems more productive of tradition-shattering novelties than work in which no similarly convergent standards are involved. How can this be so? I think it is because no other sort of work is nearly so well suited to isolate for continuing and concentrated attention those loci of trouble or causes of crisis upon whose recognition the most fundamental advances in basic science depend. (ET 234)

Or as he put it later in SSR: "Anomaly appears only against the background provided by the paradigm" (SSR-2 65).

Once an anomaly is identified and not readily resolved by the standard resources and techniques familiar in the ruling paradigm, Kuhn says that the normal scientists will start thinking creatively, out of desperation, as people who hit dead-ends are forced to think laterally. We now enter into the realm of extraordinary science: "The proliferation of competing articulations, the willingness to try anything, the expression of explicit discontent, the recourse to philosophy and to debate over fundamentals, all these are symptoms of a transition from normal to extraordinary research" (SSR-2 91). There must be a good deal of truth to this idea. However, it is not clear how this lateral thinking actually happens, and there is also no clear answer in Kuhn's writings as to how some people are capable of making extraordinary departures from the current paradigm in a crisis situation, and some people are not. Kuhn (SSR-2 90) gave two clues, noting that the creators of new paradigms are often "very young, or very new to the field whose paradigm they change." These are plausible insights,[2] but they are not sufficient.

I believe that Kuhn has taken us to an important problem, without providing a convincing solution. We can accept that innovation must be possible, because it does happen; figuring out the "how" is another issue. I agree with the assessment he gives of the situation: "We are, I think, more likely to exploit our potential scientific talent if we recognize the extent to which the basic scientist must also be a firm traditionalist, or ... a convergent thinker," as well as an innovator and a divergent thinker. I also agree that the following is an urgent problem to solve: "Most important of all, we must seek to understand how these two superficially

[2] However, Brad Wray (2003) has empirically challenged Kuhn's claim about the youth of innovators.

discordant modes of problem-solving can be reconciled both within the individual and within the group" (ET 237).

Before I make my attempt to solve this problem, I must note that behind Kuhn's formulation of the problem is the premise that normal science as he describes it really is the best way to promote the kinds of innovation that are desired in science. This premise can plausibly be challenged, as Feyerabend (1975) has done emphatically. While Kuhn has demonstrated well enough that the pursuit of normal science has been a rather effective way of generating innovation, he has not shown that it is more effective than all other alternatives. I do think that there are actually conceivable methods of scientific research and education that would be more effective. Here history is not an unambiguous guide. We can agree with Kuhn that "no part of science has progressed very far or very rapidly before this convergent education and correspondingly convergent normal practice became possible" (ET 237). However, the conditions that governed scientific research and its limitations at the origins of our current scientific disciplines may be different from the conditions under which they currently exist, not to mention those that they will face going into the future. An interesting contrast to Kuhn's historical view is provided by Joseph J. Schwab, who published *The Teaching of Science as Enquiry* in 1962, the same year as the first edition of Kuhn's SSR. Harvey Siegel (1990, 99–102), among others, has noted a close parallel between Kuhn's distinction of normal/extraordinary science and Schwab's distinction of stable/fluid inquiry. But Schwab's view was that as science develops more and more research is devoted to fluid inquiry, so it becomes more and more necessary to train scientists for fluid inquiry – in other words, to equip them for critical thinking.

I would like to propose that science can be organized in a less dogmatic way without causing scientists to lose the kind of focus that is characteristic of research within a paradigm. In short, the answer is pluralism, and pluralism need not be flabbiness. To put my point in Kuhn's idiom: a scientific field ought to be able to maintain more than one paradigm simultaneously, allowing scientists within each paradigm to carry out focused and disciplined research, yet allowing the whole field to reap the benefits of diversity that I have called "benefits of toleration" in my advocacy of pluralism in science (Chang 2012a, chapter 5). I believe that the maintenance of multiple paradigms is possible not just briefly during periods of extraordinary science, but during normal periods as well. This is one place where the changing historical circumstances may matter greatly. I can understand that in the early days of science there may not have been sufficient material and human resources to support more than one tradition. In our current situation there is a great deal of investment in science, and we can afford to maintain multiple subcommunities within a field, or at least some maverick individuals, each pursuing a distinct and rigorous (even rigid) program of research.

I think Kuhn had an unduly pessimistic view of human psychology, in maintaining that only those people who have confidence in the rightness of their paradigm would tackle difficult esoteric puzzles:

> Problems of this sort are undertaken only by men assured that there is a solution which ingenuity can disclose, and only current theory could possibly provide assurance of that sort…. Who, for example, would have developed the elaborate mathematical techniques required for the study of the effects of interplanetary attractions upon basic Keplerian orbits if he had not assumed that Newtonian dynamics … would explain the last details of astronomical observation? (ET 235)

Observing the habits of affluent humans in the twenty-first century might have changed Kuhn's mind on this point. People put in enormous concentrated efforts into difficult technical puzzles without any faith that the system of thought or practice within which they work represents some unique and ultimate truth. Consider all the specialist technical effort that goes into designing information-technology systems, or even online games. In the realm of science, too, people nowadays routinely engage in very difficult formal modeling without much realist faith in the rightness of their approach, or any real assurance that their approach will yield good results for sure. And believing one system to be capable of providing solutions to some difficult problems does not mean believing that it is the only one that can solve problems. It certainly does not preclude one from learning to work with multiple problem-solving systems. Kuhn himself said the same about scientists who learn competing paradigms and make a choice between them during periods of extraordinary science. Strictly obeying the rules of chess while playing chess does not mean denying that there are other games played by other rules, or committing that one will never play any other game.

6.5   Pluralist Scientific Training

Having challenged Kuhn's notion that monism (or "paradigm-monopoly") is a required accompaniment to effective scientific work, I can return to the question of education in a different way. If scientific research is to be organized in a pluralistic way, how should scientists be trained? Could each individual scientist be educated in a pluralist way? I believe so. My view is that science students should be capable of learning multiple systems at an introductory level, even if they can only afford to be trained to an advanced level in one system. Training can properly include exposure to multiple ways of knowing, and such pluralistic exposure can be carried up to a fairly high level.

What I am advocating here may sound like a pipe dream when it is put in such an abstract way, but it is in fact something that happens routinely in the typical education of scientists. And paradoxically, it is physics, the science that formed Kuhn's thinking most strongly, that most clearly exhibits the plurality of education. Any well-educated physicist today will have learned classical mechanics in three different formulations, special and general relativity, classical thermodynamics and statistical mechanics, wave and matrix versions of quantum mechanics, and some more advanced version of quantum theory including quantum electro-dynamics and quantum field theory. These are all very different ways of understanding the physical world, and they provide very different modes of problem-solving. And each physicist going through basic training up to graduate school is forced to learn *all* of these modes of thinking, not just some of them. That is to say: narrow and restricted as the mainstream education of physicists may appear, it actually embodies a good deal more plurality and diversity than meets the eye. I believe that a look into the state of education in other sciences would also reveal the same kind of pattern.

This observation should help us answer the descriptive puzzle concerning the innovativeness of science carried out by those trained through normal-science education. Scientific training as it stands is not as monistic as it might appear, dogmatic as its tenor may be. In an important sense, Watkins was right: the demonstrated innovativeness of science means that it cannot be as regimented as Kuhn suggested. Having learned to think in such a diverse variety of ways and having learned to solve problems with such a diverse range of tools, it is no wonder that scientists can be versatile and adaptable when they are searching for ways to tackle a recalcitrant anomaly. The same point can be made about the broad range of material, instrumental, and computational tools, as well as conceptual ones, at the disposal of a well-trained scientist. If one adds to the mix the other resources that a scientist would have from a broad mathematical education or some training in other fields of science, then the innovative capabilities of a normal scientist begin to look less and less mysterious.

One interesting thing about this plurality in actual science education is how hidden it is – not just in Kuhn's description of normal science but in the scientists' own perceptions. I believe that this is because the plurality that is allowed is very carefully controlled. A great deal of pedagogical effort goes into assuring students that there are no contradictions between the different theories that they are taught. Quasi-philosophical notions of equivalence and reduction play a key role here: physics students are given hand-waving explanations on how the Heisenberg and Schrödinger formulations of quantum mechanics say really the same thing; after the conceptual shock of first learning special relativity, they are told to calm down because the formulas of Newtonian mechanics are still good in the low-velocity limit despite the glaring conceptual incompatibilities between the two theories. So the students can go on believing that there is just one harmonious body of theories.

Pondering about this situation prompts three thoughts on how the benefits of plurality in education may be enhanced. First, taking the different theories and frameworks that they already learn, students can be encouraged to notice and wrestle with the interesting differences (and, yes, incommensurability) between them. Students can be encouraged to marvel at the fact that the same phenomena can be understood from such different perspectives as macroscopic thermodynamics and statistical mechanics, or wave and particle optics, or Hamiltonian and Lagrangian mechanics. They can also learn to treat the different theories or formulations as different tools for problem-solving, and learn to recognize that there is a choice they can make concerning the choice of tools they employ in tackling a given problem, and a possibility they can explore in combining the use of different tools.

The second thought is more challenging, but certainly not impossible: it is to extend the pluralism beyond the boundary of what is currently considered as standard. How would science education be if it consciously introduced students to certain cogent viewpoints that are currently dismissed by the majority of mainstream scientists without convincing reasons? For example, most physicists would admit that Bohmian quantum mechanics is not clearly wrong or nonsensical, yet they do not entertain it as a serious possibility. As James Cushing (1994) has argued, the neglect of Bohmian mechanics has owed a great deal to historical contingency. Léna Soler (2022) imagines a physics education that would teach students Bohmian mechanics as well as the standard quantum mechanics as formulated by von Neumann and Dirac. What would such education, rigorous on the side of each theory, do for the imaginative capacities of well-trained physicists? They could also benefit from learning Hertzian classical mechanics, Hamiltonian general relativity, or other alternatives to mainstream theories. One might question if any more content can plausibly be fitted into the already overflowing science curricula, but in order to reap the benefits of exposure to plurality one does not have to go very deep into the alternative theories. Sometimes even the awareness that alternatives exist is sufficient to open one's eyes.

Finally, the most daring layer of pluralist education would be to reach into frameworks of thought that in some ways contradict what is taken as accepted knowledge or orthodox beliefs in current science. This would open the students' minds in a maximal way, but the risk is that of opening the door to unsubstantiated ways of thinking. This is where historical work can be helpful, for precisely the reasons for which Kuhn says science textbooks keep real history out. A careful look at the history of science reveals that even the thinking of those who are recognized as great scientists of the past usually had elements that are incompatible with our current thinking. Going a little bit further, one can identify past systems of thought that are now considered quite wrong, but were respected for a time and for good reasons. One excellent example is the work of Johann Wolfgang von Goethe on optics; various modern scholars, including the philosopher Olaf Müller, have done very careful historical and physical work demonstrating that Goethe's optics, consciously going against Newton's then-orthodox optics, embodied a great deal of good experimental and theoretical knowledge (see Klug et al. 2021). Another inspiring example, this time from the realm of biology, comes from Annie Jamieson and Greg Radick's work in rehabilitating the genetics of W. F. R. Weldon, who opposed Mendelian genetics at the start of the twentieth century, and lost. Jamieson and Radick (2017) even designed a genetics course framed by Weldon's ideas, emphasizing the role of development and environment from the start (see also Radick 2016). Scientists who learn to reach a sympathetic understanding of such bygone systems of knowledge, and even to solve problems using them, would certainly gain an extraordinary level of insight and adaptability of thought that would equip them well for the challenge of tackling problems requiring novel solutions. By looking to history, we can reap the benefits of Feyerabendian proliferation without going to the anarchistic extremes of taking ideas "from outside science, from religion, from mythology, from the ideas of incompetents, or the ramblings of madmen" (Feyerabend 1975, 68). We can get quite far by listening to Goethe or Weldon instead. It is possible to imagine a liberal and pluralist science education, while respecting Kuhn's recognition of the importance of disciplined training.

Part III

Incommensurability, Progress, and Revolutions

7   Kuhn on Translation

Alex Levine

7.1   Introduction

In 1970 *The Structure of Scientific Revolutions* (SSR) had been in print for some eight years, inspiring new approaches to many of the scholarly fields on which the book had touched, while also serving as a lightning rod for trenchant criticism. One such criticism, that accepting Kuhn's view of revolutionary change in science would force us to take the scientific process as ultimately irrational, had been voiced frequently enough that Kuhn may perhaps be forgiven if some of his rejoinders display a hint of impatience by the end of the decade, as, for example, his reply to Imre Lakatos in the 1970 meeting of the *Philosophy of Science Association*:

> As I have said before, both here and elsewhere, I do not for a moment believe that science is an intrinsically irrational enterprise. What I have perhaps not made sufficiently clear, however, is that I take that assertion not as a matter of fact, but rather of principle. Scientific behavior, taken as a whole, is the best example we have of rationality. Our view of what it is to be rational depends in significant ways, though of course not exclusively, on what we take to be the essential aspects of scientific behavior. (Kuhn 1971, 143–145)

Before I even introduce the primary thesis of this chapter, let alone provide an overview of its argument, I would like to draw attention to two significant inferences a careful reader of SSR might draw from this passage. The first requires that we recall that even in the first edition of SSR, one of the senses of "paradigm" was an *exemplar*: an exemplary model, an exemplary (textbook) solution to an exemplary (textbook) problem, an exemplary repeatable observation, or even an exemplary specimen. Implicitly throughout SSR, and most explicitly in Chapter V, "The Priority of Paradigms," Kuhn insists that rather than associating a period of normal science with scientific consensus about a *theory*, let alone a theory understood in the technical sense of the closure under logical implicature of a set of axioms or other primitive presuppositions, we ought to identify such periods with shared *paradigms*, understood as *exemplars,* among other things. Theories in the narrow sense might themselves be exemplars or might contribute to paradigms in one of the other senses of "paradigm," but they were not the primary desideratum of normal science.

Second, as the reply to Lakatos makes clear, Kuhn saw paradigms, in the sense of exemplars, as constitutive not just of periods of normal science in astronomy, physics, chemistry, or any of the other fields discussed in SSR but also of his own conception of science – of the account of scientific change he had provided in SSR. To put the point slightly differently, Kuhnian history and philosophy of science (hereafter Kuhnian HPS) is itself within the scope of Kuhnian HPS. Its goals include discerning characteristic features of scientific inquiry, both those characteristic of the normal-scientific inquiries of particular disciplines within specific historical contexts and those characteristic of, or remaining invariant across, periods of revolutionary science. Rationality, Kuhn suggests, is one of the invariants, at least to the extent that while scientists, like other human beings, may occasionally lapse from it, scientific revolutions do not by themselves undermine rationality. The scientific process, punctuated as it might be by periods of revolutionary change with all the disruption they entail, is not irrational. This is not itself an empirical claim, but rather a function of the fact that the scientific process is itself our best exemplar of rationality and must be treated as such if Kuhnian HPS is to pursue the goal Kuhn articulates for it in this passage, that of shedding light on rationality.

Although I lead with my reading of this passage from Kuhn's "Notes on Lakatos," this chapter is not about either rationality or the prospects for discovering anything about rationality itself by means of Kuhnian HPS. I wish instead to provide an analogous reading of Kuhn's understanding of the relationship between *meaning* and *translation,* and of the prospects for discovering features of that relationship by means of Kuhnian HPS. I will argue that in his references to translation, translatability, and untranslatability Kuhn does *not* understand translation as the "diffusion" (to gloss the Latin *translatare*), as the transfer or "carrying across" (to gloss the Latin *transferre* along with the German *übersetzen*), as the "leading across" (to gloss the Latin *transducere* along with the French *traduire* or the Spanish *traducir*), or as any other operation performed on an antecedently fixed meaning. Instead, I will try to show that he understands *translation as constitutive of meaning* – not its only constituent, but one significant enough that, to borrow a phrase from Hilary Putnam (1975), there is no "meaning of meaning" prior to translation. I will also argue that beyond Kuhn's well-known rejection of the semantic externalism that came to characterize Putnam's philosophy of language by the middle of the 1970s (see RSS 312–313), this view of the

relationship between translation and meaning has far-reaching consequences for our understanding of both the process of scientific revolutions and its broader significance.

7.2   The Translation Metaphor

Having come to SSR after the publication of the second edition, I have,
since my first encounter, been fascinated by a contentious metaphor for
incommensurability absent from the first edition but developed over the
1960s and included in the Postscript (SSR-2, SSR-3, SSR-4).
Incommensurability is a significant obstacle to communication across a
revolutionary divide but not, contrary to what some readers of the first
edition had thought, an insuperable one: "Briefly put, what the participants in a communication breakdown can do is recognize each other as
members of different language communities and then become translators" (SSR-2 202).

Incommensurability thus consists in a full or partial failure of translation of varying tractability and duration. What it means to become a
translator, how precisely the translator bridges language communities,
and how, in turn, this process illuminates communication under conditions of revolutionary science are left unstated, though Kuhn does add
the following footnote:

The already classic source of most of the relevant aspects of translation is W. V.
O. Quine, *Word and Object*, chaps. I and II.... But Quine seems to assume that
two men receiving the same stimulus must have the same sensation and therefore
has little to say about the extent to which a translator must be able to *describe* the
world to which the language being translated applies. (SSR-2 202 n. 17)

Brief and suggestive though these passages are, they serve as the focal
point for some of the most influential critiques of Kuhn's account of
incommensurability, perhaps most famously that of Donald Davidson's
paper "On the Very Idea of a Conceptual Scheme," delivered as the
presidential address to the Eastern Division of the American
Philosophical Association in December 1973. Davidson, having greatly
enhanced the principle of charity briefly invoked in Quine's *Word and
Object* (1960), argued that substantive conceptual differences between
speakers of any languages were an a priori impossibility; the idea of a
conceptual scheme, and with it the scheme/content distinction, were thus
essentially vacuous. Consequently, given that a sentence can have truth-
value only relative to a holistic interpretation, or what Davidson calls a "T-
theory," of the entire language in which the sentence is couched, incommensurability understood as failure of translation becomes utterly banal:

So what sounded at first like a thrilling discovery – that truth is relative to a conceptual scheme – has not so far been shown to be anything more than the pedestrian and familiar fact that the truth of a sentence is relative to (among other things) the language to which it belongs. Instead of living in different worlds, Kuhn's scientists may, like those who need Webster's dictionary, be only words apart. (Davidson 1974/1984, 189)

But being "only words apart" is not, in Davidson's view, a serious obstacle to communication. One need only, as his passage implies, "pick up the dictionary." If no such dictionary has been written, no matter; charity guarantees a priori the possibility of constructing one by producing a T-theory of the other's language, a task whose ultimate success is ensured by the fact that every natural language is suited to use as metalanguage for any other.[1]

There are numerous rejoinders to Davidson's critique, many of which set out to demonstrate that Davidson and his fellow-travelers had missed something important about Kuhn's semantics, an oversight only partially excused by Kuhn's casual allusion to Quine's account of radical translation in the Postscript. Most recently, several scholars have challenged the idea that the Postscript and related essays represent a "linguistic turn" in Kuhn's philosophy of science, one undertaken in response to criticisms of the first edition of SSR following some remedial study in the philosophy of language. In fact, Quine himself acknowledged Kuhn's assistance on what would become *Word and Object* in 1959 (Quine 1960, xi; q.v. Wray 2021a, 73–74). Hufbauer (2012), Mayoral (2017), Reisch (2021b), and Giri and Melongo (2023) have established the depth of Kuhn's background in the philosophy of language as early as 1951, as well as demonstrating continuity in Kuhnian semantics throughout his career. I have elsewhere (Levine 1999) tried to situate Kuhn's semantics within the broader context of post-Fregean antipsychologism, arguing that it has the resources to account both for incommensurability and for the possibility of scientific progress across revolutions. These arguments need not be rehearsed here. What interests me in this chapter is not the technical niceties of Kuhn's semantics, but rather the significance, for his account of scientific change, of the relationship he envisions between *translation* and *meaning*. This is best discerned by reference to his invocation of translators, translation, and translatability as *metaphors* for aspects of this account of scientific change. It is in large measure thanks to Kuhn that the philosophy of science has become attuned to the

---

[1] Davidson takes the interpretive totipotency of natural languages to have been established by Alfred Tarski in his "The Concept of Truth in Formalized Languages" (1956), an essay quite clearly *not* about natural languages at all.

importance of metaphor and analogy in scientific language and reasoning.

I began this chapter by arguing, on the basis of Kuhn's discussion of scientific rationality in his reply to Lakatos (see Kuhn 1971), that Kuhnian HPS itself falls within the scope of Kuhnian HPS. Now I propose to apply post-Kuhnian work on scientific metaphor and analogy to Kuhn's own account of revolutionary science and incommensurability.[2] One consequence of this effort will be a revival of Kuhn's translation metaphor, a metaphor that, when taken seriously, still yields substantive insights into scientific change.

7.3   Metaphor as Constitutive

Analogies and metaphors operate in significant ways throughout science, philosophy, and the philosophy of science. In my collaborations with the historian of science Adriana Novoa (Novoa and Levine 2010; Levine and Novoa 2012), I focused on the reception of the central analogies of Darwinian evolutionary theory in nineteenth-century Latin America. Evolution by natural selection constituted a significant break from the Enlightenment understanding of science on which the scientific and political elites of Latin America had been weaned. Members of those elites thought of events governed by scientific laws as unfolding by necessity, but Darwinian evolution is contingent. Furthermore, for Alexander von Humboldt and many of his followers in the first half of the nineteenth century, the cosmic process was itself progressive. Belief in the inevitability of progress – in science, in nature, and in society – also ran throughout the positivist oeuvre of Auguste Comte, like Humboldt, one of the guiding lights of scientists and politicians in the newly independent Latin American republics of the first half of the nineteenth century. But Darwinian evolution, despite vestiges of teleological language found even in Darwin, is *not* necessarily progressive, and if human societies transform under Darwinian principles, their evolution need not be progressive, either. As Novoa and I have argued, in their reception of Darwinism, members of the Latin American elites were compelled to recast what Richard Boyd (Boyd 1979) has called the "theory-constitutive metaphors" of Darwinian science, accommodating them, to the extent possible, to their prior understanding of natural philosophy, and, yes, *translating* central concepts for deployment in a linguistic and cultural milieu far removed from Darwin's England. If there is such a thing

---

[2] For a parallel treatment of another of Kuhn's analogies, see Levine (2000).

as incommensurability, then in coming to grips with the Darwinian Revolution, these figures were surely contending with it.

Their efforts gave rise to some fascinating interventions in social policy, some of them quite dire, and to some long-lived and highly productive research programs in science, such as Florentino Ameghino's statistical phylogeny, or José Ingenieros' evolutionary psychology. Such successes, we have argued, owed themselves in large measure to what Harold Bloom (Bloom 1973) calls the "creative misreading" of Darwin and other European sources. I suspect that translation almost always involves creative misreading. If there is anything to Thomas Kuhn's translation metaphor, such processes must be taken into account in understanding the ways in which scientists bridge the gaps between incommensurable theories, or fail to do so.

One characteristic of Boyd's "theory-constitutive metaphors" is that they do not operate in only one privileged direction, shedding light from a familiar domain onto a domain less well understood. Quite often, some of the obscurities of the less familiar domain end up obfuscating, or calling into question, the unexamined presuppositions of the supposedly familiar domain. So it is not only with Darwin's analogy between domestic and natural selection, or with Rutherford's analogy between the solar system and the atom, but also with translation metaphors, including Kuhn's. For such metaphors to be illuminating we ought ideally to have some fairly clear prior notion of what translation is, how it is performed, and what distinguishes a successful from an unsuccessful translation. When we consider the root of the English word "translation" or the roots of related terms in related languages, it indeed appears that in naming the process, we may have had such a prior notion in mind. Translation is the diffusion, transferal, or carrying over, of meaning. Meaning is the cargo, and translation is its shipment. If we understood what the cargo was, we might well have a clear idea of how to ship it, and thus have a clear enough prior notion of translation to allow Kuhn's translation metaphor to fully explicate incommensurability. However, there is no agreed-upon understanding of meaning, nor does Kuhn to my knowledge ever make any claim to the contrary. Recent scholarship has refuted the misattribution to Kuhn of a description theory of reference, and Kuhn himself unequivocally rejects a Kripkean theory of reference, and with it meaning externalism (see Kuhn's rejection of Putnam's turn toward externalism in Kuhn RSS 312–313). As a philosopher of language the Kuhn of SSR was already sophisticated – but he also knew his limits (see Giri and Melongo, 2023).

Instead of supposing that in deploying his translation metaphor Kuhn meant to invoke a clear prior conception of both translation and meaning, despite the lack of evidence for either, I propose to take the metaphor, like other theory-constitutive metaphors, as much more symmetrical. As such, the potential of an appeal to translation to illuminate scientific communication in general and incommensurability in particular comes with the complementary potential of a Kuhnian account of scientific communication to shed light on translation and, with it, on meaning.

7.4   Communicative Breakdown

I have performed many translations, including the English translation of Paul Hoyningen-Huene's *Reconstructing Scientific Revolutions* (1989/1993). I take pride in my own work, and read translations performed by others with a practiced eye, sometimes admiringly, sometimes with disappointment. I know when a translation has met or exceeded my standards, and I know when it has failed to do so. But despite considerable effort, I am at a loss to actually *articulate* those standards. I know *how* to translate competently, but I am not entirely clear on *what* I am doing. Kuhn thought that skill in translation was weakly correlated with fluency in both source and target languages, but only weakly. As Kuhn observes in an important use of the translation metaphor predating the Postscript, translation

can present grave consequences to even the most adept bilingual. [The translator] must find the best available compromises between incompatible objectives. Nuances must be preserved but not at the price of sentences so long that communication breaks down. Literalness is desirable but not if it demands introducing too many foreign words which must be separately discussed in a glossary or appendix. People deeply committed both to accuracy and to felicity of expression find translation painful. (Kuhn 1970b, 267)

On reading this passage, every experienced translator feels a sense of validation as satisfying as it is rare. We so often pass beneath notice, even though scholars, in particular Anglophone scholars, could not do their jobs if we had not done ours. Kuhn both notices what we do and appreciates its difficulty. Significantly, though he is not himself a translator, Kuhn is speaking from experience – specifically, from the experience of communicative frustration and breakdown. Such experiences feature prominently in his 1995 interview with Baltas, Gavroglu, and Kindi (RSS 253–323), in which he both enumerates salient instances and remarks on their significance to his intellectual trajectory. Reflecting on the schools in which he completed his secondary education, Kuhn recalls that

those schools gave me more formal training, more language – although I was never any good, I've never been any good really at foreign languages. I can read

French, I can read German, if I'm dropped into one of those countries I can stammer along for a while, but my command of foreign languages is not good, and never has been, which makes it somewhat ironic that much of my thought these days goes to language. (RSS 258–259)

The irony would be easy to miss if Kuhn had not explicitly drawn our attention to it. Note, however, that for there to be any irony at all, what Kuhn has in mind when he says "much of my thought these days goes to language" cannot be language in general, or semantics. Given the context, the irony must be specifically linked with Kuhn's poor "command of foreign languages" and with the experience of struggling to overcome what this limitation entails. Nor are Kuhn's allusions to such experience confined to recollections of adolescence. Despite his best efforts in high school, when he was sent to France during the Second World War, he found his command of the spoken language profoundly inadequate, so much so that his recounting of the experience half a century later is still tinged with embarrassment (RSS 270–271). Reading German was at least as frustrating, and similar embarrassment colors his reminiscence of encounters with the seminal texts of Ludwik Fleck (RSS 283) and Wolfgang Stegmüller (RSS 318). In fact, Kuhn reports having needed a year and a half to read Stegmüller's book.

For an American scholar of Kuhn's generation to have struggled with languages other than English is hardly unusual, though the intellectual humility to admit the difficulty is perhaps somewhat more so.[3] Of primary interest, for our purposes, are his conclusions on the significance of translation, to which the communicative challenges he faced, together with his work in the history and philosophy of science, may have led him. Equipped with a sense for these challenges, we may view with suspicion a "merely" semantic reading of the untranslatability metaphor for incommensurability. For all the breezy tone of footnote 17 to the Postscript of the second edition of SSR, thus armed, an ironic reading presents itself.

---

[3] Curiously, another American scholar to make this admission is none other than Kuhn's close contemporary and fellow Harvard alumnus Donald Davidson, who, despite having written his dissertation on Plato's *Philebus*, reports, decades later, that "being a product of American education, neither my German nor my Greek was very good (they still aren't)" (Davidson 1997, 422). Yet as evidenced by this (Davidson 1974/1984), for Davidson, unlike for Kuhn, the success of translation is ultimately ensured. Speculation on any experiential ground for this difference is beyond the scope of this chapter, though it is perhaps significant that Davidson, unlike Kuhn, spent the Second World War in the United States Navy (Malpas 2021) and may have missed out on the kind of communicative challenge Kuhn relates (RSS).

The already classic source of most of the relevant aspects of translation is W. V. O. Quine, *Word and Object*, chaps. I and II.... But Quine seems to assume that two men receiving the same stimulus must have the same sensation and therefore has little to say about the extent to which a translator must be able to *describe* the world to which the language being translated applies. (SSR-2 202 n. 17)

By 1970, Chapters I and II of *Word and Object* may have already been the "classic" source of "most of the relevant aspects of translation," but "classic" is not a synonym for "informative" or "accurate," especially given the subsequent qualifiers. On the assumption Kuhn attributes to him, Quine "has little to say about the extent to which a translator must be able to *describe* the world to which the language applies," and thus little to say about how translators might validate the fruits of their labors. Any text in a target language may, without much effort, be passed off to the uninitiated as a translation of a source language text; the hard work consists in the validation, correction, and refinement of such. *Mutatis mutandis*, communication across a revolutionary divide may be possible, but neither party to such communication has any a priori guarantee of its possibility, let alone its eventual success. Work is required. That work consists in the parties coming at least partially to share a description of the world, which in turn demands both that they learn to talk to each other and that they do science.

As Davidson would realize, Quine's radical translation, along with Davidson's own radical interpretation, is inimical to what Kuhn was trying to do. However, Kuhn's translation metaphors, far from entailing the conceptual trap Davidson foresaw, constitute an invitation to what I shall call a *literary* interpretation of revolutionary science. We can embrace the translation metaphor self-consciously, with all the baggage that involves. One way of unpacking this option would involve considering renowned translator David Bellos' claim that "translation *is* meaning" (Bellos 2012). Translation, as a metaphor for the ways in which scientists bridge the gaps between incommensurable paradigms, suggests that in doing so scientists thereby *construct the meanings* of those same paradigms. Philosophical work on incommensurability might then, in turn, illuminate the notion of translation. This is the symmetry inherent in Boyd's notion of a theory-constitutive metaphor.

7.5 Conclusion: Becoming Translators

Davidson's argument is addressed to the notion of mutually untranslatable languages as a symptom of mutually incompatible conceptual schemes, with Kuhn as one of its targets. I hope I have shown that it misses its mark when it comes to Kuhn's translation metaphor. There remains one aspect of that metaphor, as introduced in the Postscript, that I have barely touched on in this chapter, though it points toward other critical themes in Kuhn's opus. In the Postscript, Kuhn does *not* simply prescribe that the disputants in a scientific revolution resolve their differences by performing translations. Instead, they must "recognize each other as members of different language communities and then become translators" (SSR-2 202). The distinction is an important one for Kuhn, who devoted nearly as much attention to scientific pedagogy and developmental cognitive psychology as he did to the history and philosophy of science (see several essays in ET, along with the long-awaited posthumous *Plurality of Worlds* [LW], published as Kuhn 2022). Famously, he credits developmental psychologist Jean Piaget with having furnished core insights for SSR (SSR-2 vi–vii; see also Levine 2000). Becoming a scientist is not, for Kuhn, a separate process from becoming a science communicator, a member of a particular language community. Being a member of two or more such communities does not automatically qualify the scientist as a translator – translation, after all, presents "grave consequences to even the most adept bilingual" (Kuhn 1970b, 267). But it is one necessary condition for translation, as is the capacity, acquired in the course of scientific training, to deploy the language of one's community to describe the world. Acquiring and employing these capacities are not trivial; the study of developmental and cognitive psychology helps to show just how difficult they are, and what profound changes they impose on the brains of scientists. If the changes are not as profound as those enacted in the brains of growing children, it is because, while both children and scientists are human animals, surviving childhood is a necessary condition, but not a sufficient one, for becoming a scientist. For a child to enter a language community is, among other things, for that child to have become a new source of *meaning*. Struggling with the worlds they strive to explain, predict, and control, and struggling with each other toward a basis for communication, scientists likewise become sources of meaning.

Acknowledgment

This chapter benefited greatly from comments on an early draft by the late Pablo Melogno (1979-2023), to whose memory this volume is dedicated.

8   Paradigm Shifts and Group Belief Change

Haixin Dang

8.1   Introduction

What a scientific community holds to be its core beliefs changes over time. In the early twentieth century, the physics community rejected ether theories. Geologists slowly came to accept plate tectonics and continental drift in the 1960s. In the early 2000s, particle physicists revised the highly confirmed Standard Model in order to account for neutrinos having mass. How should we understand such belief changes within science?

A natural starting point for thinking about this question is Thomas Kuhn's seminal *The Structure of Scientific Revolutions* (SSR). In it, Kuhn lays out an account of how research communities come to accept a core set of theories and how these theories are supplanted by new theories over time. He describes these revolutions in science as cycles of normal and extraordinary science. It is hard to overstate how much Kuhn's ideas have influenced philosophy of science for the last sixty years. Practically every aspect of Kuhn's account has since been debated: from the concept of a research paradigm to theories of reference to whether Kuhnian revolutions occur as theorized. However, only recently has there been philosophical work done on how Kuhnian accounts of scientific change are a contribution to social epistemology. SSR not only highlights the social dimensions of scientific knowledge and the training of scientists as social epistemic agents embedded in complex social networks but also offers a convincing account of how scientific knowledge is *collective* knowledge. In particular, Wray (2011b) has offered a positive reading of Kuhn for his insights into the social epistemology of science. Wray points out that Kuhn later retracted the metaphor of the "gestalt switch" as a metaphor for paradigm shifts because a gestalt switch is an individual experience, whereas, for Kuhn, a paradigm shift is experienced by the scientific community as a whole (Wray 2011b, 15). Following this key observation, this chapter will explore how to understand Kuhnian paradigm shifts as a genuine social epistemic phenomenon, that is, as *group* belief change.

Among social epistemologists, what a group belief is, or is not, and how it is formed have been well studied and debated (Gilbert 1987; Tuomela 1992; Mathiesen 2006; Bird 2010, 2014; List and Pettit 2011). However, much less is known about how group beliefs, once formed, may change over time (see Pettigrove 2016). There has been much philosophical debate over the nature of group belief in general (Hakli 2006). But there exist metaphysical worries over whether groups can have mental states and whether talk of a group belief is just metaphorical. Wray (2001) argues that group beliefs as studied by social ontologists like Gilbert are better characterized as group acceptance. For the purposes of this chapter, I will set aside the question of which cognitive attitudes can be properly attributed to groups. If one is particularly averse to theorizing about group belief, we can also make use of "group views." Belief changes are simply changes in views (see Harman 1986).[1]

Margaret Gilbert (2000) gave an account of group belief change by specifically picking out scientific change as the phenomenon of interest. She asks: "Is scientific change a matter of individual scientists changing their minds, or is it a matter of scientific communities changing their minds?" (p. 37) Gilbert has long defended an inflationary and non-summativist view of group belief. She believes that groups are agents in their own right, over and above the individuals (see Gilbert 1992). Group belief change cannot be just a simple matter of all or most of the individuals within that group changing their mind. Rather, group beliefs require something more, a joint commitment among the members of a group to believe or to change their beliefs *as a body*. Specifically, Gilbert argues that the joint commitment model of group belief can account for changes in scientific consensus. Weatherall and Gilbert (2016) have extended the joint commitment model of group belief to explicitly make sense of Kuhn's theory of scientific change. On Gilbert's view, a group belief is difficult to change as it would require a new joint commitment to be formed to replace the old one. A group's resistance to group belief change, as predicted by Gilbert, corresponds with Kuhn's observation that normal science is conservative and slow to process new evidence in ways that require a change of paradigm.

---

[1] It bears noting that "belief" may be a particularly reductive way to frame scientific attitudes. Scientists work with models and have complex methodological commitments that are difficult to capture as beliefs. While I will mostly be discussing how group beliefs change, these changes are also often entangled with changes in methodologies and modeling.

There have been several critical assessments of the application of Gilbert's joint commitment framework to science (Mathiesen 2006; Wray 2007; Rolin 2008; Fagan 2011). Wray (2001) and Mathiesen (2006) have argued that groups holding group beliefs as a joint commitment may not be epistemically rational, as group beliefs may not be updated appropriately in light of new evidence. For Gilbert, group beliefs are not strictly answerable to evidence, but rather what constrains group belief change is the joint commitment itself – it is a sense of obligation to the other members of the group to uphold what the group has already been committed to. This analysis threatens to render group belief change to be an arational or irrational process.

In this chapter, I will evaluate the account of scientific change given by Gilbert (2000) and Weatherall and Gilbert (2016). In particular, I argue that the *primary* normative constraint on group belief revision is the weight of the evidence being considered by the group, and not the normative constraints that arise from joint commitments. To understand group belief change, we need a better theory of how epistemic groups respond to evidence. I will only briefly sketch a positive view here, as working out a full-fledged account of group belief revision is beyond the scope of the present chapter. Instead, I will look back to Kuhn in SSR where he highlights the role of accumulating new evidence in the process of paradigm changes. While normal science tries to explain away new countervailing evidence and account for persistent anomalies, ultimately, a new paradigm is accepted when it can better account for the evidence and anomalies. Resistance to scientific change cannot be completely explained by the joint commitments among group members alone. Rather, Kuhn's account suggests that epistemic groups can be directly responsive to evidence.

This chapter will proceed as follows. I will first reconstruct the joint commitment model of group belief in more detail. Then I will critically assess the mechanisms proposed in Gilbert (2000) and Weatherall and Gilbert (2016) on how joint commitments constrain group belief change in science. Finally, I will sketch a positive view of how epistemic groups may respond to new countervailing evidence by looking to Kuhn's account of how anomalies are resolved in science.

8.2   Group Beliefs as Joint Commitments

Gilbert defends one of the classic inflationary accounts of group belief, which in other work she also terms "collective belief." In short, on Gilbert's non-summativist view, there is a collective belief that p if some persons are jointly committed to believing as a body that p. In contrast, a classic deflationary account of group belief is summative: a group believes that p when some or all of its members believe that p. Gilbert rejects simple summativism because she believes that there is an import-ant sense in which a group can have beliefs over and above its members. That is, a group can have beliefs that are not held *personally* by the group members. Gilbert (1987) gives an example of a poetry reading group where a collective interpretation of a poem emerges from the discussions of the reading group's members. However, while individual members take the collective interpretation to be the view held by the group, it could very well be the case that no one individual member *personally* believes the collective interpretation. Divergence cases like this are taken by Gilbert to be evidence against various versions of summativism. Group beliefs are more than mere aggregations of member beliefs, more than just members voting.

According to Gilbert, "a group G believes that p" should be inter-preted by the following conditions:

(i) A group G believes that p if and only if the members of G jointly accept that p.
(ii) Members of a group G *jointly accept* that p if and only if it is common knowledge in G that the individual members of G have openly expressed a conditional commitment jointly to accept that p together with the other members of G. (Gilbert 1987, 195, original emphasis)

Weatherall and Gilbert (2016) give a more concise definition of a group belief: "A group G believes that p if and only if the members of G are *jointly committed* to believing that p as a body" (p. 196, original emphasis). The authors also clarify that the idea of acting "as a body" is that "the object of the commitment, the thing being committed to, is to emulate as far as possible a single believer of the proposition in question, by virtue of the combined actions and utterances of the parties" (ibid.).

Central to both definitions is the technical concept of joint commit-ment. A joint commitment is, at the most basic level, a commitment of the will (Gilbert 1992). Gilbert (1990) specifically explains that a joint commitment is not a promise between parties to each do their part. Rather, a joint commitment is a basic notion that is at the very heart and foundation of human social relations. A joint commitment is created when each of the parties has expressed their personal willingness to be a party to it under conditions of common knowledge. Given a joint com-mitment, the parties to it are bound together and are subject to it until *all parties* concur in rescinding it. Importantly, while the parties are subject to a joint commitment, they have obligations and rights in relation to one another. When parties enter a joint commitment, they are entering a particular normative situation where each is committed to act in a certain way in accordance with the content of the joint commitment.

Joint commitments ought to be distinguished from personal commitments. I may have a personal commitment to leave my office door open whenever I am inside to signal to students and colleagues that I am free to have discussions and answer questions. As a personal commitment, I can unilaterally decide to rescind my open-door policy at will. The philosophy department, as a whole, can decide to jointly commit to a department-wide open-door policy, perhaps as a result of a faculty meeting where the members openly express willingness to institute such a rule. This joint commitment would require that each member of the department leave their respective office door open whenever possible. On Gilbert's view, this newly formed joint commitment is not composed of personal commitments. The joint commitment is, rather, a normative situation where it is common knowledge between members that each of them has expressed their personal willingness to be committed in a specific way and that they all are so committed as a body. Under such conditions, if I fail to conform to a joint commitment, absent certain background understandings, then, I risk offending all of the other parties to the joint commitment. That is, if I personally rescinded from the collective open-door policy, I open myself up to rebuke from the other members of the group. Unlike in the case of a personal commitment, the joint commitment ties me to conform to the expectations of the group.

This is fundamental to understanding Gilbert's account. Reneging on a joint commitment is an offense because it is a violation of an *obligation* to the other parties, who have rights to one's conformity to the joint commitment. "This is a function of the joint commitment itself: a joint commitment obligates each party to the others to conform to it, at the same time endowing them with rights to such conformity" (Weatherall and Gilbert 2016, 197).[2]

In the case of collective belief, members of a group are jointly committed to believing that p as a body, under conditions of common knowledge. Once a collective belief has been formed, on Gilbert's account, it becomes extremely difficult for members who are party to it to disown it. It also becomes difficult for members to express *personal* skepticism toward the collective belief. Gilbert (2000) explains this in more detail:

If they speak contrary to the collective belief without a preamble, they will be regarded by both themselves and the others involved as acting out of line. More

---

[2] The normative force of a joint commitment can be quite strong. If one were to renege on a joint commitment, then the other members of the group can *rebuke* one. The members of the group can *demand* that one follow through with the joint commitment.

specifically, they will have violated obligations they had to the others, and the others have the standing to take other measures responsive to this situation. These could take the form of anything from a mild rebuke to complete ostracism. Even if they do not speak without a preamble but make it clear that they are expressing what they personally believe, doing this only makes it clear that they are not personally in sympathy with the collective belief. That is liable to lay them open to suspicion. Hence *there is a significant initial cost* in every case of mooting an idea that runs contrary to a collective belief. (p. 42, original emphasis)

Note here that even in cases where a party of the collective belief expresses that they *personally* believe something contrary to the collective belief, this would open them up to suspicion by the group. There is a strong sense in which the collective belief *constrains and restricts* the personal speech of members.

A joint commitment to believe that p as a body does not strictly require a personal commitment from each member to believe that p, or for each member to personally act as though they believe that p. Rather, the members of the group are only required to act as separate mouthpieces of the group, upholding the belief in any context in which they are acting in their capacity as group members. This is why, for Gilbert, members must be careful to use a preamble to distinguish between personal speech and speech as a member of the group. Failing to use the preamble properly is a violation of one's obligations to the rest of the group. However, collective beliefs can be quite tricky, in that openly expressing that one does not personally believe the collective belief can still be costly. While it may not earn one a rebuke or exclusion, it still makes one vulnerable to suspicion from others.

Given this background on Gilbert's account, we can see now that once a group belief is formed, it is very difficult to change. It is even difficult for a member to openly speak against a group belief without some initial cost. A group belief change, in Gilbert's sense, will involve getting all parties to jointly commit to rejecting the previous belief and then to jointly commit to forming a new one. It is not enough that members personally reject the group view. Gilbert (2000) writes: "for a body of which I am a member to change its beliefs requires something akin to an agreement to stop believing that p together and to start believing that q instead, and this is not easy to achieve" (p. 46).

It would be useful here to contrast Gilbert's view with a deflationary account. On the summativist view, group belief change is much more straightforward. When enough members individually change their beliefs, the group belief change automatically follows. Group belief change would not require more than changing the minds of some suffi-cient number of individual members – a majority, a supermajority, or perhaps everyone. On the other hand, for Gilbert, if a majority or even all of the individual members of a group believe something, it does not follow that there is a group belief, as it could be the case that they have not yet formed a joint commitment to believe it as a body. However, once the group belief is formed and the members are jointly committed in the right way, the group belief can remain stable no matter how individual beliefs may vary.

8.3   Group Belief Change and Scientific Change

Gilbert (2000) and Weatherall and Gilbert (2016) argue that scientific consensus can be properly interpreted as a group holding a collective belief through a joint commitment. Gilbert even argues that the difficulty of group belief change is a *virtue* of her view because she can successfully account for important sociological aspects of scientific communities. She notes two particular facts to be explained: (1) that collective beliefs act as "brakes" on scientific change and (2) that scientific change is often initiated by neophytes or outsiders.

Let us address (2) first. There is a prevalent public perception that major epistemic breakthroughs are made by young researchers. The Fields Medal, the most prestigious award in mathematics, is limited to recipients under the age of forty. Isaac Newton was twenty-three when he had his *annus mirabilis* and discovered – among many things – calculus and the law of universal gravitation. Albert Einstein was sixteen when he came up with one of his most important thought experiments: What would happen if he could chase a beam of light?

The idea that revolutionary change comes from either young scientists or those new to a field is also remarked on in Chapter VIII of SSR. Gilbert argues that young scientists are more likely to contribute revolutionary changes to science because they are not subject to the joint commitment in the same way as someone who is entrenched in the community – not being party to the scientific consensus allows one to go against it without fear of rebuke from members of the community. While Kuhn does not employ the technical notion of joint commitment, Kuhn (SSR-4) does echo a similar line of reasoning: "these are the men who, being little committed by prior practice to the traditional rules of normal science, are particularly likely to see that those rules no longer define a playable game and to conceive another set that can replace them" (SSR-4 90). However, Kuhn notes that while there exists a common perception that young scientists or outsiders play an outsized role in scientific change, whether or not this is true is ultimately an empirical question.

Wray (2003, 2004), through a survey of the historical record, showed that it is not the case that young scientists contribute more revolutionary research. Wray (2003) reports that a larger portion of significant scientific discoveries is, in fact, attributable to middle-aged researchers. Further, Line Andersen (2017) examines Gilbert's assertion that a scientist's outsider status can allow the scientist to more easily buck the consensus within an epistemic community. Andersen examines a case where outsiders to the mathematics community were able to make an important mathematical contribution. However, Andersen concludes that the facts of the case can neither confirm nor refute Gilbert's theory. We cannot make a judgment about whether the community's joint commitments played any role in preventing or bringing out the discovery. Given these empirical studies, it appears that Gilbert's claim – that scientific change is often initiated by neophytes or outsiders and this fact can be explained by the joint commitment model of group belief – has not been borne out.

Weatherall and Gilbert (2016), instead, focus on defending (1), the claim that collective beliefs act as brakes on scientific change. They argue that collective beliefs in Gilbert's sense can be analogous to a Kuhnian paradigm. Collective beliefs can regulate how the group functions, evaluates new evidence, and determines which projects are deemed appropriate for its members to pursue. Theoretical and methodological commitments associated with a Kuhnian paradigm may best be understood, they argue, as a set of foundational *collective* beliefs of the scientist within a paradigm. Becoming a full-fledged member of a scientific community requires one to jointly commit to what the community holds to be its core beliefs – to assent to the prevailing consensus and to work within the accepted paradigm. By entering a joint commitment, the scientist is then obliged to defend the consensus or otherwise face public rebuke or ostracism. Speaking out against the consensus – even with the preamble to signify that one is not speaking as a mouthpiece of the group – will still be personally costly. It casts doubt on one's commitment to the group and may signal unreliability to other members.

This analysis *appears* to echo much of Kuhn's analysis of normal science. Kuhn characterizes normal science as "puzzle-solving," that is, research based on established sets of scientific practice, with accepted exemplars and methods that can be fruitfully applied to new problems. When a scientist is working within normal science, their day-to-day research does not contradict the theories and methods that are foundational to that paradigm (see SSR-4 11). Even when faced with anomalies, or facts that seem not to fit the paradigm, scientists first try to explain away the anomaly with the tools available to them from within that paradigm.

Scientists do not immediately abandon a paradigm due to there being anomalies. Kuhn notes that "to be accepted as a paradigm, a theory must seem better than its competitors, but it need not, and in fact never does, explain all the facts with which it can be confronted" (18).

Normal science, then, tends to be conservative and thus slow to admit paradigm change. Countervailing evidence is often either explained away or ignored. Weatherall and Gilbert argue that the underlying explanation of this conservatism could be the fact that scientists within a paradigm are bound together by a joint commitment. They write:

Even when evidence against a consensus is found by an individual scientist, one might expect it to be ignored, suppressed, or explained away by its discoverer, since such evidence would force a *psychologically unsustainable conflict* between the scientist's commitment to act in a certain way and his (or her) beliefs concerning the epistemic warrant for those actions. In the case of evidence contrary to a consensus that is made public, the scientific community, also, may ignore, suppress, or explain it away. For instance, it may be collectively affirmed that a crucial experiment cannot have been properly done, or it may be assumed that facts that are not at odds with the consensus can explain it, however implausible such an assumption really is. For these reasons, scientific communities *should be expected* to hold certain beliefs in the face of considerable conflicting evidence, to the point of being, or at least appearing, irrationally dogmatic or epistemically irresponsible. (Weatherall and Gilbert 2016, 201, my emphasis)

Note the mechanism the authors have posited as responsible for the group not being responsive to new evidence: members of the group are psychologically constrained by the normative force of the joint commitment itself. How does this mechanism work? Gilbert (2000) explains in more detail:

It is important to see how participation in a collective belief can have consequences even for one's private thoughts, inhibiting one from pursuing spontaneous doubts about the group view, inclining one to ignore evidence that suggests the falsity of that view, and so on. Fulfilling the relevant commitment may not require that one personally believes what one's group believes. Nonetheless, it is awkward to say one thing and think another. This is so for a number of reasons. One may then resist following up certain ideas. One may – consciously or unconsciously – calculate that it is not worth doing so, since by making public the likely conclusions, even characterized as one's personal conclusions, one would risk a host of negative reactions to oneself and possibly even the breakdown of one's professional group. Knowing that this is something one is not prepared to do, one may avoid pursuing any relevant thoughts, even those one suspects of having something to them. (p. 45)

So, being party to a joint commitment will constrain what one can personally believe and personally do. Changing the collective belief would require all the members to jointly commit to a new belief.

However, individual members are, in a sense, psychologically constrained by the original joint commitment, so much so that one often discounts and resists the countervailing evidence, preventing members from coming forward to question the original joint commitment.

While Gilbert has always highlighted the importance of cases where members' personal beliefs may diverge from the group belief, there is tension in her account here. Even though an agent can form personal commitments, when their personal commitments clash with a joint commitment they are a party to, the agent is left in a psychologically "awkward" position. On Gilbert's account, the agent's obligation to uphold the joint commitment may be so strong that it overrides the agent's personal epistemic assessments. It is a consequence of Gilbert's view that even though a group member may personally recognize that new evidence ought to be accounted for, the group as a whole cannot respond to the new evidence. Surprisingly, once the agent is committed to being part of a body that believes that p, there appears to be some constraint on whether the agent can personally doubt that p.

Are epistemic groups this resistant to revising their beliefs, even in the face of convincing countervailing evidence? It bears mentioning that there is some intuitive appeal behind the idea that groups can be, or at least appear to be, irrationally dogmatic or epistemically irresponsible. For example, corporations often fail to change or adapt to evolving market conditions (Wright et al. 2004). There is now a growing literature on group polarization, a phenomenon where members of groups become more extreme in their beliefs after discussion with other like-minded members (Sunstein 2002; Del Vicario et al. 2016). While the dynamics behind these social phenomena are complex, these cases do show that groups often do fail, or at least appear to fail, to account for evidence in an epistemically responsible way.

For the remainder of this chapter, I will offer an alternative view of group belief change. In particular, I argue that the normative force of joint commitment in constraining group belief change is not primary. Ultimately, scientific groups are answerable to evidence. While scientific groups do seem resistant to certain kinds of new evidence, this resistance is not necessarily epistemically irresponsible. Rather, we need a theory of how a group can process and respond to new evidence, and we can look to SSR and Kuhn's analysis of revolutionary change in science as a starting point.

8.4   Kuhn and Group Belief Revision

Given Gilbert's arguments for the difficulty of group belief revision, it is surprising that groups can and do change their beliefs at all. While normal science is conservative, periods of revolutionary science still occur. Gilbert (2000) acknowledges that the details of how a group changes its beliefs, especially a scientific group, are "worthy of study" (2000, 46). Weatherall and Gilbert (2016) apply the joint commitment model of group belief to explain how the string theory community is resistant to change and new evidence,[3] but the authors do not give an account of how an established scientific consensus eventually can be overturned. Because Gilbert claims that the normative force of the joint commitment will prevent group members from individually questioning the established consensus, this mechanism makes it mysterious how one may escape the joint commitment. I will argue here that the strength of the available evidence explains how both individual and group beliefs change.

Take a simple case: Mara is a member of a scientific group that has a collective belief that p, where all the members have openly expressed willingness to let p stand as the view of the group under conditions of common knowledge. Mara is committed to upholding p as a member of the group. However, she starts to encounter evidence against p during her research. Mara at first discounts the evidence, then she tries to explain it away in a manner that is consistent with p. She does this because of her commitment to holding p as a member of the group. Over time, Mara encounters more and more evidence against p. Mara begins to suspect that perhaps p ought to be rejected. Now, Mara is in what Gilbert had called a "psychologically unsustainable conflict" between her personal beliefs and the group belief. If she speaks out against p, she risks rebuke from her group members or, worse, complete ostracism. She might be prevented from getting more funding for her research or her papers may be rejected by journals. What should Mara do?

In Gilbert's telling, it seems that Mara has an obligation to keep believing that p. That is a consequence of being a party to a joint commitment. Mara may find it difficult to declare that she rejects that p even with a clear preamble to mark that she is speaking only for herself. If she does so, other members of the group will suspect her loyalty and exclude her from their future collaborations. So, Mara does not argue against her group and lets p continue to stand as the group belief. Gilbert

---

[3] The main claim in Weatherall and Gilbert (2016) is a conditional: *if* Gilbert's theory of group beliefs is correct, *then* the idiosyncratic features of the string theory community could be explained. Thanks to James Weatherall for highlighting this for me. In this chapter, I focus on assessing Gilbert's (2000) theory of group belief as applied to scientific change, but not to the string theory community in particular.

predicts that a widely held belief within a community will be difficult to change in this way.

But the evidence against p remains! Mara cannot ignore it indefinitely. Whether Mara ought to keep her obligation to her group has to be weighed against how strong the evidence is. Eventually, the evidence against p will be so strong and so persistent that Mara will be forced to speak out. She may have to argue with the other members of her group for them to also see the evidence against p. Mara's obligation to uphold and defend the joint commitment can go only so far. Her personal commitment to responding to the evidence in an epistemically appropriate way may be stronger than her joint commitment to hold p as the view of the group.

The most plausible way to understand the role of the joint commitment in constraining what agents can or cannot do is to view these special obligations that arise from a joint commitment to be only *pro tanto* obligations.[4] Mara lacks, all-things-considered, obligations to uphold the collective belief. She is also subject to other normative obligations, such as the norms of rationality or epistemic norms. Ultimately, if the evidence against p is strong and convincing, Mara ought to reject p, regardless of whether or not she has a joint commitment to uphold p. Once Mara has rejected the group view, she can then perhaps try to change the mind of the other group members. Epistemic norms must play some role in regulating how beliefs are formed and changed, at least, among members of scientific groups. A joint commitment can be broken, as it were, when the agents have other, stronger reasons to not follow through.[5]

---

[4] Elsewhere, in her work on joint action, Gilbert (1992) has claimed the agents have *pro tanto* obligations grounded in their joint commitments to perform their part of a joint action. However, agents may lack, all-things-considered, obligations to perform the joint actions, due to extenuating circumstances like side agreements or moral considerations (see also Kopec and Miller 2018). This should apply analogously to cases of collective belief as well.

[5] Analogously, for joint action, consider the following example. David and Nancy jointly commit to researching the efficacy of a drug together. This requires that David and Nancy coordinate their actions to bring about the study outcome as a group. Both Nancy and David at the beginning of their study believe that the drug is promising and deserves to be studied – perhaps they even jointly accepted that as a body. If Nancy fails to produce data according to their agreed-upon experimental protocol, David has the right to rebuke her for failing her obligations, and vice versa. Over the course of the collaboration, Nancy begins to suspect that the drug is harmful to patients and stops doing her part of the study. David still believes that the drug is promising and wants to continue the research as planned. Now, does David have the right to rebuke Nancy for not following through with their joint commitment to run the study? The normative force of the joint commitment does not automatically trump other competing normative considerations in this case. Agents are subject to many competing norms at the same time – both epistemic and moral. While a joint commitment gives rise to some specific obligations, agents can and do choose to act on other normative considerations. Nancy may fail the specific obligations she had toward the collaboration to follow through with a research plan.

If joint commitments are only *pro tanto* obligations that can be overridden by epistemic obligations, then members of scientific groups that have a group belief can still be responsive to evidence. Gilbert's view on scientific change deemphasizes how epistemic norms that govern how one ought to rationally respond to evidence can compel the members of a group to reject its original belief and adopt a new one. While normal science may be conservative in its evaluation of new evidence, scientists who are engaged in normal science are still responsive to evidence. When a paradigm has a demonstrated record of success, however, the strength of the countervailing evidence must be correspondingly strong – after all, extraordinary claims require extraordinary evidence. Extraordinary evidence is difficult to come by and alternative explanations must be evaluated before extraordinary evidence can be accepted as such. The fact that a scientific group resists revising their core theories in response to new evidence does not need to be primarily explained by the joint commitments of its members. Instead, the fact can be explained simply by the very nature of extraordinary evidence. What constrains the group from revising its core theories is not obligations to each other, but, rather, the group is constrained by evidence that is difficult to gather and difficult to interpret.

Another gap in Gilbert's account is how to explain why the members of a group jointly commit to upholding a group belief to begin with. How does a paradigm get established in the first place? Kuhn tells us that a paradigm becomes established because it is *successful*: it is good at generating new problems and good at giving a framework to study new problems. Scientists engaged in normal science are trained to solve outstanding puzzles using the tools and methods of the paradigm. A paradigm gains wide acceptance and becomes established because it is more successful than its competitors:

Paradigms gain their status because they are more successful than their competitors in solving a few problems that the group of practitioners has come to recognize as acute. To be more successful is not, however, to be either completely successful with a single problem or notably successful with any large number. The success of a paradigm … is at the start largely a promise of success discoverable in selected and still incomplete examples. Normal science consists in the actualization of that promise, an actualization achieved by extending the knowledge of those facts that the paradigm displays as particularly revealing, by increasing the extent of the match between those facts and the paradigm's predictions, and by further articulation of the paradigm itself. (SSR-4 24)

Instead, Nancy can prioritize fulfilling a different obligation, the moral obligation to conduct ethical research that minimizes harm to patients.

Paradigms persist because they generate new problems for practitioners. A paradigm is successful not because it is complete, but rather because it invites practitioners to work out more of its details and to account for anomalies within its framework. Practitioners then adopt a paradigm and take up its commitments because of *what the paradigm can do.* That is, scientists have rational reasons to accept and work within a paradigm.

In contrast, for Gilbert, the reasons why agents choose to express openly a willingness to be jointly committed as a body are incidental. The joint commitment itself is the foundation of the social group, not the reason why the agents decided to form the joint commitment in the first place. Weatherall and Gilbert (2016) argue that theoretical and methodological commitments associated with a paradigm can best be understood as a set of foundational collective beliefs of a community of scientists. But why did *these* particular collective beliefs become paradigmatic and not another distinct set of collective beliefs? Gilbert is generally uninterested in giving an analysis of this question.[6]

Kuhn, however, does have a distinct analysis of why one paradigm triumphs over a competitor. The successful paradigm is better at solving puzzles and can account for more anomalies. When the paradigm comes to be accepted in a community, it seems that scientists sign on to the paradigm because they believe that the paradigm is good at solving problems that they have "come to recognize as acute." On Kuhn's description, adherents to a paradigm often have rational reasons to accept one paradigm over a competitor. Kuhn discusses this at length at the end of SSR Chapter XII, "The Resolution of Revolutions." In fact, in a later essay (ET 320–339), Kuhn explicitly lays out how competing theories can be rationally assessed. Theories should be accurate in their predictions within their domain, be consistent, be broad in scope, present phenomena in an orderly and coherent way, and be fruitful in suggesting new phenomena or relationships between phenomena.[7]

---

[6] For Gilbert, it does not matter what the reasons are because different members will have different and competing reasons to make themselves party to a joint commitment. Joint commitment is a general and basic fact about social groups. Members of one group can jointly commit to believing that p. Members of a competing group can jointly commit to believing that ¬ p. Gilbert is not interested in analyzing why some people believe p and while others do not. Instead, she is interested in analyzing how the joint commitment itself regulates the group and its members.

[7] Critics of Kuhn's account of paradigms often accuse Kuhn of introducing irrationality and relativism to scientific changes. But this is a misinterpretation of Kuhn's project (see Wray 2011b; Hacking 2012). While Kuhn's account of rationality in science may not satisfy those with more foundationalist leanings, Kuhn does not think scientists choose theories irrationally.

Once a paradigm becomes widely established and scientists are trained within it, scientists become very reluctant to reject it, even when faced with strong countervailing evidence. However, Kuhn's analysis of why practitioners of a paradigm take a long time to abandon an established paradigm diverges from the one given by Gilbert. No paradigm explains all anomalies. Much of the work done within normal science is trying to understand anomalies and account for them within the accepted methods and theories of the paradigm. However, members of the community openly discuss these anomalies as being unaccounted for within the paradigm. This is a signal that there is an open problem that would be a suitable research project within the bounds of the paradigm.

One of Kuhn's examples of how anomalies are handled within a paradigm is the discrepancy between Newton's original computation and the observed motion of the Moon's apogee, the point where the Moon is farthest from the Earth. This is a case of how an anomaly is resolved within normal science. For over sixty years after the publication of the *Principia*, physicists tried to resolve the discrepancy, without success. Newton's failed prediction regarding the motion of the Moon's apogee was well known and discussed. Kuhn even notes that during the time that the discrepancy remained unresolved, "there were occasional proposals for a modification of Newton's inverse-square law. But no one took these proposals very seriously, and in practice, this patience with a major anomaly proved *justified*" (SSR-4 81, my emphasis). Newton's inverse-square law proved too successful elsewhere to be abandoned, even when there existed important evidence against it. The community, in a sense, had faith that it could be explained. In 1749, Clairaut showed that the mathematics being used was wrong and that with simplified mathematics Newton's laws did accurately predict the observed motion. Kuhn uses this example to show that a persistent and recognized anomaly does not always induce crisis and cause the community to abandon a highly successful paradigm.

If the community had a collective belief in Gilbert's sense in Newton's laws, then we would expect those who speak out against it to be rebuked and ostracized. Perhaps we would expect that doubts cast on the Newtonian theory be formulated very cautiously and carefully, with a clear preamble and disclaimers. Yet, in this historical episode, it was the leading lights of the Enlightenment who openly discussed the problems facing the Newtonian paradigm.

For instance, d'Alembert had once remarked:

The [Newtonian] system of gravitation can be regarded as true only after it has been demonstrated by precise calculations that it agrees exactly with the phenomena of nature; otherwise the Newtonian hypothesis does not merit any preference over the [Cartesian] theory of vortices by which the movement of the planets can be very well explained, but in a manner which is so incomplete, so loose, that if the phenomena were completely different, they could very often be explained just as well in the same way, and sometimes even better. The [Newtonian] system of gravitation does not permit any illusion of this sort; a single article or observation which disproves the calculations will bring down the entire edifice and relegate the Newtonian theory to the class of so many others that imagination has created and analysis has destroyed. (quoted in Hankins 1985, 37)

In fact, Clairaut, d'Alembert, and Euler all publicly discussed that calculating the Moon's motion can only be approximated, which posed an important test for Newton's laws. In the summer of 1747, Euler explicitly acknowledges the need to correct the inverse-square law to account for the motion of the Moon in a paper read before the Berlin Academy of Science. Around the same time, d'Alembert also presents a paper to the Paris Academy and the Berlin Academy on the problem of the Moon's orbit. Clairaut, in late 1747, bolstered by Euler's and d'Alembert's earlier papers, openly suggests adding additional terms to the inverse-square law. This created much uproar among his colleagues. Note that in suggesting a modification to Newton's laws, Clairaut is openly questioning a core commitment of Newtonianism. Newton's laws had been so well confirmed on other matters that modifying the law just to account for lunar motion seemed unjustified and ad hoc. In particular, his peers criticized him because "Clairaut's additional term, specifically introduced to explain the motion of the Moon's apsides, implies that the laws that rule the motion of Earth around the Sun, the fall of an apple to the ground, and the orbit of the Moon are no longer one and the same" (Bodenmann 2010, 28). Euler rejected Clairaut's modifications on mathematical grounds (Bodenmann 2010, 29) – not because he took Clairaut to be a bad Newtonian. Surprisingly, in 1749, Clairaut finally found a way to reconcile the Moon's observed motions with Newton's laws, not through the addition of new terms but through a simplification of the mathematics used, thus confirming that Newton's laws were correct after all.[8] Clairaut turned himself from a heretic to the great defender of Newtonianism in the span of just two years.

Gilbert predicted that within normal science, anomalies would be ignored or explained away because members have jointly committed to upholding the core tenets of the paradigm. In this case, we see that

---

[8] For a full accounting of this historical episode, see Hankins (1985) and Bodenmann (2010).

anomalies were openly discussed and were actively worked on as open problems – a successful paradigm leaves some anomalies open for exploration. But as the Moon problem persisted, Clairaut, d'Alembert, and Euler all started to publicly question the inverse-square law in the 1740s. They did not suffer, as far as we can tell, a "psychologically unsustainable conflict" between their personal beliefs and the group belief. Rather Clairaut, d'Alembert, and Euler were eager to give a novel solution to the Moon problem to make their name. The result of their speaking out is not ostracism. Much of the rebuke Clairaut received challenged him on the merits of his proposal. It was, indeed, difficult for Clairaut to convince his community that Newton's laws needed to be modified in 1747. But the reason the community resisted can be explained by the previous success of the Newtonian paradigm. The resistance, in this case, to Clairaut's modifications was justified, Kuhn suggests, because there was a way to reconcile the calculated and observed motion of the Moon that confirms Newton's laws. The true mark of a successful paradigm is its ability to, eventually, account for the trickiest of anomalies. Clairaut's colleagues did not resist his modifications because they felt a sense of obligation to uphold what they have jointly committed to as a community.[9] Rather, they were skeptical that a paradigm with a long and proven track record ought to be abandoned without further investigation. The mathematical challenge of calculating the Moon's orbit – which essentially requires one to solve a three-body problem – is immense, so it was a rational response for the community to demand more proof and exploration of alternatives.

The reason scientists do not readily form new beliefs in response to new evidence is not that the previous beliefs arationally constrain members' ability to evaluate new evidence. Rather, epistemic groups are faced with several deep problems regarding the nature of extraordinary evidence, possible alternatives, and how to evaluate them. What it takes to rationally resolve these deep problems requires time and careful consideration by many group members. This accounts for why changing a scientific consensus is slow and difficult.

Gilbert's work on scientific change has been provocative and a fruitful starting point for thinking about the role of collective beliefs and evidence

[9] One may object to my characterization of the case here by arguing that Clairaut, d'Alembert, and Euler are all not British scientists, giving them a kind of outsider status. They were perhaps not the most committed to upholding the core paradigm. Defenders of Newton were much more dogmatic in Britain as opposed to the Continent. But even in Britain, the problem of the Moon's apsides was a well-known problem for Newton's theory. While not British, Clairaut, d'Alembert, and Euler all viewed themselves as Newtonians and working with the framework of Newtonian mechanics.

in theory change. While there are many striking similarities between Kuhn's observations about how normal science proceeds and Gilbert's theory about how collective beliefs work, there are several key features of normal science that the Gilbertian view cannot account for. In particular, Gilbert's view elides the important role evidence plays in group belief change. I have argued here that the *primary* normative constraint on group belief revision is the weight of the evidence being considered by the group, and not the normative constraints that arise from joint commitments. However, to fully understand group belief change, we will need a better theory of how epistemic groups respond to evidence. Kuhn's observations about the nature of normal science and anomalies are suggestive of a social epistemic account of how an epistemic group and its members can evaluate new evidence. Developing this account will be important for future work.

9 The Puzzle of Promise, aka "Kuhn's Problem"

Chris Haufe

> The issue is which paradigm should in the future guide research on problems many of which neither competitor can yet claim to resolve completely. A decision between alternate ways of practicing science is called for, and in the circumstances that decision must be based less on past achievement than on future promise.
> —Thomas Kuhn, *The Structure of Scientific Revolutions*, 4th ed., 157–158

9.1 Introduction

I have two main objectives in this chapter. The first is to articulate and refine our understanding of the puzzle of promise and the challenges it raises for the epistemology of science. The second is to lay the groundwork for a line of philosophical inquiry that might terminate in a solution to the puzzle. I pursue each of these objectives broadly within the constraints formulated by Kuhn himself. Before moving onto the main business, I would like to spend a little while developing a perspective on the significance of the puzzle of promise, both in general terms and as it relates to Kuhn's peculiar intellectual orientation.

9.2 Kuhn's Rhetoric and the Significance of the Puzzle

I think the first thing one needs to appreciate about the puzzle of promise is just how unabashedly Kuhnian it is. It is a classic illustration of what I will call *Kuhn's gambit*, a rhetorical device that appears over and over again in *The Structure of Scientific Revolutions* (SSR) as well as in many other writings. Kuhn's gambit involves three signature steps. First, Kuhn uses examples from sociology or history of science to affirm the widely recognized significance of a practice or event. For example, in his 1961 "The Function of Measurement in Modern Physical Science" (ET 178–224), Kuhn affirms the universally acknowledged importance of measurement, referencing Lord Kelvin's famous dictum concerning measurement as a prerequisite for knowledge. For his next trick, Kuhn provides an account of *why* the practice is significant that cuts directly against the most intuitive or received view. Did you think the reason measurement is important is because it is the only way to test whether the match between theory and observation is exact? Because if that is what you thought, boy, were you wrong. The real reason measurement is important is because it gives the practitioner a sense of what "reasonable agreement" between theory and data looks like. That is step two – yanking the epistemological rug out from underneath his hapless victim. For his closing move, Kuhn will argue on philosophical grounds that his particular interpretation of a practice's significance is logically unavoidable, once one appreciates the epistemic situation of the practicing scientist. "Reasonable agreement" has to be what practicing scientists are after because it is all a practitioner can hope for. "Exact agreement" cannot be achieved, due to the fact that the mathematical consequences of theories can be calculated to arbitrary degrees of precision.

The puzzle of promise unfurls in like fashion. Kuhn begins by reminding us of all those revolutionary moments in the history of science in which some great work transcended all others to define a new era of investigation. We all remember those, don't we? Of course; they are the stuff of scientific legend. Their overnight success, as we all know, derived from the fact that they were just able to solve more problems than "competing modes of scientific activity." Psych! No it didn't. The instantaneous takeover by the likes of Lavoisier, Newton, and Darwin was due to the *promise* their works appeared to possess in the eyes of a sizable portion of the relevant scientific community. The reason we know this is because we know it *could not* have been a matter of which theory solves more problems – the incumbent has typically solved a great many; the challenger, only a few. Checkmate.

Kuhn's gambit is symptomatic of the general rhetorical tenor of SSR. Part of the great power of the book lies in how it is provocative without being outlandish. Taking on "the image of science with which we are now possessed" is no easy task, because the epistemological catechism is so deeply engrained that it has become the very lens through which we understand the elements of scientific inquiry. Kuhn's gambit works by inverting that lens. Instead of looking *back* at the track record of success laid down by a paradigm, or an element of scientific practice, Kuhn consistently directs our attention to the *future* that paradigms and practices portend. Appearing toward the very end of the book, Kuhn uses the puzzle of promise as a crescendo to mark the culmination of a series of gambits, all of which have been designed to lead us to the fundamental question facing a scientist: Which paradigm should in the future guide research? The future-directed focus of scientific decision-making is the master theme that propels the shift in vision effected by SSR. The puzzle of promise is a statement of that theme in the clearest possible terms.

The second feature of the puzzle's significance as it relates specifically to Kuhn's thought is the central role that it plays in criticism of his epistemology of science. By insisting that (1) scientists are laser-focused on how to guide research in the future, and (2) problem-solving track records do not play an important role in these decisions (because new paradigms would always lose to incumbents), Kuhn left very little epistemological space for arguing that paradigm choice could be made on any basis other than "faith." Anyone who has followed the discussion of SSR from 1962 until relatively recently is painfully aware of this claim's legacy. His focus on the search for future promise and his explicit invocation of faith in guiding that search invited a torrent of different forms of abuse, generally oriented toward the perception that Kuhn's image of science was as of an arational process. Philosophers of science were scandalized. Historians and sociologists found it oddly arousing.

It is worth noting that the specific threat posed to the rationality of science by the puzzle of promise differs from the one posed by incommensurability. In Kuhn's framework, incommensurability threatens the rationality of science by making it impossible to appraise the merits of any paradigm other than the one currently in place (and not really even then). But suppose, contrary to Kuhn, that we could do a side-by-side comparison with respect to some epistemological desideratum, and base our paradigm choice on which paradigm best satisfied it. The rationality of paradigm choice is no longer threatened by incommensurability. Yet it remains under threat nevertheless, because no one has told us how to rationally assess the promise of a paradigm, and (according to Kuhn) these choices are being made on the basis of future promise. In fact, forget other paradigms altogether. Questions about future promise do not just stop once we have decided on a paradigm. They are ever-present. They apply to all the elements within a paradigm. The arationality of these judgments does not emanate from a communication breakdown. Rather, it flows from deeper questions about what could, even in principle, constitute evidence for future promise. Kuhn's appeal to "faith" indicates where his sympathies resided.

This points to a further, deeper sense in which the puzzle of promise is significant for understanding the legacy of SSR and of Thomas Kuhn. Kuhn was well aware that he was breaking with traditional epistemological frameworks. But we are never given a statement regarding what the fundamental problem is; and we are told even less about what, if anything, he is offering in its stead. Certainly, we are presented with specific reasons for rejecting this or that philosophical view, such as his direct attack on falsificationism in Chapter 9 and his direct indictment of positivism in Chapter 12 (and indirectly in several other places). None of these attacks, however, move on to articulate a general epistemological misorientation of the sort to which Kuhn alludes in his introduction. The puzzle of promise discharges this burden in the way it suggested that dominant forms of theory appraisal were asking the wrong question (and probably still are). The epistemology of science during the twentieth century had been overwhelmingly concerned with the question of "To what degree *has* the theory *been* confirmed?" Running counter to this cultural mainstay, Kuhn asks, "What promise does the theory hold for the *future*?" These two questions bring into focus completely distinct epistemological concerns. The puzzle's propensity to raise fundamentally novel epistemological questions lies at the center of Kuhn's divergence from pretty much all extant epistemology of science.

There are a couple of issues raised by the puzzle whose intellectual significance is broader than its relation to Kuhn. One of them comes to the fore in a deceptively subtle comment that Kuhn makes on behalf of promise-based concerns in his famous paper "Objectivity, Value Judgment, and Theory Choice" (ET 320–339). In the midst of ticking off all the traditional epistemic values – simplicity, scope, and so on – Kuhn singles out promise-based concerns as being "of special importance to actual scientific decisions." *Actual* decisions? What other kind of scientific decisions are we interested in? There is a strong temptation to believe that the inclusion of "actual" here is deliberate allusion to the nearly universal tendency among philosophers of science to not really pay any attention to the way scientific decisions-making *actually* works. SSR is often credited with issuing in an era of practice-oriented studies of science. The puzzle of promise plays no small role in this legacy, because one simply cannot ignore the fact that promise – unlike, say, truth – *has* to be an important factor in actual scientific decisions. Scientists cannot live in a theory's glorious past. They carry their investigative tools into the future, and any decision they make needs to keep that future firmly in mind. The glory days matter only to the extent that they inform our estimates of a theory's promise. In this way, the puzzle compels attempts to understand the nature of scientific inquiry to pay closer and closer attention to the details of how scientific practice actually works.

Last, the orientation toward promise-based considerations provides a fundamentally new perspective on traditional epistemological questions. We touched briefly on how the notion that paradigm choice is made on the basis of future promise raises concerns about rationality (further discussion below). But even something as basic as the notion of *evidence* becomes reinvigorated through the lens of promise. What counts as "evidence for future promise"? Kuhn's direct answer in SSR is that nothing does; that is why we have to go on faith. But he also says a lot of other stuff, stuff about the kinds of properties to which practitioners are paying attention when they choose between paradigms. He clearly does not think that the choices are arbitrary. And yet he clearly does not think that those choices are geared solely toward the acquisition of truths about nature. Reflection on what might constitute evidence for future promise raises questions that simply do not arise when reflecting on what might count as evidence for truth. Ditto for other paradigmatically epistemic concerns. Our views, for instance, about what makes a scientific inference inductively valid when the goal of inductive inference is to maximize future promise may prove to be inconsistent with what we think is required to maximize probable truth. These are the sorts of really deep philosophical challenges that arise when we try to take seriously Kuhn's rhetorical question of "How could the history of science fail to be a source of phenomena for which theories about knowledge are legitimately asked to apply?"

9.3   Historical Development of the Puzzle

I now want to look at the evolution of the puzzle of promise over three different phases in Kuhn's writing, starting with a more careful discussion of the contours as it appears in SSR. There he asserts that a decision directed toward future promise "can only be made on faith." Kuhn's very deliberate language here is designed to evoke the distinctively antirational, or at least arational, resonances of the term "faith." Probably Kuhn liked it also because of its contrast with cultural conventions that depicted science as being opposed to faith by its very nature. In this iteration of the puzzle, Kuhn is clearly enamored of the notion that promise-oriented decisions cannot be reconciled with an image of science as a rational process, let alone a *paradigmatically* rational one.

Admitting that the use of the term "faith" is strictly for its rhetorical power, however, does not explain why Kuhn *thinks* that the decisions must be made on faith, or some suitably arational avatar for faith. The primary reason he cites is that any new paradigm will have a poorer track record than an incumbent when it comes to sheer number of problems solved. Is Kuhn suggesting that, were the challenger to have solved more problems than the incumbent, the choice would be rational? Part of me wants to say, "Yes." But a couple of important considerations tug in a different direction. First, it is not immediately obvious why a superior track record of problem-solving should translate into *future* promise. After all, history would seem to indicate that that inference is far from watertight. Second, much of what is said elsewhere in SSR clearly cuts against the idea that comparing problem-solving track records is a viable means by which to rationally assess the relative virtues of competing paradigms. For one thing, the meaning of "solved problem" should differ across paradigms, which characteristically place different demands on what a solution to a scientific puzzle needs to look like. When we compare the record of solved problems for one paradigm to the record for another, we are comparing incommensurables. Kuhn argues for this exact point in the Postscript (SSR-2, SSR-3, SSR-4). In point of fact, however, he does not think practitioners are normally in this position anyway, since challenger paradigms more or less lack track records altogether. The only reason the challenger has appeared in our radar is because it does a good job with one or two high-value problems. So the part of me that wants to say "Yes" is correct: the essence of the puzzle lies in the fact that there would seem to be no evidentiary basis for projecting the future promise of a paradigm. Choosing on the basis of formal or aesthetic criteria, as practitioners in crisis mode will frequently do (SSR-4 154–158), apparently amounts to faith-based decision making in Kuhn's view as articulated in SSR. These criteria are not the sort that can anchor the *rationality* of a choice.

Phase II of the puzzle is his surprising embrace of some traditional philosophical views in the Postscript to the 1970 edition of SSR. Postscript-Kuhn is a considerably more measured Kuhn when it comes to fundamental questions about epistemology. He is also much less enthusiastic in his efforts to emphasize the arationality of science. Notably absent is any appeal to faith or other enemies of reason. Moreover, what appeared to be a relatively minor point about the use of formal and aesthetic criteria in SSR (the purpose of which was to underscore the extent to which paradigm choice could not be rationally sanctioned) is elevated to a central feature of Kuhn's thinking about the competition between paradigms. The Postscript insists there might be "good reasons for choice" and they need not differ "from those usually listed by philosophers of science – accuracy, simplicity, fruitfulness, and the like" (SSR-4 198). Yet here again we see the irresistible pull of Kuhn's gambit, as he coyly reminds us that just because practitioners agree on the importance of, say, fruitfulness does not mean they agree on what fruitfulness *is*.

No cause is advanced by denying that Kuhn is flatly going against the thrust of the puzzle of promise as it appears in SSR, where it was supposed to show that paradigm choice is not governed by reason. What happened between 1962 and 1970? Basically, he got served. I think the reference to faith was the one place in SSR where Kuhn was more radical than he needed to be, and he was rightly criticized for it. Obviously, this was not the only thing for which he was criticized; the incommensurability thesis, for instance, endured more than its share of vitriol. However, there he stuck to his guns. He *never* backed down from his claims about world changes (see RSS 224–252, his last published article). And, indeed, that element of his thought became increasingly plausible over time (Hoyningen-Huene 1989/1993; Wray 2021a, part IV). But in the case of faith-based decisions, he seemed to recognize that he had gone too far. In place of faith, he supplies the well-known epistemic values, which he even deigns to concede might be "good reasons for choice." How far we are from faith! For all that, though, there is no sign that he has moved beyond the idea that paradigm choice is made on the basis of future promise. The Phase II iteration of the puzzle, then, is not about whether paradigm choice can be motivated by reasons but about how the admittedly good reasons like accuracy, simplicity, and fruitfulness relate to future promise.

Phase III of the puzzle is where things really get cooking. The year 1977 saw the publication of "Objective, Value Judgment, and Theory Choice" (ET 320–339) and with it the reemergence of direct discussion of promise. There Kuhn specifies that the kind of promise that matters in theory choice is "the promise of concrete successes" (ET 322, n. 6), which since SSR had meant solved problems. (Some readers will recall a lovely passage in Chapter 3 of SSR in which Kuhn describes normal science as "the actualization of ... a promise of success"; see SSR-4 24). Interestingly, this reference to the promise of concrete success takes place amid Kuhn's explication of the term *fruitfulness*. His explication reads like a refinement of the various references to fruitfulness he had made over the years, where that notion was never further articulated. Because of this intimate connection between fruitfulness and promise, I frame the final version of the puzzle of promise in terms of fruitfulness: *How are judgments of fruitfulness potential rationally justified?* The puzzle of promise, in my view, amounts to understanding how practitioners discern a potential for fruitfulness.

9.4   Solving the Puzzle

Let us now turn to thinking about how we might solve the puzzle. To set things up, I am going to outline its major features and constraints as given to us by Kuhn, as well as some philosophical demands that a solution to the puzzle would need to satisfy. I then look in detail at Kuhn's attempt to solve the puzzle, which I regard as, at best, a bizarre lapse in judgment. I conclude with some brief suggestions for how we can do better.

The constraints I have in mind are the major features of the paradigm decision-making environment as described by Kuhn. We have seen a number of these already, such as (1) that the decisions are not based on track record, (2) that debates are not settled by rational argument (though often depicted as if they are), and (3) that decisions are frequently governed by "aesthetic" or "formal" criteria (later "values"). These had been in place since SSR in one form or another. But there is another constraint that, I think, adds some much-needed clarity to the question of whether practitioners can pursue a choice-making strategy that they can reasonably expect to advance inquiry – namely, Kuhn's assertion in the Postscript that "the demonstrated ability to set up and to solve puzzles presented by nature is ... the dominant criterion for most members of a scientific group" (SSR-4 204).

The "ability to set up" problems does not receive much attention in SSR. Given how interested Kuhn became in it later, I am inclined to believe that he came to realize that something important had evaded his notice when he was thinking through SSR. This can be seen in the evolution of his thinking about Newton's second law, $f = ma$. Recall that in SSR Kuhn draws on Norwood Hanson's discussion of the law to make the claim that it "behaves for those committed to Newton's theory very much like a purely logical statement that no amount of observation could refute" (SSR-4 78). Hanson's section on $f = ma$ does include some discussion of this point. But it is part of a broader context aimed at understanding *why* the law is treated that way by practitioners. And there much of what Hanson says focuses on the *use* of the law as a "schema" or a "technique" – in other words, a way of setting up problems (Hanson 1958, 99–105). A technique cannot be proven false, because a technique cannot be true or false. A schema is an ordering principle that, like a statement in sentential logic, can be relatively indiscriminate with respect to the content of the elements to be arranged. It is precisely this organizing or structuring function of the law in scientific practice that comes to occupy Kuhn's interest in $f = ma$ as a "shared example" in the Postscript. The reason it "behaves ... like a statement which no amount of observation could refute" is because it is not really conceived of as a statement at all. It is a rule for structuring problems in order to do calculations. You put your "force" stuff on the left side, you put your "mass" and "acceleration" stuff on the other side, and voila! Kuhn's focus on the "ability to set up" problems would appear to undergo significant development in the following years, reaching its full maturity in his essay, "Second Thoughts on Paradigms" (ET 293–319), which was published for the first time in Frederick Suppe's edited volume, *The Structure of Scientific Theories* (1974). "Second Thoughts" shows a Kuhn deeply immersed in the realization that paradigm choice is largely about exemplars, and that what exemplars give practitioners is largely the intuitive ability to structure problems.

Then there is the issue of which properties a constraint-obeying solution to the puzzle of promise ought to exhibit if it is to be philosophically satisfying. The first of these is the demand that *a solution should explain the salient empirical facts concerning how paradigm choice appears to operate.* If, for example, scientific communities tend to adopt a new theory or paradigm only when it has achieved novel predictive successes, then our solution to the puzzle should explain why novel predictive success is generally considered relevant to the acceptance of a theory. Or, if novel predictive success does not figure significantly in paradigm choice, that might also require explanation, it being so intuitive (and therefore salient) that novel predictive success *would* feature prominently in paradigm choice. Of course, Kuhn made no secret of the fact that then-dominant philosophical models could not explain the facts concerning how, historically, paradigm choice has worked. The puzzle of promise encapsulates one such fact: historically, practitioners have chosen paradigms with an aim toward future promise. Kuhn's view in SSR was that the choices are based on faith, which is not explanatory. Why, if you were aiming toward future promise, would you choose your paradigm on the basis of faith? It makes no sense. Now, if, as Kuhn asserted, you *had* to choose on the basis of faith, then that would explain why you do in fact choose on the basis of faith. But, as we saw, that model of paradigm choice did not survive beyond SSR, despite being explanatory in its own way. I do not suppose anyone was sorry to see it go.

The second criterion, chosen by a panel of experts consisting of myself and no one else, is that *any solution to the puzzle is going to need to offer fresh insight into the nature of epistemic justification.* This is really the heart of the puzzle itself, for it was precisely Kuhn's assertion in SSR that choices aiming toward future promise could not be epistemically justified that got us into this whole mess in the first place; "faith" is just another word for "nonepistemic considerations." In fairness, though, there is one thing that Kuhn got right with his appeal to faith: if these choices are epistemically justified, the sense of justification has to be fundamentally unlike any conception of which we are currently aware. Any account of paradigm choice that successfully validates the epistemic rationality of the practitioners in Kuhn's narratives will *per force* reveal something about the nature of epistemic justification that is quite distinctive.

The strange thing is that, like the demand for explaining the actual behavior of choice, Kuhn clearly embraced in SSR the idea that epistemic justification was nothing like we thought it was. And, indeed, there are various points throughout the book, mostly concentrated in the first couple of chapters, in which Kuhn seems to be preparing to take epistemic justification in an entirely new direction. He liberally heaps abuse on dominant epistemologies. He tells us how much the history of science is going to change our image of science. He all but mocks the idea that the history of science might be irrelevant to theories of knowledge. But we never end up getting any statement. Kuhn brings the ball to the one-yard line and then punts it – hard – sending it sailing out of the stadium along with his credibility.

Another possibility is that Kuhn *does* offer an alternative conception of justification in SSR, but that it is so radically different from traditional views that it becomes difficult to recognize it as a species of justification. That is actually pretty plausible. The cornerstone of Kuhn's account of paradigm change is that of a shift in the way practitioners see the world. It is not really a "choice" made on the basis of "reasons," because it is not really a choice at all. Any sense of epistemic justification that is to be salvaged from SSR is going to have to accept the fact that new paradigms are not chosen; they choose us. Kuhn, aware of the corner into which he had painted himself by the end of the book, cedes all the remaining reason-giving ground to faith. But perhaps he has not thereby given up the epistemic ghost entirely. Insisting that any workable conception of epistemic justification must be partially grounded in an ability to articulate reasons or in a set of conscious motivations for a decision fails to do justice to the various species of knowledge – Polanyi's tacit knowledge, for instance – that simply do not work that way. If scientific thought is as deeply intertwined with perception as Kuhn believed in SSR, it is probable that the only kind of epistemic justification that would have made sense to him is one that has more to do with cognitive frames and reflexes than it does reasons for justified belief. Of course, that would be a sense of justification with which the philosophy of scientific knowledge has yet to come to terms. But it certainly would, if successful, offer fresh insight into the nature of epistemic justification.

The third demand I would place on a solution to the puzzle of promise is *that it shed light on why the pursuit of promise works so well*. Apparently, basing decisions concerning how in the future to guide scientific research on what would maximize future promise is a really good idea, because scientific knowledge increases perpetually in depth and in breadth. Why is that? As with the previous two demands, Kuhn is very much aware of the importance of discharging a philosophical burden of this variety. In fact, this one is particularly close to his heart. For recall that a common version of Kuhn's gambit involves revealing the shocking truth concerning the epistemic function of a practice generally known to be important already. His preferred explanation for the practice's significance typically elaborates a causal chain leading from his version of the practice's function to an end that seems plainly to be important for the development of scientific knowledge. This convention is observed, for example, in his explanation of why the acquisition of a sense of "reasonable agreement" is important (see above). In many ways, SSR is a giant causal model of paradigm change, composed of smaller scale causal models, each of which occupies a node in the process by which revolutions occur. However, when it comes to the pursuit of *promise*, SSR is silent. It is not trivial to ask why decisions made on the basis of future promise should so routinely result in the growth of knowledge. Those kinds of decisions might just as easily have resulted in utter epistemic devastation. But they do not. They are *productive*.

9.5   Kuhn's Do-Over

We have seen how faith drops out of the story after SSR, replaced by philosopher-approved epistemic values as the basis for paradigm choice. Kuhn's capitulation on this issue seems strongly to indicate that he was sensitive to the fact that damage had been done to the cognitive authority of science by insisting that paradigm choice was arational. But how helpful is his turn to epistemic values, really? Does it fortify the citadel of scientific reason? Does it move us toward a solution to the puzzle of promise, along the lines I have articulated above? I will argue that it does not. Although the puzzle remains unsolved, some of what Kuhn says contains the seeds of a potential solution.

Before we begin assessing the ways in which Kuhn's mature model of paradigm choice fails to solve the puzzle, I want to look at a subtle but important shift in its development that bears on the comparative worth of different stages of his view. When the epistemic values appear in the Postscript, they are held up by Kuhn as potentially "good reasons" for paradigm choice. As the years went by, however, Kuhn seems to have increasingly come to see choosing on the basis of accuracy, simplicity, scope, and so on as an *analysis* of rationality itself. From about 1977 onward – after the publication of "Objective, Values, and Theory Choice" (ET 320–339) – he consistently adheres to the custom of asserting that to choose on the basis of accuracy, simplicity, scope, and so on was just what *rationality* means in the context of "evaluating scientific belief" (RSS 251).

This is not that surprising; it is a very Kuhnian thing to do. It is what one would have expected Kuhn to say at the end of SSR, rather than claiming that decisions were made on the basis of faith. It is what one would have expected Kuhn to say in the Postscript, rather than the more modest claim that the canonical epistemic values were candidates for "good reasons" for paradigm choice. And, to be fair, the Postscript discussion of values does start trending toward that mature post-1977 view, as I suggested above with regard to his reference to practitioners' differing interpretations of fruitfulness, as well as of the epistemic worth of fruitfulness. But after 1977, the overt Kuhnianness of his view on the connection between rationality and the epistemic values is explicit. For a Kuhnian of sufficient purity, different paradigms will necessarily involve different criteria for what constitutes a rational decision, or what constitutes a solved puzzle. In the epistemological paradigm that governs modern science, "Accuracy, precision, scope, simplicity, fruitfulness, consistency, and so on, *simply are the criteria which puzzle solvers must weigh* in deciding whether or not a given puzzle about the match between phenomena and belief has been solved.... [T]hey are the 'defining' characteristics of the solved puzzle" (RSS 251–252; emphasis added). This comes from Kuhn's last published essay. But the tenor is not substantially different from his discussion in "Objectivity, Value Judgment, and Theory Choice." For those accustomed to reading Kuhn, it will not escape notice just how out of character this statement is.

Without articulating exactly what is so weird about it, my hypothesis is that its awkwardness stems from the fact that Kuhn was trying to have his cake and eat it too. In his heart, he never really abandoned the idea that decisions aiming at future promise could be made only on faith. While he ultimately was forced by cogent philosophical criticism to admit that scientists can, in principle, base their decisions on "good reasons," he still longed to retain the anti-epistemic spirit that had tempted him in SSR. Eventually, he was able to reach a detente with epistemology by claiming that, although accuracy and so on are the criteria for rational decision making, there is nothing special about those criteria necessarily; that is just what "rationality" means. Modern science could have developed based on entirely different choice criteria. If modern science had evolved the use of criteria other than the ones with which it is actually associated, we would have used the term "rationality" to refer to those alternative criteria instead. Thus, while he abandons the appeal to faith, it amounts to the same thing – he deprives us of any means by which to argue for the cognitive authority of science over other forms of inquiry or belief formation.

Now then, let us see how Kuhn's mature position for paradigm choice fares along the dimensions that matter for a solution to the puzzle of promise. The first question is, does Kuhn's view explain the empirical facts concerning what grounds paradigm choice? The evidence is equivocal, but it equivocates between "not really," and "definitely not" – so, not exactly a win for Kuhn. The issue turns on whether or not defining rationality in terms of the canonical choice criteria amounts to an explanation of the facts about paradigm choice or not. Here is a sense in which it could explain those facts: pre-theoretically, we assert (reasonably) that modern science exemplifies what it means for inquiry to be rational. Based on that pre-theoretic designation, we predict that, historically, paradigm choice can be expected to conform to the criteria we pre-theoretically associate with rationality (accuracy, scope, etc.). When we look at the history of science, we find (with Kuhn) that this is in fact how the choices are governed. The fact that science is a rational process explains why actual paradigm choices exhibit the particular behavior that they do. In this way, Kuhn's mature view explains the empirical facts concerning what grounds paradigm choice. This is the "not really" version.

An alternative rendition goes like this: pre-theoretically, science exemplifies rationality. Therefore, whatever criteria govern scientific decision making are the criteria for rational decision making. Therefore, accuracy and so on are (at least some of) the criteria for rational decision making, because that is how scientific decision making in particular is governed. On this rendition, Kuhn's mature view explains the empirical facts concerning what grounds paradigm choice because it implies those facts. This is the "definitely not" version, because it is trivially true that Kuhn's view explains those facts. It was designed to accommodate those facts: find out how decisions are actually made, and then define scientific rationality in relation to those findings. Neither this nor the "not really" version are very attractive. The only thing that prevents the "not really" version from being a complete disaster is the fact that there are good pre-theoretic motivations (more on this below) for designating modern science as the exemplification of rational inquiry. But they are not the motivations to which Kuhn appeals, and in no way does he ever attempt to link the epistemic values to those motivations.

The second question is whether Kuhn's mature view offers fresh insight into the nature of epistemic justification. I think it is obvious that it does not do this at all. He states explicitly that he will take on board the canonical rational choice criteria as articulated by philosophers of science. We get no new perspective on how epistemology might be transformed by the image of science that its actual history reflects. It essentially dials back one of the most radical advances in SSR; certainly, it is the most radical advance on a distinctively philosophical matter.

The last issue has to do with whether Kuhn's mature view explains – or is able to explain – *why* choices that are based on the canonical epistemic criteria work so well for the advancement of science. Why do choices based on these criteria so routinely result in solved puzzles, that is, in the realization of the promise of concrete successes? Nowhere in his published work does Kuhn address this question. He addresses analogous questions. Take, for example, the issue of why normal science results in solved puzzles. Kuhn offers a plethora of causal mechanisms that contribute toward this end. Or how about his causal explanation for why it is pedagogically "unexceptionable" to suppress the history of science in our efforts to train new scientists (SSR-4 142)? He is happy to explain why "novelty for its own sake is not a desideratum in the sciences" (SSR-4 169). There are so many examples of this kind of discussion in Kuhn's work. Yet when it comes to the criteria for paradigm choice, the issue of their causal efficacy never comes up. I find it impossible to persuade myself that these questions never occurred to him. For instance, he would have been well aware of the fact that philosophers of science had long been interested in explaining what was epistemically efficacious about simplicity. He would have been cognizant of the fact that philosophers of science working in the dominant mode would have naturally assumed that the reason accuracy is epistemically valuable is because it is a guide to truth; this would have been a perfect venue in which to whip out Kuhn's gambit and mercilessly thrash them with it.

So why does he never address these kinds of questions in the context of paradigm choice? I think that, deep down, we all know the reason: it is another reflection of the fact that he never really abandoned faith in the arationality of paradigm choice. Kuhn's decision to designate the epistemic values as an analysis of rationality has the same effect on our ability to do epistemology that his appeal to faith does. In the framework provided by his mature view, it is not even possible to ask *why*, for example, accuracy is an epistemic ideal capable of exemplifying rationality, or why it works well as a governing ideal; it is just one of reason's "'defining' characteristics." This approach keeps paradigm choice as far out of epistemological reach as does faith. In neither case does it make sense to ask for a causal story about the relationship between the mechanics of choice and their effects – or really any explanation of any kind. What makes accuracy a "good reason"? That is not a question for which candidate answers are in short supply. And Kuhn *never* suffered from a shortage of insight into the causal processes that facilitate the solving of scientific puzzles. But his framework precludes us from asking: "Why accuracy instead of, say, inaccuracy?" He has no problem explaining why novelty for its own sake is undesirable as a governing value (it is a natural byproduct of his model), or why falsification is not psychologically plausible from the perspective of actual scientific decision making. It is not plausible to think that his silence on the explanatory power of the epistemic values was an oversight. By framing them as an analysis of rationality, he hoped to render them explanatorily inert.

9.6   The Salience of the Setup

In this last section I refocus our thinking about the puzzle of promise in the light of two considerations that Kuhn made explicit only after SSR: (1) the status of the "ability to set up" problems as the "dominant criterion" for puzzle-solving and (2) the elaboration of fruitfulness, "the promise of concrete success," and its status as a criterion "of special importance in actual scientific decisions." I do not think it is accidental that these two considerations are independently singled out as occupying their own special stratum in the decision-making hierarchy.

Consider the paradigm-defining scientific achievements to which Kuhn appeals: Newton's *Principia* and *Opticks*, Franklin's *Electricity*, Ptolemy's *Almagest*, and so forth. In SSR Kuhn is very focused on the fact that new paradigms are embraced after solving an important outstanding empirical problem or two. To be honest, he is overly focused on it. So much so that, when he finally turns to epistemological questions surrounding paradigm choice in Chapter 12, he can only assign epistemic value to the problems that the challenger paradigm has solved. But that is not all that these new theories give us. When is a solved problem more than a solved problem? When it is treated as an exemplar of how to solve problems. Because Kuhn had been insufficiently appreciative in SSR of the importance of the ability to set up problems, he was not quite able to see that the absence of a long track record of success did not necessarily preclude practitioners from perceiving within a single solved problem a whole possible future for inquiry – or, at least, a possible future for certain problems under whose recalcitrance they had long suffered. They can see in Newton's derivation of Kepler's first law more than the fact that a body in uniform rectilinear motion will define an ellipse when subject to continuous force directed at a stationary point. They see a whole strategy for setting up and solving problems of motion. They can see in his geometric derivation of the inverse-square law a general program for calculating magnitudes associated with force. They see what I. Bernard Cohen called "the Newtonian style" (1980). Yes, Newton solves a number of important outstanding problems. But those solutions are not of much value to most practitioners unless they suggest something about solutions to other problems. And if they succeed in *that*, then the absence of a track record is easily forgiven.

Seen in this way, it makes sense why "the ability to set up and solve problems" would be the dominant choice criterion. If the "promise" that attracts "an enduring group of adherents away from competing modes of scientific activity" is the promise of solved problems, then we should expect ability to set up problems – to set them up for inevitable solution – to ascend to a position of special importance. And if that is right, then we can understand why fruitfulness is *also* singled out as being of special importance: if the promise of concrete success rests in the ability to set up problems, and fruitfulness is the promise of concrete success, then a paradigm's fruitfulness must rest in its ability to set up problems. In other words, an ability to set up problems is what grounds attributions of fruitfulness. As the ability to set up problems is the dominant choice criterion, the other criteria – accuracy and the like – become subordinate to problem setup; they are epistemically valuable to the extent that they facilitate the setting up and solving of problems.

Here is one significant virtue that this account has over both Kuhn's SSR view and his mature view: it actually explains the empirical facts concerning theory choice. Accuracy and so on are not valued simply because they are conceptually associated with rationality. They are valued because of their causal relationship to problem-solving; *ceteris paribus*, increasing accuracy helps with the setting up and solving of problems. Similarly for the other epistemic values Kuhn mentions. Or, if they do not help with the setting up and solving of problems, that is something we could discover and then revise our conception of rationality accordingly. There is no room in Kuhn's mature view to discover that accuracy is not a defining criterion for rational choice making in scientific contexts. That would be a little like discovering that all bachelors are actually married. Essentially, this account opens the door to learn that we were wrong about what is epistemically valuable.

The account also promises new insight into the nature of epistemic justification in a couple of ways. First, it raises the possibility that the canonical picture of epistemic justification as determined by the use and satisfaction of the canonical epistemic criteria could be overturned. It all depends on what we find has, historically, genuinely contributed to the setup and solving of problems. More significantly, though, it frees us entirely from the reliance of epistemic justification on a noteworthy track record of success, or on any sense of "confirmation." All that matters is what sorts of properties contribute to practitioners' ability to set up problems for solution. It is easy to see how properties like simplicity could feature in such a process; judgments of simplicity do not depend on track records. Similar causal stories could be told for the other canonical epistemic criteria (or, failing that, alternative criteria). Aside from the sheer excitement this prospect holds, there are good historical reasons why we would want to be freed in this way. Because, for each of the problem-solving revolutionary scientific works that Kuhn cites, there are scores more that had similar effects on the scientific community but *which did not actually solve any problems.* How can something attract an enduring group of adherents without actually solving any problems? As I suggested above, and as Kuhn states explicitly ("paradigm debates are not really about relative problem-solving ability"; see SSR-4 156), it is never really about the specific solved problems. It is about the ability to set up problems. Practitioners can see a strategy for structuring problems even in a failed attempt to solve problems. Exactly how they can do this is the heart of the puzzle of promise. I provide a painstaking yet philosophically edifying sketch in my forthcoming book *Fruitfulness* (Haufe 2024).

The foregoing suggests an alternative conception of scientific rationality, potentially at odds with the one upon which Kuhn eventually settled: for a scientific decision to be rational is for it to be based on criteria that contribute to the ability to set up problems for solution. We do not need faith here (well, maybe a little). More than anything, we need to know what sorts of structural properties make it easy for us to solve problems. My conjecture is that the canonical epistemic virtues can be convincingly tied to some cognitive need or other that, when satisfied, streamlines the problem-solving process. It also allows for the kind of diversity in epistemic values that Kuhn briefly mentions in the Postscript. However, it is not a diversity that depends on membership in divergent disciplinary language communities (despite being consistent with that). Rather, it acknowledges that people differ psychologically from one another to varying degrees, and those differences might easily translate into differences in what individuals find most helpful to their ability to structure and solve problems.

10 The Orwellian Dimension of Scientific Progress

K. Brad Wray

10.1 Introduction

In Chapter XIII of *The Structure of Scientific Revolutions* (SSR), where Kuhn discusses progress in science, he notes that after a revolutionary change of theory, scientists need to write new textbooks to incorporate the new theoretical perspective. The revisions to the textbooks, though, do not merely involve the addition of the new discoveries that led to the change of theory. Rather, the task inevitably involves some "rewriting" of the history of the discipline. Kuhn suggests that there are parallels to George Orwell's *1984*, where history is constantly being rewritten. According to Kuhn, after a revolutionary change of theory, the "new" history of a scientific field is written to emphasize the continuity through the change, thus supporting a widely held conception of scientific progress, one that Kuhn believes misrepresents the facts.

My aim in this chapter is to examine (1) the role Kuhn's comparison to Orwell's *1984* plays in his argument and (2) the significance of the rewriting of a discipline's history after a change of theory. The process, I argue, tells us something about both (1) how scientists are trained to work effectively and (2) the nature of the changes that occur during a scientific revolution, changes that Kuhn provocatively compared to *world changes*. I want to begin, though, by examining some hitherto unknown background regarding this discussion in SSR.

10.2 Some Background

Kuhn's passing reference to Orwell's *1984*, where he discusses scientific progress, is striking for at least two reasons. First, it is interesting that Kuhn decided to make a reference to a popular novel, especially one that was published so recently. Orwell's novel was published in 1949, only thirteen years before SSR was published. Second, this reference was a late addition to SSR. There was no reference to Orwell's *1984* in the manuscript version of SSR that Paul Hoyningen-Huene christened as

"*Proto-Structure*" (TSK Archives–MC240 1961b). *Proto-Structure* is the draft of SSR that was completed around April 1961 and circulated among some trusted colleagues, as Kuhn solicited feedback (see Hoyningen-Huene 2006b, 611).

There are some significant differences between SSR and *Proto-Structure*, some of which have been noted before. Most notably, *Proto-Structure* does not contain Chapter V of SSR, "The Priority of Paradigms." Sometime after April 1961, the chapter on normal science, Chapter IV in *Proto-Structure*, was developed further and divided into two chapters. Thus, the chapter in *Proto-Structure* titled "Normal Science as Rule Determined" becomes "Normal Science as Puzzle-Solving" and "The Priority of Paradigms" in SSR (see Hoyningen-Huene 2015a, 187). Clearly, this suggests that as late as spring 1961 Kuhn was still working out the exact relationship between normal science and paradigms. This is not particularly surprising, since, as I have argued elsewhere, Kuhn really did not work out his view on paradigms until the early 1970s, about a decade *after* the publication of SSR (see Wray 2011a).

George Reisch has suggested that Kuhn added this extra chapter on paradigms in an effort to address criticisms that James B. Conant raised about the paradigm concept after reading *Proto-Structure* (see Reisch 2016, 15). Specifically, Reisch claims, the new chapter is intended to show that "paradigms are indispensable for understanding science because they are prior to and *more* fundamental than the theories, conceptual schemes, or climates of opinion recognized in extant accounts of how science works" (Reisch 2016, 15; emphasis in original). Further, in the published book Kuhn emphatically claims that the application of a paradigm to a scientific problem is *not* a rule-governed activity. Clearly, this view is at odds with the chapter title in *Proto-Structure*: "Normal Science as *Rule Determined.*"[1]

But there are other noteworthy differences between SSR and *Proto-Structure* that have not yet been discussed. For example, Kuhn notes in *Proto-Structure* that the final chapter on progress and science is less finished than the preceding chapters. This note about the state of the manuscript is embedded in the manuscript, marked off in brackets, like this [...], at the beginning of the final chapter. Kuhn claims that "this section ... is still far from right, particularly in its later pages" (TSK

---

[1] Ian Hacking provides the best analysis of the use of paradigms, insofar as their application is not rule governed (see Hacking 2016). The application of a paradigm, Kuhn suggests, involves analogical reasoning, seeing one sort of scientific problem as like another in important respects. To my mind, Alexander Bird is right to argue that "the exemplar idea is ripe for renewed investigation and development ... in a climate that is more receptive than that in which Kuhn himself faced" (Bird 2012b, 880).

Archives–MC240 1961b, 166). One assumes these remarks in the brackets were directed either to his colleagues, who were reading the draft manuscript, or to the editor at the University of Chicago Press.

In the bracketed remarks, Kuhn gives some sense of what he will need to add to the existing manuscript to complete the chapter. Specifically, he claims that he needs to describe the nature of scientific communities. More precisely, Kuhn notes that "the 'special' character of the authority exercised by the scientific community must be considered in more detail" (TSK Archives–MC240 1961b, 166). Clearly, he was aware of the sociological dimensions of his theory of science (see Wray 2011b, chapter 10; also Wray 2015).

Interestingly, Kuhn also cut some material from the manuscript, specifically material in the final chapter, where the discussion of Orwell would be placed. Most notably, he ended the manuscript with a few remarks about Hitler! Kuhn makes these remarks in an effort to address an objection that he anticipated. Specifically, the objection is that the account of science that he develops may seem to suggest that "might makes right" in science. He then proceeds to address this anticipated criticism.

Kuhn raises the Hitler remark immediately after he presented the Darwinian dimension of his view of the development of science. Kuhn suggests that the selection between competing theories is subject to a process like evolution by natural selection in the biological world, a comparison he retained in the published book. This image is invoked as part of his argument aiming to show (1) that we do not need to assume that science is aiming at the truth in order to explain its success and (2) that the developmental process of science is not aptly characterized as an evolution toward *anything*.

Having invoked the Darwinian image, Kuhn then suggests that some readers may find this dimension of his account "distasteful," as it may suggest that he believes that "in the sciences ... might makes right" (TSK Archives–MC240 1961b, 176). Kuhn notes, "I have, in fact, been told that on my view of science Hitler would have been right if only he had succeeded in subjugating the world" (TSK Archives–MC240 1961b, 176). Elaborating, he explains, "I may seem to have said that it makes no difference how an argument among scientists is settled, that Hitler, for example, could have been allowed to settle them" (TSK Archives–MC240 1961b, 177).[2]

---

[2] It is unclear who suggested this criticism to Kuhn in these terms. Cavell mentions discussing this objection with Kuhn, but he does not mention Kuhn's manuscript in this discussion (see Cavell 2010, 355). Cavell suggests that Kuhn raised the objection and

Kuhn then addresses this anticipated misinterpretation of his view. He begins by noting that the view that this criticism ascribes to him, specifically, the claim that "it makes no difference how arguments among scientists are settled," is "totally absurd" (TSK Archives–MC240 1961b, 177). Thus, in no way is Kuhn meaning to suggest that "might makes right" in science. Rather, according to Kuhn, "the continued existence of science depends upon the continued existence of a very special sort of community, one that ... is committed to solving problems about nature and knows very well what problems it has solved" (TSK Archives–MC240 1961b, 177). He then explains that "when authority alone, and particularly nonprofessional authority, is the arbiter of paradigm debates the process will cease to be science" (TSK Archives–MC240 1961b, 177). Thus, he emphasizes the great difference between (1) a group of scientists resolving their differences and settling on which of two competing theories is superior and (2) a political leader making the decision. These are, as he rightly notes, radically different processes. If they were not, so his critic seems to suggest, it would not matter if Hitler, or any other authoritarian figure, determined what scientists believe. As noted above, this discussion of Hitler does not appear in SSR.

The discussion of Orwell in SSR is not quite a replacement for the discussion about Hitler in the manuscript. Instead, the Orwell discussion emerges against the background of his comments on the rewriting of textbooks after a scientific revolution. Here his concern is with the representation of the history of a scientific field in textbooks, and the rewriting of that history as a response to a change of paradigm or theory.[3] In the course of this discussion, Kuhn again uses the phrase "might

---

expressed the concern that he could not adequately address it (see Cavell 2010, 355). Cavell suggests that Kuhn's concern was that the view he was developing in SSR seemed to "[destroy] the truth of science," and implied that "instruction and agreement are the essence of the matter" (Cavell 2010, 355). Cavell recalls Kuhn framing the problem this way: "Hitler could instruct me that a theory is true and get me to agree" (Cavell 2010, 355). Cavell recalls responding as follows: "No [Hitler] could not; he could not *educate* you in, *convince* you of, *show* you, its truth. Hitler could declare a theory to be true, as an edict. He could effectively threaten to kill you if you refuse to, or fail to, believe it. But all that means is that he is going to kill you; or perhaps kill you if you do not convince him, show him, that you accept and will follow the edict" (Cavell 2010, 355; emphasis in original).

[3] Though earlier in the book, Kuhn used the term "paradigm" to refer to what he later came to call exemplars, by the end of the book he is often using the term "paradigm" interchangeably with theory. Thus, a paradigm change is a change of theory. Kuhn's sliding between different uses of the term "paradigm" is the basis of Conant's concern, discussed above. See also Conant in Cedarbaum (1983), as well as concerns raised by Dudley Shapere (1964) and Margaret Masterman (1970).

makes right" (see SSR-4 166).[4] But here, Kuhn's point is that one must say more about progress through changes of paradigm than that new textbooks are written after a revolution. The new textbooks emphasize the continuity through changes of paradigm, which in turn supports the view of scientific progress Kuhn is reacting against, a view that suggests that the growth of scientific knowledge is strictly cumulative. The rest of his discussion of progress through changes of paradigm is meant to show that it is inappropriate to suggest that "might makes right" in science.

This is still quite a dramatic image, but not nearly as dramatic as invoking Hitler.[5] As others have noted, through the process of writing the manuscript, Kuhn used a variety of metaphors. As Kuhn revised the manuscript for publication, he discarded some of the more provocative metaphors.[6]

10.3   More on Textbooks

The remarks in Chapter XIII related to Orwell are not the first place Kuhn discusses how science is represented in textbooks. Kuhn devoted a whole chapter to the issue of textbook science, Chapter XI, titled "The Invisibility of Revolutions."[7] It is set between Chapter X, "Revolutions as Changes of World View," a rather controversial chapter,[8] and Chapter XII, "The Resolution of Revolutions," where Kuhn explains how

[4] Incidentally, Kuhn also uses the phrases "might makes right" in his reply to his critics at the Bedford College conference organized by Imre Lakatos (see Kuhn 1970b, 260). I thank Markus Seidel for drawing this to my attention.

[5] It is probably fortunate that Kuhn did not include the discussion of Hitler in SSR. Errol Morris, who has a rather critical view of Kuhn's SSR, notes that his wife said that "you can't use Hitler as a spice, nor as a flavoring. The minute you put Hitler in the soup, it becomes Hitler Soup" (Morris 2018, 176).

[6] Reisch notes that in an earlier manuscript Kuhn invoked the notions of ideology, indoctrination, and propaganda (Reisch 2016, 23 and 25–26). These terms were widely used in Cold War America, especially in discussions of Cold War politics. But they are absent from the published book. J. C. Pinto de Oliveira notes that in a 1958 manuscript of SSR, Kuhn discusses similarities between science and art; these too are largely eliminated in the printed book (see Oliveira 2021, 72).

[7] Kuhn notes in the Foreword to the English translation of Ludwik Fleck's *Genesis and Development of a Scientific Fact* that he was "much impressed by Fleck's discussion ... of the relation between journal science and vademecum science. The latter may conceivably be the point of origin for my own remarks about textbook science" (Kuhn 1979, ix). I thank Markus Seidel for reminding me of this.

[8] Many critics have objected to Kuhn's remark that a revolutionary change of theory involves something like a *world change*. See, for example, Peter Godfrey-Smith, who suggests that Chapter X is the worst chapter in SSR (Godfrey-Smith 2003, 96). Wray and Andersen (2020) try to give some sense to the claim that a scientific revolution involves something like a world change.

revolutions in science come to a close, that is, how a new normal science tradition comes to be accepted.[9]

Chapter XI is intended to provide "a last attempt to reinforce conviction about [the] existence and nature" of scientific revolutions (SSR-4 135). Kuhn notes that though he has illustrated numerous cases of revolutions in science, he believes that scientists and laypeople have been primed not to see them. This is so, even though, as far as Kuhn is concerned, scientists work in a *different world* after a revolutionary change of theory. What Kuhn finds remarkable is that such significant changes seem to be rendered invisible. This is something Kuhn learned from studying the history of science. He could not reconcile what he learned about the history of science with what was presented in science textbooks.

In fact, Kuhn claims that the revolutionary changes of theory are rendered invisible in (1) popularized accounts of science, (2) philosophies of science, and (3) science textbooks. Common to all three genres, Kuhn notes, is a focus on "an already articulated body of problems, data, and theory" (SSR-4 136). That is, these sources focus on ready-made science, rather than science in-the-making, to use Bruno Latour's terms (see Latour 1987). As Kuhn explains, "all three record the stable *outcome* of past revolutions and thus display the bases of the current normal-scientific tradition" (SSR-4 136; emphasis in original). There is no concern in these various sources with conveying a sense of the research frontier where problems, data, and theory are less stable and less well articulated.

Kuhn is quick to acknowledge that, at least in the case of science *textbooks*, there is a rationale for the distorted image of science that is created. Textbooks are "vehicles for the perpetuation of normal science" (SSR-4 136). Specifically, they "aim to communicate the vocabulary and syntax of a contemporary scientific language" (SSR-4 136). That is why they "have to be rewritten in whole or in part whenever the language, problem-structure, or standards of normal science change" (136). And that is why they are selective about the history they include in their accounts of past scientific discoveries. Their ability to effectively communicate the vocabulary and syntax of contemporary scientific language

---

[9] Paulo Pirozelli (2021) provides an insightful analysis of Kuhn's views on how revolutions are resolved. Pirozelli rightly argues that Kuhn distinguishes between (1) the factors that explain "how … scientists choose new theories" and (2) the factors that explain how "consensus is formed" (Pirozelli 2021, 1). Philosophers need to see that even if various "socio-epistemic mechanisms" are responsible for consensus formation, as Kuhn suggests, this does not mean that theory-choice is not resolved on the basis of a consideration of "epistemic criteria" (see Pirozelli 2021, 1).

would not be enhanced by a *more accurate* account of the history of the field. In fact, such a history might very well be a significant distraction, undermining the effectiveness of textbooks in serving the purposes for which they are created.[10]

Kuhn notes that the various sciences come to rely on textbooks only once they settle on their first paradigm, that is, once they settle into the pattern of development that he presents in SSR (see SSR-4 136).[11] Fields such as philosophy, for example, are very different in this respect. Consequently, such fields have little or no need for textbooks, and insofar as they use them, they are very different in structure from science textbooks. Even textbooks in the social sciences are noticeably different from textbooks in the natural sciences. In a sociology or political science text, for example, a student is typically exposed to a number of different theoretical frameworks, even in the presentation of a single topic, such as crime, for example (see Wray 2016a, 69 and 70). This is not what one finds in a chemistry or physics textbook.

What is it that textbooks do in their presentations of the history of science? According to Kuhn, "textbooks ... begin by truncating the scientist's sense of his discipline's history and then proceed to supply a substitute for what they have eliminated" (SSR-4 137). As Kuhn notes, "characteristically, textbooks of science contain just a bit of history ... often ... in scattered references to the great heroes of an earlier age" (137). Thus, in physics textbooks there are likely to be references to Galileo, Newton, and Maxwell. In chemistry, passing references to Boyle, Lavoisier, Dalton, and Mendeleev are not uncommon. Kuhn claims that "from such references both students and professionals come to feel like participants in a long-standing historical tradition" (137). The

[10] Ernst Mach had a different attitude about science students learning the history of their discipline. He argued that when science students are taught the history of science they learn to be modest, for they see that others in the past felt as strongly about the theories they worked with as scientists do today about their theories, *even* those theories that we now regard as false (see Mach 1911, 17–18). Thus, the student can see that their own strong convictions about the truth of the theories they accept are not reliable indicators of the truth of those theories.

[11] Even in the middle ages, astronomy was taught by means of documents explicitly written for students. John North explains that, "although Ptolemy's *Almagest* lay behind the curriculum [at medieval universities], it was rarely used, but was replaced by digests such as that by Alfraganus ... or by one of several derivative works with the generic title *De sphere*" (North 1992, 348). Paul Grendler provides a list of "Statutory Texts for 'Astrology' at Bologna" in 1405. Students read parts of Ptolemy's *Almagest* only in their fourth year, after reading a number of more basic texts (see Grendler 2002, 410, table 12.1). Kuhn notes elsewhere that "astronomy, statics, and optics ... are the only parts of physical science which, during antiquity, became the objects of research traditions characterized by vocabularies and techniques inaccessible to laymen and thus bodies of literature directed exclusively to practitioners" (ET 36).

young chemist in training, for example, can feel themself as a contributor to the same enterprise to which Boyle, Lavoisier, Dalton, and Mendeleev contributed. But, according to Kuhn, this is a tradition "that ... never existed" (137). The tradition that the student identifies with is merely a construction, a product of the rewriting of the history.

It is worth clarifying the respects in which the history of science presented in textbooks is a misrepresentation. According to Kuhn, "science textbooks ... refer only to that part of the work of past scientists that can easily be viewed as contributing to the statement and solution of the text's paradigm problems" (SSR-4 137). That is, the textbook leaves out the parts of the past that cannot be easily reconciled with a story of continuity, a story of the cumulative growth of scientific knowledge: "Partly by selection and partly by distortion, the scientists of earlier ages are implicitly represented as having worked upon the same set of fixed problems and in accordance with the same set of fixed canons that the most recent revolution in scientific theory and method has made seem scientific" (SSR-4 137). This is the key to understanding Kuhn's larger point. Selection and distortion are the means by which a history is constructed that gives scientists-in-training the sense that the field is essentially the same throughout its history. And this is what prevents scientists and laypeople (as well as many philosophers of science) from seeing the world-changing nature of scientific revolutions. The student is left with the impression that the problems that occupy scientists today are essentially the same problems that occupied their predecessors. Their predecessors are thus rendered near-contemporaries of sorts.

As Kuhn notes, "no wonder that textbooks ... have to be rewritten after each scientific revolution. And no wonder that, as they are rewritten, science once again comes to seem largely cumulative" (SSR-4 137). In this way, revolutions are effectively erased from the history of science. The new history makes the research world of earlier scientists continuous and seamless with our own.

Kuhn claims that this practice of making "the history of science look linear or cumulative" even "affects scientists looking back at their *own* research" (SSR-4 138; emphasis added). For example, Kuhn notes that John Dalton's own "accounts of the development of his chemical atomism make it appear that he was interested from an early date in just those chemical problems of combining proportions that he was later famous for having solved" (138). But Kuhn insists that "those problems seem only to have occurred to [Dalton] with their solutions" (138). In looking back at his own research in this distorted manner, Dalton has inadvertently erased the revolutionary dimension of his early work, specifically, "the revolutionary effects of applying to chemistry a set of questions and concepts, previously restricted to physics and meteorology" (138).[12]

Kuhn suggests that Newton did something similar in his treatment of Galileo (see SSR-4 138). Contrary to what Newton implies, Galileo's questions and concerns were not the same as Newton's own questions and concerns. But in presenting them as similar, Newton effectively diminishes the significance of his own contribution. Kuhn, though, is insistent: *Galileo was not working in Newton's world.* Concepts wholly unknown to Galileo had to be developed before Newton could make his contributions.

Why would a scientist of the calibre of either Dalton or Newton behave in such a manner, diminishing the magnitude of their *own* contributions to science? Kuhn does not address this question. But it seems that the behaviour may be due to cognitive dissonance. Dalton, for example, might have found it difficult to reconcile the beliefs he had after he made his important discovery about the law of fixed proportions with his beliefs before. That is, he may have found it challenging to think clearly about what he had thought before, once he had made the revolutionary transition afforded by his own contributions. This gives some sense to Kuhn's notion of working in a new world after a revolution. Scientists are not at liberty to return to the state they were in before they comprehended a significant scientific breakthrough, even if they were the ones responsible for the breakthrough. Indeed, Kuhn thought that this was one of the noteworthy differences between the scientist's experience during a revolutionary change of theory and the experience one has when one looks at a gestalt image, like the duck/rabbit. As Kuhn explains, "the scientist does not preserve the gestalt subject's freedom to switch back and forth between ways of seeing" (SSR-4 85). Thus, Dalton cannot recapture his impressions and thoughts once he has experienced the transition to the new way of seeing the world.

In fact, there is some evidence that scientists are not readily able to retrieve the past, once they have experienced a revolutionary change of theory. Kuhn's own experience interviewing the scientists involved in the Quantum Revolution in physics in the early twentieth century seems to support this. Kuhn was surprised that many physicists were unable to recall the past (see RSS 303). As John Heilbron, who was working with

---

[12] Kuhn's analysis of Dalton's understanding and misunderstanding of his own discovery is based largely on Leonard Nash's analysis (see Nash 1956). Nash argues that "*the creation of the chemical atomic theory was the direct outgrowth of Dalton's search for an explanation of the differences in the solubility of various gases in water. The theory was not just 'first applied' to the gas-solubility problem: it was devised first to solve this very problem*" (Nash 1956, 108, emphasis in original). Kuhn taught the general education history of science course at Harvard with Nash, and they corresponded even after Kuhn left Harvard in 1956 (see Wray 2021a, 56–59).

Kuhn on the History of Quantum Physics Project, explains: "few inter-
viewees were reliable informants about events then as much as fifty years
in the past" (Heilbron 1998, 510). In a progress report prepared during
the History of Quantum Physics Project, Kuhn provides a candid assessment of the value of the interview material collected from Peter Debye, a
Nobel Laureate in chemistry. "Professor Debye's remarks on the reception of the Bohr atom are quite explicitly a reconstruction of how things
must have been or ought to have been. As such, they are entirely lacking
in the sort of circumstantial detail which would make them either useful
or entirely credible" (Kuhn, cited in Heesen 2020, 95).

This reconstruction of the past that scientists such as Dalton, Newton,
and Debye engaged in is the same sort of process that occurs when
textbooks are rewritten after a scientific revolution. The impression one
gets from a textbook presentation is that "from the beginning of the
scientific enterprise ... scientists have striven for the particular objectives
that are embodied in today's paradigms" (SSR-4 139–140). That is, a
scientific field is treated in an ahistorical manner and assumed to be
concerned with the same sorts of research problems.

But this is precisely what Kuhn challenges in his theory of the development of scientific knowledge. He insists that the objectives of a scientific field change with each change of theory, and some of these changes
are quite profound. As he explains, "many of the puzzles of contemporary normal science did not exist until after the most recent scientific
revolution" (SSR-4 140). Some of the puzzles are suggested, for the first
time, by the new theory. For example, prior to the Copernican
Revolution in astronomy, the problem of explaining what keeps the
Earth in its orbit did not exist. Similarly, geoscientists began reflecting
on the mechanisms that move, or could move, the continents laterally
only once they began considering the possibility of continental drift. This
was not a problem for geoscientists before the early 1900s, when Alfred
Wegener first suggested that the continents drifted.

Kuhn is thus setting scientific fields in their history. "Earlier generations pursued their own problems with their own instruments and their
own canons of solution" (SSR-4 140). In fact, Kuhn notes that it is not
"just the problems that have changed. Rather the whole network of fact
and theory that the textbook paradigm fits to nature has shifted" (140).
That is, with each revolution in a field, the focus of the field shifts.[13] But
much of this is rendered invisible with the rewriting of textbooks.

---

[13] The most obvious example of this shift in focus that Kuhn discusses is the change from
Aristotelian physics to Galilean physics. Aristotelian physics encompassed such things as

The textbook image of science is thus irreconcilable with the notion of *world changes*.

10.4   The Value of the Orwellian Approach

Are scientists rewriting the textbooks in their field after a revolutionary change of theory just to cover over aspects of the past that now seem unscientific and, thus, threaten to undermine the cognitive authority of scientists? Are the authors of textbooks set on saving scientists from appearing prone to mistakes?

When Kuhn presented the Orwellian concern, his point was, in part, to explain where scientists and the public get the idea that the growth of scientific knowledge is strictly cumulative. Kuhn notes that he had uncritically assumed that the growth of science was strictly cumulative before *he* studied the history of science. His Aristotle epiphany, where he learned to read Aristotle's writings on physics in a new way, and came to understand that Aristotle's project was fundamentally different from Galileo's and Newton's projects, played an important role in his own realization that the growth of scientific knowledge is not as it is commonly understood to be. In fact, Kuhn's motive for writing a book on scientific progress and scientific revolutions was, to a large extent, aimed at correcting this misconception.

Kuhn rightly notes that the rewritten histories of science in science textbooks tend to emphasize continuities and downplay the discontinuities through changes of theory. But this practice is not the consequence of nefarious motivations on the part of the scientists writing the "new history" of the discipline. That is, the scientists writing the new textbooks do not have any explicit motive to prevent the students who read the books from knowing the truth about the past. In this respect the comparison with *1984* is a bit misleading. But there is another respect in which the comparison with *1984* is illuminating. Specifically, there is a sense in which the authors of science textbooks aim to control the thoughts of the students reading the textbooks, much like the rewriting of history in *1984* was intended as a means of thought control.

The above remark requires some clarification. Textbooks aim to focus the attention of scientists-in-training on specific features of the world, the features that are relevant to the theories and models presented in the text. That is, a large part of what a textbook seeks to do is to direct the scientists-in-training to what is most relevant in the world. After all,

---

the change of an acorn to an oak tree. This was no part of Galileo's project, nor is it a part of contemporary physics (see ET xi).

without some sort of theoretical assumptions, directing the attention of scientists-in-training to some features and not others, a young scientist would not know what to look for. So the selective nature of the history of science presented in textbooks serves to reinforce the focus, or myopia, that helps make scientists effective at addressing the sorts of problems they are expected to address in *normal* scientific research, the sort of research that most scientists should expect to do, most of the time, throughout their careers. In this respect, the textbook is like the paradigm, or accepted theory, intentionally directing scientists to certain features in the world.[14]

But directing the attention of scientists in this way not only leads scientists to take notice of some things that others who lack their training might overlook; it also leads them to not see other features in the world. These are two dimensions of the theory-ladenness of observation. It is both an aid for scientists, in particular, as they conduct normal scientific research, assuming the accepted theory is adequate for the tasks at hand, and, at times, a hindrance to them. Specifically, it becomes a hindrance when the field is in crisis, and the long-accepted theory is failing to meet the needs of scientists. At these times scientists need to be able to break out of the framework that has been instilled in them during their training and that they have relied on to date.

Kuhn did not think that scientists needed to learn a more complete history of science in order to work effectively. Though such a history may enlighten students on what *really* happened, say, between 1774 and 1776, with respect to the discovery of oxygen, it may not help them see and attend to the sorts of things they need to attend to if they are to be effective researchers working in the current research tradition. The textbook needs to help them see which features of the world are relevant to the sorts of research they will be expected to conduct at the end of their training, not the sorts of things that illuminate the events of the past. In this respect, science education is much like most sorts of education, consciously concerned with shaping how students think and behave. Some measure of thought-control is thus called for.

And because textbooks play a significant role in priming scientists-in-training, directing their attention to some things and not others, the old textbooks, with the "old" history, can be an impediment, drawing attention to features of the world that are now regarded as irrelevant.

---

[14] In fact, when Kuhn first introduced the paradigm concept, at the beginning of Chapter II in SSR, he explicitly connects textbooks to paradigms, defining "the legitimate problems and methods of a research field for successive generations of practitioners" (SSR-4 10).

10.5   The Aftermath

Surprisingly, few readers have taken issue with Kuhn's claim that textbooks are rewritten after a scientific revolution to emphasize the continuities and downplay the discontinuities between the old theory and its successor. Even though many aspects of Kuhn's philosophy of science were subjected to criticism, this particular dimension of his view was not.[15]

More recently, a few scholars have commented on it. It is mentioned briefly in a 2010 article by the political scientist Thomas Walker (see Walker 2010). This is the same article that also mentions the pet monkeys in Kuhn's office at Princeton.[16]

And Steve Fuller discusses it in two books: *Thomas Kuhn: A Philosophical History for Our Times*, where he compares Kuhn to Chance Gardiner, the protagonist of Jerzy Koziński's *Being There*; and *Kuhn vs. Popper: The Struggle for the Soul of Science*, which addresses the Kuhn-Popper debate. In the first book, Fuller states that Kuhn "argued ... that normal science cannot proceed without scientists' [sic] having an Orwellian sense of their own history" (Fuller 2000b, 126). Fuller elaborates, noting that as far as Kuhn is concerned, history of science *should* "be rated 'X' for scientists" (Fuller 2000b, 303).[17] In the latter book, Fuller states that Kuhn's "ultimate significance lies in the attention he drew to the role of historical revisionism in the establishment of a new paradigm" (Fuller 2004, 22). Fuller is concerned that, on Kuhn's view, arguments, evidence, and reason have little role in the establishment of a new paradigm. As Fuller puts the point, Kuhn believes that "scientific revolutions succeed not because the same people are persuaded of a new way of seeing things (*á la Popper*) but because different people's views start to count" (Fuller 2004, 23).

Interestingly, some of the critics of SSR have raised a form of the Hitler concern discussed in *Proto-Structure*, though no one appeals to Hitler in framing the criticism. Often when philosophers of science discuss the relationship between Kuhn's view and the Strong Programme in the

[15] The early critics were concerned with (1) the implications Kuhn's view had for the rationality of science, (2) the threats and implications posed by the perpetual cycle of revolutionary changes of theory, and (3) the dogmatic nature of normal science, as Kuhn characterized it (see Wray 2021a, chapter 9).

[16] I have dispelled the myth that Kuhn had *monkeys* in his office at Princeton; I discovered that he only had *one monkey*, and it is not clear that the monkey was in his office much. Later the monkey went to live with a woman who claimed to be a relative of Wittgenstein (see Wray 2018).

[17] The phrase "should the history of science be rated X?" originated with Stephen Brush (see Brush 1974).

Sociology of Scientific Knowledge, or at least a view commonly attributed to the Strong Programme, concerns have been raised about whether Kuhn thinks that theory choice is determined by factors such as power, rather than evidence.

Michael Friedman, for example, discusses what he calls "the post-Kuhnian predicament" (Friedman 2001, 48). Friedman acknowledges that Kuhn has attempted "to distance himself from relativism", but he believes that "some of his followers have willingly embraced" the relativist implications of Kuhn's view (p. 49). Friedman has in mind the Strong Programme, Barry Barnes, David Bloor, Steven Shapin, and Simon Schaffer (p. 49).

Thus, it appears that the unnamed early critic who alerted Kuhn to this concern was quite prescient. At least *some* readers were apt to think that Kuhn was arguing that the competition between two theories is resolved on the basis of power. Kuhn would have to confront this charge again and again, over the years (see, e.g., RSS 224–252 and 317).

The Hitler concern alerts us to the fact that we must come to terms with the processes by which scientists are able to rebuild consensus after the community has fragmented, following the breakdown of a long-successful normal scientific tradition. Formulated this way, it should not surprise us that various social processes are at work. In fact, the social dimensions of this issue are captured in the title of Barry Barnes' paper "Thomas Kuhn and the Problem of Social Order in Science" (2003). But Barnes insists that the problem that Kuhn is grappling with affects normal science as well. We really need to understand what keeps a research community together.

Some philosophers of science are now taking seriously the need to reflect on this issue, to explain how a new consensus is constituted in a research community after a long-accepted theory has been deemed inadequate (see Wray 2011b; but also Pirozelli 2021). Philosophers now know that the influence of social factors on science need not imply that the process is determined by non-epistemic factors. That is, a consensus forming process can be both social *and* epistemic.

Kuhn, though, had a very specific idea of the sorts of social factors and interests that influence scientists, at least when science is functioning well. He was most explicit about this in his speech upon receiving the John Desmond Bernal Award by the Society for the Social Studies of Science. He was especially concerned to distinguish his own view from the view of the Strong Programme, a view with which he became increasingly uncomfortable as time passed. Speaking to the society, Kuhn insisted that scientists are "inculcated by training" with "special cognitive interests" (Kuhn 1983, 30). Specifically, these include "the love of truth … [and] fascination with puzzle-solving" (Kuhn 1983, 30). And, as far as Kuhn was concerned, "we shall not … account at all for scientific knowledge without recognizing the determining role in scientific practices of just such special interests" (Kuhn 1983, 30).

Kuhn was concerned that the then-current sociological studies of science were predominantly concerned with socioeconomic interests (see Kuhn 1983, 30). I do not think this is a fair assessment of sociological studies of science in the early and mid-1980s. But Kuhn's reading was not unique to him. Rather, it was the standard reading among philosophers of science (see, e.g., Laudan 1984, 21).

10.6   Concluding Remarks

In summary, the issue concerning the threat to the rationality of science, real or not, is quite distinct from issues that concerned Kuhn in his analyses of textbook history of science. Kuhn's concern in highlighting this Orwellian practice was to understand (1) how science textbooks function and (2) why scientific revolutions are often rendered invisible. Textbooks are effective means for conveying the current theory and practices because they are written with the aim of drawing the attention of scientists-in-training to the aspects of the world that are relevant, given the theoretical understanding of the world expressed in the book. That the "history" they recap is a distortion matters little. But it does leave scientists with a misconception of how scientific knowledge develops, effectively erasing the radical nature of the changes that inevitably occur in the development of science. Further, the distorted histories in science textbooks *may* be responsible, to some degree, for the gap between (1) the expectations that the public have of science and (2) what scientists are able to deliver.

Acknowledgements

I thank Lori Nash and Markus Seidel for constructive feedback on earlier drafts. I also thank my audience when I presented the chapter at a conference in celebration of the hundredth year of Thomas Kuhn's birth, in Aarhus, in June 2022.

11 Essential Tensions in Twenty-First-Century Science

Hanne Andersen

11.1 Introduction

Kuhn's 1962 monograph, *The Structure of Scientific Revolutions* (SSR), is one of the most influential works in philosophy of science from the twentieth century. Impressively for a book in philosophy of science, SSR has sold more than a million copies.[1] According to Andrew Abbott (2016), SSR has received more than 15,000 citations in peer-reviewed literature in the humanities and social sciences during the fifty years from 1962 to 2012. However, the peer-reviewed scholarly literature in these fields represents only a fraction of the total attention that SSR has received. Not only has it been a significant contribution to history, philosophy, and sociology of science,[2] but it has also been read and discussed by many scientists, science educators, science managers, and policy makers. Thus, Abbott's analysis also showed that while the citations from the core audience in the humanities and social sciences first increased linearly for about two decades and then leveled out, citations from applied fields such as education, law, management, and business show a steady increase over time, without any levelling (ibid., 168f.). Similarly, extending the search for citations to SSR outside the peer-reviewed literature in the humanities and social sciences, a search in Google Scholar on September 2022 using Harzing's Publish or Perish reported almost 150,000 citations to the English editions alone.

Through this wide readership, SSR has had an important influence on how science has been viewed over the last decades. For example, a grant agency such as the National Science Foundation emphasizes its aim of supporting transformative research that "leads to the creation of a new paradigm" (https://beta.nsf.gov/funding/learn/research-types/transformative-research#definition). Similarly, searching through the project

---

[1] For a recent estimate of the number of sold copies, see Wray (2021a, 1). For an earlier estimate together with a list of translations, see Hoyningen-Huene (1992, xv).
[2] See Wray (2021a) for a detailed exposition of the influence that SSR has had on social science, history of science, and philosophy of science.

highlights from projects supported by the European Research Council, the notions of paradigms and paradigm shifts figure prominently (see https://erc .europa.eu/search?keys=paradigm).

Kuhn's phase model was based on an argument about an essential tension between tradition and innovation that would drive both incremental progress within a paradigm and transformative progress across a scientific revolution. Revisiting this argument, in this chapter I shall argue that the argument builds on a number of assumptions about the practices of science that hold for science conducted by individuals within a discipline, but not necessarily for the increasingly collaborative and interdisciplinary science we see today. On this basis, I shall focus on the way in which scientists cluster together in teams – what I shall refer to as the granularity of science – and discuss how changes in this granularity affect Kuhn's essential tension argument.

11.2   Kuhn's Essential Tension

Summarized very briefly, Kuhn described scientific progress as a development that takes place in specialized fields. Within such fields, research is conducted by a community of researchers who have all been trained to share a set of received beliefs, what Kuhn termed a paradigm or, later, a disciplinary matrix. Through this training, all members of the community have acquired equivalent abilities: first, to identify new research problems that other members of the community will also see as relevant and, second, to assess whether a proposed solution is acceptable within the paradigm. On this basis, researchers in the community pursue the various puzzles that the reigning paradigm defines. By conscientiously and truthfully drawing on the shared set of tools that the paradigm provides, and by continually keeping track of new results produced, it is clear to the members of the community which problems have been solved, which problems still need to be solved, and which new problems have become apparent.

On the one hand, normal science is immensely effective in solving the problems that the paradigm defines. This faithful work conducted within the boundaries of the paradigm enables the scientific community to move faster and penetrate deeper than if its members kept spending time examining all sorts of alternatives. In this sense, normal science is a tradition-preserving activity. On the other hand, the stable, focused effort to investigate each question and solution that a paradigm presents in excruciating detail at the same time also ensures that any limitations of the paradigm will be discovered eventually. As Kuhn argued:

At least for the scientific community as a whole, work within a well-defined and deeply ingrained tradition seems more productive of tradition-shattering novelties than work in which no similar convergent standards are involved. How can this be so? I think it is because no other sort of work is nearly so well-suited to isolate for continuing and concentrated attention those loci of trouble or causes of crises upon whose recognition the most fundamental advances in basic science depend. (ET 234)

Hence, again and again, the ultimate effect of this tradition-bound work is the shattering of the same tradition. This is what Kuhn described as an essential tension between tradition and innovation, or between convergent and divergent thought. On Kuhn's model, the ultimate effect of the tradition-bound work is to change the tradition itself. But this process of changing the tradition also has a particular structure. When recalcitrant problems are discovered, the previous consensus in the community on (1) what the relevant problems are, (2) which tools to use to solve them, and (3) what an acceptable solution looks like begins to dissolve, and different members of the scientific community start suggesting different modifications or replacements of the received beliefs. This distribution of conservative and innovative dispositions among the members of the community ensures a spreading of risk in the scientific community that is crucial. As Kuhn argued: "If all members of a community responded to each anomaly as a source of crises or embraced each new theory advanced by a colleague, science would cease. If, on the other hand, no one reacted to anomalies or brand-new theories in high-risk ways, there would be few or no revolutions" (SSR-4 185f.). Thus, for the essential tension between tradition and innovation to work as a driver of change, a distribution between conservative and innovative dispositions among the members of the scientific community is also needed. If no one reacted to problems in a high-risk way, there would be no possibility of change. Vice versa, if everybody reacted to each little deviation by suggesting radical changes, science would become a mess.

In his 1959 article "The Essential Tension" (ET 225–239), Kuhn argued that it was a particular kind of education that assured that, although students were trained in convergent thought, divergence could nevertheless emerge when loci of trouble had been identified. Thus, Kuhn argued that, because students were trained to recognize and classify new research puzzles as samples of known types by similarity rather than by necessary and sufficient conditions, they could differ on what they considered more or less important characteristics while still agreeing on the same classification. Only when loci of trouble were discovered would these latent differences manifest themselves as different reactions to the discovered loci of trouble.[3]

When such a divergence has become apparent, the various alternatives will compete in offering better solutions to the loci of trouble, and

---

[3] See Andersen (2000) for a detailed exposition of Kuhn's view of science education.

eventually a new consensus will be established among members of the scientific community on which one to adopt.

In summary, on Kuhn's phase model of the development of science, normal science is a convergent activity based on a settled consensus, but at the same time the ultimate effect of this tradition-bound work is to change the tradition. The resources from which to start building alternatives to the tradition are the latent differences established by a special kind of training, and while the development of new alternatives is initiated by individuals, settling on a new consensus is the act of the community.

11.3   Hidden Assumptions in Kuhn's Analysis

As described above, Kuhn's analysis of the development of science explained why science conducted within disciplinary boundaries, and using the tools that the discipline agrees on, is extremely efficient at both solving the puzzles that the paradigm defines and identifying any loci of trouble. Similarly, Kuhn's analysis of science education explained how the identification of loci of trouble could cause members of the same community to diverge in their approaches, despite their similar training. However, Kuhn's analysis of how convergent and divergent thought most effectively contribute to the progress of science builds on a number of assumptions about the practice of science that were not fully spelled out in Kuhn's work.

The effectiveness of normal science in identifying and solving all the problems that the paradigm defines, as well as in identifying any loci of trouble, depends on community members being able to effectively keep track of the development of their field. And they can do this only if there is open communication between a manageable number of peers on a manageable volume of results within a shared endeavor, where community members can divide labor among them selves rather than having to perform and master everything individually. While Kuhn explicitly reflected on the number of peers in a scientific community, he did not explicitly discuss how labor was to be divided within such a community. Thus, in the Postscript to SSR, Kuhn described scientific communities as communities of scientists who would attend the same special conferences, exchange drafts and preprints, communicate closely, and cite each other, and his intuition was that these would be "communities of perhaps one hundred members, occasionally significantly fewer" (SSR-4 177).[4]

---

[4] Kuhn did not provide an argument for what he saw as an optimal size of a community. However, elsewhere in the Postscript he referred to work by Price, Crane, Hagstrom, and others on scientific communities. In his famous monograph *Little Science, Big Science*, Price (1963) had argued that if scientists produce around a hundred papers during a

Thus, his analysis of communities as the "producers and validators of scientific knowledge" (ibid.) was directed not at entire disciplines such as physics or chemistry, nor at subdisciplines such as organic chemistry or solid-state physics, but at the narrow research field within which members communicate closely.[5]

However, how efficient such a community will be in identifying and solving puzzles as well as in identifying any loci of trouble will depend on, for example, whether community members devote their time and energy to working in parallel on solving the same puzzle while competing to be the first to solve it, or if instead they work in parallel on different puzzles to which each contributes in a different way to the overall progress of the field. This again may depend, for example, on how community members diverge in their preference for solving either only the hardest and most prestigious puzzles or the many, less challenging puzzles. Similarly, the efficiency with which the community identifies and solves puzzles and identifies any loci of trouble also depends on whether community members tend to share results with each other at an early stage or keep them secret until as many personal achievements as possible have been secured.[6]

Further, the efficiency by which a community will be able to identify and solve puzzles, as well as identify any loci of trouble, will depend on several things. First, it will depend on whether members of the community can establish sufficient warrant for trusting one another to be competent, conscientious, and truthful.[7] Second, it will depend on whether individual community members are in a position that allows them to meticulously assess new results and to raise questions and voice concerns whenever something "seems wrong."

career, it would be feasible to keep track of the production of a few hundred colleagues, but not more. This estimate has later been criticized by Wray, who argues, first, that many scientists produce much fewer than a hundred papers and, second, that scientists also read literature from neighboring fields. On this basis, he estimates the size of scientific communities to be between 250 and 600 scientists. Recent empirical studies indicate that communities should not become too big. Thus, based on their examination of 1.8 billion papers, Chu and Evans (2021) have found that in fields with a very high annual production of papers, citations tend to focus on already highly cited work. This means that new ideas do not diffuse in the field, and that, instead, there is an "ossification of the canon" (Chu and Evans 2021, 1).

[5] In this respect, Kuhn's view on the structure of disciplines resembles the fractal views presented by Abbott (2001) or Collins (2011; 2018); Collins et al. (2020).

[6] For philosophers as well as sociologists of science addressing these issues, see, e.g., Hagstrom (1974), Hull (1989), Kitcher (1990), or Poulsen (2001).

[7] See Hardwig (1985; 1991) for a detailed analysis. See further Andersen (2014) for a detailed application of this framework on trust among team members and coauthors.

Similar considerations hold for the risk-spreading mechanism. Although latent differences may exist in the community, these will manifest themselves only if individual community members are in a position that allows them to pursue alternative directions even though others do not find these alternatives particularly promising. This may depend, for example, on whether resources are available for research pursuing this alternative direction and/or whether they have the freedom to pursue their own ideas or are subordinates to a scientist who defines research in a particular direction.

The historical examples that Kuhn drew on in SSR were primarily examples of scientists working individually, and questions of funding or hierarchical relations did not enter his analysis. Admittedly, he did state that fundamental innovations were often made by scientists who were new to the field. He argued that this was because newcomers to a field would be less committed by prior practice to the paradigm that defined normal science and, consequently, would be more "likely to see that [the traditional rules] no longer define a playable game and to conceive another set that can replace them" (SSR-4 90). However, this argument still treats newcomers as autonomous agents. The argument does not consider if the situation would be different for researchers working under the direction of a strong principal investigator (PI). Hence, what needs to be added to Kuhn's analysis is how such factors as hierarchical relations between the members of the scientific community, funding availability, and so on affect the ability to identify loci of trouble as well as the possibilities for spreading risks within the scientific community.

As I shall argue in the following, these qualifications to Kuhn's analysis require us to attend to important elements of scientific practice that have changed over time, and I shall discuss how these changes may prompt us to revisit the analysis of the essential tension and the lessons that we draw from it.

11.4   How Science Has Changed since *Structure*

Kuhn's work was directed at how science had developed over the centuries prior to the publication of SSR. However, during the more than half a century that has passed since the publication of SSR, science has changed considerably. Thus, while Kuhn's historical studies were focused on individual scientists working alone[8] and within disciplinary

---

[8] For a detailed list of the individuals whose contributions Kuhn discussed in SSR, see, e.g., Wray (2003).

boundaries, science today has become increasingly collaborative and interdisciplinary.

The steady increase in collaboration has been documented by sciento-metric studies that, from analyses of the millions of papers indexed in Web of Science, have shown that more and more research, especially in the natural, health, and technical sciences, is published by teams of coauthors, and that the mean size of such teams is steadily increasing (Wuchty et al. 2007). Similarly, it has been documented that while the most frequent size of authorship groups before 2000 was two or three, it has now grown to six to ten (National Science Board 2007).[9]

Similarly, the increase in interdisciplinarity has been documented by scientometric studies of how often coauthors have different disciplinary affiliations, or how much citations cross disciplinary boundaries (Porter et al. 2007; Porter and Rafols 2009; Abramo et al. 2012).

11.5   The Implications of Granularity for the Essential Tension Argument

As described above, scientists no longer primarily work alone, as was the case for most of the examples that Kuhn drew on in SSR. Instead, they increasingly work in small- to medium-sized teams.[10] In other words, over the last decades, there has been a gradual change in what I shall here call the granularity of the scientific community, that is, the way in which the efforts of scientists are distributed within the community and the way in which scientists cluster together in teams of different sizes.

The question is how this change affects the efficiency with which science solves puzzles, identifies loci of trouble, and spreads the risks associated with suggesting alternative approaches. To investigate this in more detail, in the following I shall discuss how different organizations of small- to medium-sized teams can affect scientists' efficiency in three respects:

---

[9] It should be noted that these studies of how the prevalence of different sizes of coauthor groups develop over time do not necessarily reveal how far the increase in coauthorship reflects an increase in actual collaboration compared with a change merely in the practices of assigning authorship. For example, if we go back a century, many junior scholars would be sole authors for work where, today, similar work would grant coauthorship to senior scientists for their supervision. The development in many European countries toward paper-based dissertations instead of monographs may further add to this effect.

[10] Admittedly, some scientists work alone, and some scientists work in very large collaborations, but the focus here will be on the most prevalent situation, namely, the small- to medium-sized teams.

1. Their ability to solve puzzles; that is, how does the increase in team-work affect scientists' ability to keep track of the puzzles solved, as well as their ability to distribute labor efficiently among them selves in identifying, specifying, and solving new puzzles?
2. Their ability to identify potential loci of trouble; that is, how does the increase in teamwork affect scientists' ability to critically scrutinize and criticize each other's contributions?
3. Their ability to spread the risks when developing and investigating alternative approaches; that is, how does the increase in teamwork affect scientists' freedom to pursue alternative directions?

11.6   The Efficiency of Different Types of Teams

Collaborating in a team enables scientists to divide labor among them selves, and often this will enable them to produce more work while spending fewer resources, compared with what would have been the case if they had each worked individually and not coordinated with each other. However, the efficiency of a collaboration depends both on how knowledge and expertise is distributed among team members and on the process by which labor is divided among team members. In the following, I shall investigate these two aspects in turn.

11.6.1   *Dividing or Sharing Labor, Knowledge, and Expertise*

The division of labor works differently depending on whether the scientists in a team

1. have contributory expertise in the same field; that is, they each have the expertise required for contributing to research in the same field
2. are in the process of training novices to achieve contributory expertise in the same field as the senior researchers in the group
3. have contributory expertise in different research areas, or
4. are in the process of acquiring contributory expertise within the same area.

In the following, I have more to say about these various types.

Same Area of Expertise When team members share the same area of expertise, they are in principle all capable of performing the same tasks. However, although in principle they could therefore progress by each performing in parallel all of the tasks required for solving a puzzle they are working on, that would create redundancy and thereby be a waste of time and resources. Instead, they can progress more quickly by dividing the tasks among themselves and working on multiple tasks in parallel. Such concerted action performed by a team can therefore be more efficient for solving puzzles compared with the case where the same number of scientists either worked in parallel on the same puzzles, because they were competing for priority, or did not coordinate their work, and sometimes happened to solve the same puzzles without knowing it.

Although the team members all share the same area of expertise, they will work on different tasks, and they will therefore each produce different pieces of new knowledge. Although these pieces of new knowledge are at first distributed among them, they can later be shared by each member testifying about the various pieces of knowledge they have produced. As argued by John Hardwig (Hardwig 1985; also 1991), this kind of trust of scientific knowledge received by testimony requires that the involved scientists trust each other's competence, conscientiousness, and trustworthiness. However, because they all share the same area of expertise, they can easily communicate with each other about each of the individual contributions, and they all have the same abilities to assess whether each partial contribution seems reasonable. Hence, trust can be established relatively easily.

A special subset of this type of team are teams in which a senior scholar trains novices to acquire contributory expertise in the same field as the senior scholar. In doing so, the senior scholar may divide work among the novices to be carried out in parallel. In this case, it is initially primarily the senior scholar who is in a position to assess whether all partial contributions seem reasonable, and it is also a standard element of supervision to do so. However, the novices in the team will gradually acquire the same expertise as the senior scholar, and therefore gradually become as capable as the senior scholar in assessing their peers and their own partial contributions. If team members were to remain the same, then this type of team would evolve over time to become a team of equal experts. However, typically, the younger team members will move out of the team after some years of training while new novices will enter the team and start training.

Different Areas of Expertise When team members have contributory expertise in different areas of research, they are not all capable of performing the same tasks. In this case, labor will typically be divided among them according to their area of expertise.[11] That enables and

---

[11] For an example, see the rich description in Wagenknecht (2016, chapter 4) of a team of physicists, chemists, microbiologists, and geologists working together on planetary science.

often requires team members to defer to each other for work that they would not have been able to carry out themselves, unless they first spent a substantial amount of time acquiring additional expertise. In this case, the issue of testimony and trust becomes more intricate. Because they have contributory expertise in different fields, they cannot as easily communicate with each other about their individual contributions, they cannot necessarily assess each other's competence, and they do not necessarily have the ability to assess whether contributions provided by other team members seem reasonable or not.

However, as argued by Peter Galison (1997; 1999), scientists involved in interdisciplinary collaborations can be seen as working in a trading zone where they gradually develop a pidgin or creole language in which to communicate. Similarly, as argued by Collins et al. (2007), although they have contributory expertise in different fields, they will gradually develop interactional expertise in each other's area, which enables them to interact in meaningful ways. This gradual development of interactional expertise and of a pidgin or creole language means that there is a trade-off between trust and learning from each other. On the one hand, the more scientists simply trust each other's testimony, the quicker they can solve puzzles and make progress in their research endeavors. On the other hand, the more they aim at justifying this trust and at making sure that each other's various contributions are critically scrutinized so that any loci of trouble will be discovered, the more they need to communicate and to spend time developing interactional expertise in additional areas.

A special subset of this type of team are teams in which all team members are in the process of acquiring contributory expertise within the same area. For example, this may happen if a group of researchers decide to enter together into a new research area that they have seen others pursue.[12] In contrast to the case in which a senior scholar trains novices, in this case none of the group members starts with contributory expertise in the field. Hence, none of the group members is initially in a position to supervise others or fully assess whether the work produced lives up to the accepted standards of the field and whether obtained results look reasonable. This assessment will therefore happen outside the group, when research results are presented to other researchers with contributory expertise in the field.

---

[12] Admittedly, this situation may be relatively rare in research. Instead, this type of team may be more frequently found in education, where students collaborate while acquiring contributory expertise in a field.

This distinction between disciplinary teams consisting of team members who share the same area of expertise and interdisciplinary teams whose team members have different areas of expertise should be seen as the ends of a continuous spectrum that varies from teams in which all team members share the same area of expertise to teams in which all team members have completely different areas of expertise.

The Processes of Sharing and Dividing The efficiency with which teams can solve well-defined puzzles, critically scrutinize results, and spread the risks in pursuing alternatives depends not only on the degree to which expertise, labor, and knowledge are distributed among the team members. It depends also on the processes through which labor is being distributed among the team members and the processes through which the pieces of knowledge that they have each produced are merged into shared results. These processes work differently depending on whether team members are equal peers or whether some have authority over others. In the following, I shall investigate these two types in turn.

Egalitarian Teams When researchers collaborate as equal peers, their collaboration can be described as a shared, cooperative activity.[13] This means, first, that all team members share the same intention of pursuing a specific research activity, and they each intend to pursue this research activity in accordance with and because of this shared intention; second, that all team members need to make sure that their individual plans for how to carry out their part of the activity mesh with the plans of the others. This may change as their work progresses. For example, some may run into problems and need assistance from others. Others may progress quicker than anticipated and need results from the rest of the team before they can continue. And still others may uncover unexpected questions that need to be resolved for any work to progress. All team members therefore need to continually check how each other is progressing and to adjust their individual plans in response to emerging deviations. This kind of concerted action performed by a team can be very efficient for providing critical scrutiny of each solved puzzle from several different angles, and thereby for identifying any potential loci of trouble.

This kind of collaboration requires that all team members have sufficient interactional expertise to be able to communicate efficiently and to track the status of the research activity and of each other's contributions.

---

[13] See Bratman (1992) for the notion of shared cooperative activities in general, and Fagan (2012a; 2012b) specifically on shared cooperative activities in science.

While this requirement is fulfilled easily in disciplinary teams, in inter-disciplinary teams it may require that team members invest substantial time in developing interactional expertise in each other's areas.

Hierarchical Teams When a principal investigator (PI) has authority over the others and dominates the collaboration by giving directives and orders, the team's collaboration is better described as a joint intentional action with authority.[14] In such a hierarchical team, the PI typically distributes tasks and collects contributions, so as to match their cognitive goal, and also intends that their orders in this process of distributing tasks and collecting contributions causes other group members to see them as reasons for the group members adopting them as subplans and adjusting their other subplans accordingly. However, in order for the PI to be able to track the status of all team members' activities, take decisions on when tasks need to be adjusted or redistributed, and facilitate that partial contributions are integrated into final group results, the PI needs sufficient expertise in all the areas covered, and this is more easily fulfilled in disciplinary than in interdisciplinary research.

This distinction between hierarchical and egalitarian teams should be seen as two opposite ends of a spectrum that varies from scientific research activities in which all involved scientists engage as equal partners to scientific research activities in which a PI acts as an authority to whom all other team members are subordinate.

11.6.2 Types of Team Organization

These analyses of the process by which labor is divided and knowledge is combined among team members yields a two-dimensional spectrum that can serve as background for investigating various team types (Figure 11.1). However, combining the two distinctions should not simply be seen as outlining four quadrants of equivalent possibilities. First, as noted above, each of the two dimensions describes a continuum rather than a dichotomy. Hence, the four categories that can be described by combining them should be seen as ideal types that allow for wide variation. Second, as also noted above, the various combinations are not equally likely. Hierarchical organizations may function more easily within a setting where contributory expertise is largely shared; vice versa, egalitarian organizations may have a larger potential within a setting where contributory expertise differs. Given

---

[14] See Shapiro (2002) on this notion.

Figure 11.1 Types of team organization

this background, in what follows I shall provide a conceptual clarification of two popular types of teams – hierarchical, disciplinary "rockets," and egalitarian, interdisciplinary "constellations of stars" – and discuss their efficiency for solving puzzles, identifying loci of trouble, and spreading risks when developing and investigating new approaches. That is, I will explore the essential tension as it is manifest in contemporary scientific research, where collaboration is the norm.

Hierarchical "Rockets" In hierarchical teams, a principal investigator is ultimately responsible for dividing tasks among team members and for bringing their partial contributions together to form an integrated result. Some scholars have used the metaphor of a rocket to describe the organization of such groups: Placed in the cockpit, the PI acts as the commander who sets the direction of the rocket, possibly assisted by a small number of officers in the form of permanent staff or postdocs who oversee daily activities and serve as liaisons to the large fuel tank of junior scholars, typically graduate students whose work drives the rocket forward in the direction set by the PI (Rasmussen 2014).

Such a fuel tank of junior scholars can produce large amounts of routine work quickly and cost-efficiently. Hence, such teams can be extremely efficient in solving puzzles within a well-defined research agenda. On current authorship practices, the PI will often be coauthor on a substantial share of the publications produced by the team. It is therefore also a very efficient organization for making the PI gain prominence. However, as noted earlier, the efficiency of the group depends on the PI's ability to direct the work, and the requirement of expertise in all areas of the group's work is more easily fulfilled for disciplinary than for interdisciplinary research.

In principle, such teams can also be very efficient in identifying loci of trouble. However, the situation is more intricate than it is for puzzle-solving. On the one hand, the fuel tank of junior scholars has a large capacity for critically scrutinizing any results produced and for investigating in detail if anything deviates from what is expected. On the other hand, this requires that the junior scholars have sufficient contributory expertise to recognize any detail that deviates from what is expected, and to have an idea about how to handle it. Further, it also requires that the hierarchical organization is open for team members at the bottom of the hierarchy to point out problems that can have the potential to challenge the direction of the research agenda defined by the PI at the top of the hierarchy. Since the junior scholars are typically under training, and therefore in the process of gaining contributory expertise in the field while being supervised and assessed by a senior colleague, often the PI, this can be difficult.

Research teams organized as rockets may particularly be challenged if loci of trouble are identified and there is a need for spreading the risks while developing and investigating a multitude of alternative approaches. This situation is very different from distributing large amounts of routine work among junior scholars while supervising and assessing whether their work complies with commonly accepted standards. Thus, attempts at investigating a multitude of alternative approaches may acquire the character of splitting the rocket into multiple, competing endeavors. For this reason, hierarchical research teams organized as rockets may promise a potential high gain in puzzle-solving but at the same time also harbor a potential high risk of failing, if the original research direction runs into problems.

Egalitarian "Constellation of Stars" In egalitarian teams, team members together organize how tasks are divided and individual contributions merged into shared results. In contrast to the "rockets," there is no commander who sets the direction; instead, all team members have an equal say. Typically, all group members have contributory expertise, just in different areas, and some scholars therefore describe this constellation of equal seniors as a "constellation of stars" (Rasmussen 2014).

In principle, such a team can produce large amounts of routine work quite quickly. However, there are several reasons why it is more common to see the kind of puzzle-solving that builds on large amounts of routine work performed by hierarchical rockets rather than egalitarian constellations of stars. First, the rocket's fuel tank of junior scholars is cost efficient. In contrast, a constellation of stars may not be very cost-efficient in its operations. Second, routine work can constitute important training for young scholars. In contrast, such routine work is often regarded as tedious by a constellation of stars.

Instead, the strength of a constellation of stars is its ability to perform interdisciplinary work. This kind of team is very efficient in solving puzzles that require multiple areas of contributory expertise. Further, due to the team members' different areas of contributory expertise, they automatically bring different perspectives to the solution of puzzles and to the critical scrutiny of the results produced. Hence, these types of teams can also be very efficient in identifying potential loci of trouble. However, this requires that they have sufficient interactional expertise to be able to communicate to each other why a result that may seem reasonable from one perspective nevertheless seems problematic from another perspective.[15]

Finally, due to the team members' different areas of contributory expertise, a constellation of stars also harbors the potential to develop different alternatives, once loci of trouble have been identified. In some cases, such different alternatives will show different potential in different areas, and it may be possible for an interdisciplinary constellation of stars to explore multiple alternatives.

11.6.3   *Empirical Studies of the Efficiency of Different Types of Teams*

So far, the analysis of the efficiency of different types of teams for solving puzzles, for identifying loci of trouble, and for spreading the risks associated with developing and investigating alternative approaches has been based on theoretical considerations. However, recent empirical investigations of the efficiency of different types of groups for incremental puzzle-solving and for presenting new approaches that disrupt research fields vindicates the analysis.

Examining papers published in the journals *Nature*, *Science*, *PNAS*, and *PLoS ONE*, Xu et al. (2022) found that publications reporting research in which a large fraction of team members had been included in defining, directing, and presenting the research were more likely to disrupt previously accepted ideas compared with teams where only a small fraction of the team members were included in these leadership roles. Similarly, they were more likely to have long-term impact, as measured by citations, and to facilitate greater productivity for the average author. However, for papers written by teams where a large fraction of team members were acting in various support roles, these members were more likely to be

---

[15] See Andersen (2010) for a detailed exposition of this argument.

recognized as contributing incrementally to the development of the field, to have a more short-term impact as measured by citations, and to primarily amplify productivity for a few, top team members.

Similarly, examining 65 million papers published over half a century, Wu et al. (2019) found that work authored by a large group of coauthors tends to build on more recent work than work authored by small groups of coauthors, and they also found that the work of a large group of coauthors tends to get more immediate attention. This indicates that these publications form part of the development of existing ideas. In contrast, work authored by a small group of coauthors tends to draw on literature from a larger time span, and is more likely to introduce new ideas that eclipse the attention to the work on which it builds. Thus, publications by small groups of coauthors tend to be more disruptive than publications by large groups of coauthors.

11.7   Conclusion

As described in the introduction to this chapter, the influence of Kuhn's phase model for the development of science can be found in many science policy documents, and much funding is devoted to research that is expected to be transformative and lead to the creation of new paradigms. However, it is important to recognize that Kuhn's phase model builds on an essential tension between tradition and innovation, and, hence, transformative research cannot be promoted alone but only as a complement and a result of research within the tradition.

Based on his analyses of disciplinary science conducted by individual researchers working alone, Kuhn argued that work within a well-defined tradition is immensely effective not only in solving the puzzles that the tradition defines but also in identifying potential loci of trouble. However, in this chapter, I have argued that this depends on how well the work is organized and how free scientists are to pursue their own agendas. Therefore, Kuhn's argument seems to be highly sensitive to changes in how scientists cluster together in teams, and to how these teams are organized.

Finally, I have argued that while hierarchically organized teams have the potential to be very efficient in solving puzzles within a focused research area, they can have difficulties dealing with loci of trouble and initiating transformations. I have also described how this analysis is vindicated by empirical studies showing that large, hierarchical teams tend to produce publications that report incremental contributions to their field and receive a large, short-term impact. Similarly, I have argued that while egalitarian teams may be less efficient in solving puzzles, they may be better equipped for identifying loci of trouble and initiating transformations. I have also described how this analysis is vindicated by empirical studies showing that small, egalitarian teams tend to produce publications that are transformative and have a long-term impact.

Recognizing that the core of Kuhn's argument is to point out the equal importance of both tradition and innovation, what can be concluded from this analysis is that both kinds of team organization are important for the progress of science.

Part IV

Kuhn's Impact on the Philosophy, Sociology, and History of Science

12   The Ambiguous Legacy of Kuhn's *Structure* for Normative Philosophy of Science

Jonathan Y. Tsou

12.1   Introduction

Six decades after its original publication, the legacy of Kuhn's *The Structure of Scientific Revolutions* (SSR) for analytic philosophy of science remains ambiguous. On the one hand, SSR was the key work – along with Quine's "Two Dogma's of Empiricism" (1951) – that contributed to the demise of logical empiricism in the 1960s and 1970s. Moreover, SSR modeled a novel methodological approach for doing philosophy of science, which spearheaded the "historical turn" in philosophy of science and the rise of the history and philosophy of science (HPS) tradition. In terms of legacy, SSR was undoubtedly successful in shifting the methodological assumptions of post-positivist philosophy of science toward historical analyses and away from logical analyses. On the other hand, the methodological assumptions of SSR introduced unclarity regarding how philosophers of science should address normative issues and what kinds of normative questions they should address. Prior to SSR, analytic philosophers of science were preoccupied with addressing normative questions, such as the demarcation problem (e.g., Popper 1935/1959; Carnap 1936, 1937, 1956; Hempel 1950). After the publication of SSR, attention shifted away from such general ("universalist") philosophy of science issues, and attempts to address normative questions in an ahistorical manner became unfashionable.

This chapter examines the legacy of Kuhn's SSR for normative philosophy of science. As an argument regarding the history of twentieth-century philosophy of science, I contend that the main legacy of SSR was destructive: SSR shifted philosophy of science away from addressing general normative philosophical issues (e.g., the demarcation problem, empirical testability) and toward more deflationary and local approaches to normative issues. This is evident in the first generation of post-SSR philosophers of science in the 1980s and 1990s, who adopted a pluralist approach to HPS. As a metaphilosophical argument regarding the methods adopted in HPS, I argue that there are a plurality of legitimate philosophical methodologies for inferring normative claims from historical cases. I frame this argument as a response to Joseph Pitt's dilemma of case studies. I reject Pitt's dilemma for its presupposition of an unrealistic and unfruitful standard (viz., epistemic certainty) for assessing HPS arguments and its analysis of philosophical methodology at the level of *individual arguments*. Pitt's dilemma is most usefully understood as identifying potential points of criticism for HPS arguments.

The chapter begins with an examination of Kuhn's normative philosophy of science in SSR and his position that historical cases provide *evidence* for philosophical claims. Kuhn's philosophical methodology is insufficiently articulated, and his utilization of case studies is subject to objections (viz., interpretative bias, hasty generalization) implied by Pitt's dilemma. I subsequently examine four post-Kuhnian methodological perspectives: (1) Ian Hacking's particularism, (2) Helen Longino's practice-based approach, (3) Michael Friedman's neo-Kantianism, and (4) Hasok Chang's complementary science. These views suggest alternative methodological strategies in HPS for addressing normative issues. I conclude by articulating some outstanding methodological challenges for the pluralist tradition of HPS – associated with the Stanford and Minnesota schools of philosophy of science – that emerged in the 1980s and remain influential.

12.2   Kuhn's Philosophical Methodology and Its Problems

The topic of this chapter is encapsulated by a question that Paul Feyerabend posed to Kuhn about whether SSR presents normative arguments. I critically examine Kuhn's response that his normative arguments (e.g., science should solve paradigm-prescribed puzzles) are supported by historical cases. I subsequently discuss problems with Kuhn's methodology with reference to the dilemma of case studies.

12.2.1   *Kuhn's Normative Philosophy of Science*

At the 1965 International Colloquium of Philosophy of Science (also see Hoyningen-Huene 2006b), Feyerabend asked whether SSR intended to present normative prescriptions or historical descriptions:

Whenever I read Kuhn, I am troubled by the following question: are we here presented with *methodological prescriptions*, which tell the scientists how to proceed; or are we given a *description*, void of any evaluative element, of those activities which are generally called "scientific"? Kuhn's writings ... do not lead to a straightforward answer. They are *ambiguous* in the sense that they are compatible with, and lend support to, both interpretations. (Feyerabend 1970, 199, emphasis in original)

In response to Feyerabend, Kuhn was characteristically nuanced and murky:

The answer is that, of course, they should be read in both ways at once. If I have a theory of how and why science works, it must necessarily have implications for the way in which scientists should behave if their field is to flourish. The structure of my argument is simple and, I think, unexceptional: scientists behave in the following ways; those modes of behavior have (here theory enters) the following functions; in the absence of an alternative mode *that would serve similar functions*, scientists should behave essentially as they do if their concern is to improve scientific knowledge. (Kuhn 1970b, 237, emphasis in original)

In the Postscript added to the second edition of SSR in 1970, Kuhn puts the point as follows:

[SSR] presents a ... theory about the nature of science, and ... the theory has consequences for the ways in which scientists should behave if their enterprise is to succeed. Though it need not be right, any more than any other theory, *it provides a legitimate basis for reiterated "oughts" and "shoulds."* Conversely, one set of reasons for taking the theory seriously is that scientists, whose methods have been developed and selected for their success, do in fact behave as the theory says they should. *My descriptive generalizations are evidence for the* [normative] *theory precisely because they can also be derived from it*, whereas on other views of the nature of science they constitute anomalous behavior.

The circularity of that argument is not ... vicious. The consequences of the viewpoint being discussed are not exhausted by the observations [i.e., case studies] upon which it rested at the start. (SSR-4 206–207, emphasis added)

Kuhn indicates that his theory of scientific change – wherein science progresses from stages of normal science to revolutionary science – is both descriptive and prescriptive.[1] As a normative theory, this implies that scientists *should* engage in puzzle-solving or "mop-up work" (i.e., dogmatic and narrow empirical research intended to support, elaborate, and expand the currently favored paradigm) characteristic of normal science (SSR Chapter 3) *until* a competing paradigm emerges that has

[1] An alternative, arguably more candid, statement of Kuhn's methodology is found in a letter that Feyerabend wrote to Kuhn in the early 1960s. Feyerabend paraphrases and quotes Kuhn's response (from an unpublished letter from Kuhn to Feyerabend) to his prescriptive-descriptive question as follows: "You say that you are not interested in a prescriptive methodology, but in a 'more realistic notion of the practice actually used successfully in physical science'" (Feyerabend to Kuhn, cited in Hoyningen-Huene 2006b, 617). On this understanding, SSR was more concerned with presenting an accurate historical description of successful scientific practices than articulating philosophical prescriptions about how science should proceed. In their correspondence, Feyerabend chides Kuhn for assuming that his description of historical facts is *value-free*.

greater puzzle-solving power.[2] For Kuhn (SSR Chapter 13), science is *rational* and genuinely scientific insofar as scientists commit themselves to the best available paradigm with respect to (current or future) *puzzle-solving power*. From a methodological standpoint, Kuhn's descriptions of (putatively successful) scientific revolutions provide the basis (and "evidence") for his normative claims. Kuhn argues that this "derivation" of normative from descriptive claims is not question-begging ("viciously circular") because his normative generalizations can be "tested" against other successful cases of paradigm change.

12.2.2   The Dilemma of Case Studies

Kuhn's favored methodology of deriving normative (philosophical) generalizations from particular case studies is limited insofar as it requires impartial selection (and accurate description) of historical cases and a sufficient number of cases to support normative generalizations. Pitt (2001, 373) presents these limitations as a dilemma:

1. Either historical cases are selected because they support a philosophical conclusion or they are randomly selected.
2. If cases are selected because they support a philosophical conclusion, then there is no assurance the historical data has not been manipulated to support the conclusion.
3. If case studies are randomly selected, then it is unclear how many cases are needed to support a philosophical conclusion.
4. Thus, the use of cases to support philosophical conclusions is either (potentially) biased or (potentially) a hasty generalization.

Pitt's dilemma implies that philosophical arguments based on case studies fall prey to the fallacies of question-begging and/or hasty generalization. Mary Hesse anticipated this problem when she remarked that "the case history approach is liable to fall into either dogmatic partisanship for a single theory of science, or into lack of generality" (Hesse 1980, 4).

Pitt's dilemma is particularly salient for SSR since Kuhn's analysis appears to violate both horns. In SSR, Kuhn supports his theory primarily through three cases: the Copernican Revolution, the Chemical Revolution, and the Einsteinian Revolution. Kuhn presents these cases to support the (descriptive) generalization that large-scale scientific

---

[2] *Pace* Kuhn, Feyerabend (1975) argues that science should be a pluralistic ("anarchistic") enterprise that constantly encourages a proliferation of inconsistent methods and theories. Feyerabend (1970) premises his argument for pluralism on the assumption that normal science, as Kuhn describes it, does not exist (Tsou 2003b).

change is discontinuous and noncumulative insofar as revolutions involve the *replacement* (in whole or in part) of an established paradigm with an incommensurable one (SSR-4 92). Regarding the question-begging charge, critics argue that Kuhn's historical descriptions are inaccurate and presented through the lens of his theory of scientific revolutions (Shapere 1964).[3] For example, Kuhn's presentation of the Einsteinian Revolution neglects continuities (e.g., Newton's laws are derivable as a limiting case in special relativity), while exaggerating discontinuities (Earman 1993).[4] Moreover, Kuhn's presentation of this revolution as a *forced choice* between "incommensurable paradigms" that would define physics is a misleading. The Einsteinian Revolution did not involve a *replacement* of Newtonian mechanics (e.g., Newton's theory is still used widely by scientists and engineers), but the recognition of the limited domain of application for Newton's theory (e.g., classical mechanics applies to objects on Earth moving slower than the speed of light). Regarding the hasty generalization charge, a larger or more random selection of scientific cases might conflict with the philosophical conclusions that Kuhn reaches. For example, analysis of the transition from views of blending inheritance to Mendelian inheritance (Olby 1966/ 1985) might reveal a more continuous and cumulative view of scientific revolutions than the narrative suggested in SSR. Similarly, examination of the shift from behaviorist psychology to cognitive psychology might yield a much more cumulative and continuous picture of scientific change (e.g., behaviorist methods of classical and operant conditioning are retained in cognitive psychology and remain central in contemporary neuroscience) than the "replacement view" suggested in SSR. The presence of such alternative historical narratives highlights the significance of the hasty generalization objection.

12.3   Post-Kuhnian Approaches to Philosophy of Science

The following section examines some post-Kuhnian methodological approaches to philosophy of science: Ian Hacking's particularism, Helen Longino's practice-based approach, Michael Friedman's neo-Kantianism,

---

[3] Pitt's question-begging objection focuses on bias in *case selection*. Herein, I extend this objection to include bias in *case interpretation*. A related but distinct issue that I do not address concerns whether Kuhn's philosophical methodology commits a naturalistic fallacy (Giere 1973; Schickore 2011; Schindler 2013).

[4] Kuhn (SSR-4 101–102) contends that these continuities are *negated* by changes in the meaning of terms (e.g., "mass") across theories. Chang (2012b) argues persuasively against the significance of meaning incommensurability, pointing out that the meanings of theoretical terms frequently change with theory revisions.

222 *Jonathan Y. Tsou*

and Hasok Chang's complementary science. These approaches fall broadly within the HPS tradition inspired by Kuhn but offer more clearly articulated methodological positions on normative questions.[5]

12.3.1 Hacking's Particularism

Ian Hacking is a central member of the Stanford School of philosophy of science (Suppes 1978; Cartwright 1983; 1999; Hacking 1983; 1992; Galison 1987; Dupré 1993), which is characterized by a commitment to scientific pluralism, disunity of science, and practice-based accounts of science (Ludwig and Ruphy 2021; Cat 2022). In their overarching commitment to disunity of science and rejection of universalist accounts, the Stanford School opposes both the logical empiricists' unity of science ideal and Kuhn's (monistic) attempt to present a global (or "universal") account of scientific change (cf. Wray 2021c).

The philosophical methodology that characterizes Hacking's various philosophical projects is a form of philosophical particularism, which eschews drawing *universal* claims about science.[6] Hacking's methodological particularism recommends drawing particular philosophical conclusions from historical cases and avoiding generalizations about science. This approach is motivated to address the second horn of Pitt's methodological dilemma. By closely examining particular case studies, philosophers can avoid making invalid inferences by drawing qualified and local normative conclusions.

Hacking's particularism is evident in his analysis of scientific objectivity ("Let's Not Talk about Objectivity"), where he urges philosophers to stop discussing objectivity *in the abstract* (Hacking 2015). Hacking distinguishes two different types of questions about objectivity:

---

[5] My survey examines work associated with three of the most prominent methodologically oriented HPS societies: *The International Society for History of Philosophy of Science* (HOPOS), *The Society for Philosophy of Science in Practice* (SPSP), and *The Committee for Integrated HPS* (&HPS). Friedman offers an example of HOPOS methodology; Hacking and Longino provide contrasting examples of methodologies adopted in SPSP; and Chang presents a prominent example of SPSP and &HPS methodology (Chang is a cofounder of both societies). SPSP and &HPS are contemporary societies that represent the pluralist/disunity of science tradition of HPS that emerged in the 1980s (Ludwig and Ruphy 2021).

[6] Hacking's particularism is inspired by J. L. Austin's ordinary language philosophy and Michel Foucault's "history of the present" (Hacking 2002). Tsou (2015) notes particularist (i.e., bottom-up) aspects of Kuhn's philosophical methodology in SSR. Kuhn's case study methodology reflects his training in James Bryant Conant's General Education of Science curriculum at Harvard (Wray 2021a, chapter 2). For a fascinating historical examination of Kuhn and Conant's collaboration in the context of the Cold War, see Reisch (2019).

1. "Ground-level questions": specific questions about particular cases that have some bearing on the objectivity of science (e.g., "Can we trust medical research when it is funded by pharmaceutical companies?").
2. "Second-story questions": general questions about objectivity that assume that objectivity is a stable epistemic ideal (e.g., "What is scientific objectivity?").

Hacking argues that philosophers should stop talking about (2) and only talk about (1), which motivates his imperative: "let's get down to work on cases, not generalities" (p. 29). Hacking regards abstract philosophical ideals (e.g., "objectivity," "truth," "facts") as "elevator words" that exist at a higher level than actual things in the world (Hacking 1999, 22–24). He objects to philosophical analysis of such words because they are circularly defined and use of them gives the false impression that there are stable meanings for such terms. His favored methodology analyzes these words *as they appear in various historical sites* (Hacking 2002).

Hacking's particularism is also salient in his defense of entity (or experimental) realism. Hacking distinguishes between realism about theories (as true or false) and realism about entities (as existing or not). Hacking eschews questions about realism about theories, but he maintains that questions about realism about entities are answerable. Hacking's argument for entity realism emphasizes the use and manipulation of theoretical entities as a basis for believing in their reality. He argues that we have good evidence for the reality of an entity under specific circumstances: "The experimenter is convinced of the reality of entities some of whose causal properties are sufficiently well understood that they can be used to interfere elsewhere in nature. One is impressed by entities that one can use to test conjectures about other more hypothetical entities" (Hacking 1982, 75). Hacking contends that we have evidence for the reality of "unobservable" entities that experimenters can *control* and *manipulate* to study other parts of nature: "We understand the effects, we understand the causes, and we use these to find out about something else" (Hacking 1983, 24). Hacking draws this normative conclusion based primarily on a single case study, namely, experimenters' manipulation of electrons in the use of electron guns to study quarks. He writes: "We are completely convinced of the reality of electrons when we regularly set out to build – and often enough succeed in building – new kinds of device[s] that use various well-understood causal properties of electrons to interfere in other more hypothetical parts of nature" (p. 265, emphasis removed). This leads to Hacking's well-known slogan for entity realism: "If you can spray them, then they are real" (p. 22).

Hacking's analysis indicates the qualified and limited normative conclusions he is willing to draw from case studies. As a methodological strategy, Hacking's particularism grabs the second horn of Pitt's dilemma by qualifying and limiting the scope of normative claims away from universal questions (e.g., when are scientific theories true?) toward more particular and local normative questions (e.g., when do we have evidence that the unobservable entities posited by theories exist?).

12.3.2 Longino's Practice-Based Methodology

Whereas Hacking eschews discussing scientific ideals in the abstract, Helen Longino (1990) articulates a general social account of objectivity that opposes Kuhn's account of scientific objectivity. Her practice-based methodology is important to examine since it represents a common methodology adopted in contemporary HPS.

Longino rejects Kuhn's "theory-choice" framework (ET 320–339), which examines questions of objectivity in terms of *choices* that scientists make between incommensurable paradigms. In Kuhn's framework, theory-choice is objective if there is a *stable* set of *epistemic values* (e.g., empirical accuracy, predictive power, explanatory scope, simplicity, fruitfulness) shared by proponents of competing paradigms, which provides an intersubjective basis for making *mechanical* ("algorithmic") *decisions* between competing paradigms. Kuhn argues that paradigm choice is never fully objective in this sense because competing paradigms are premised on incommensurable value-sets (e.g., specific values are interpreted differently or weighted more heavily within a paradigm).[7] Longino rejects Kuhn's (rational-choice) framework for its individualistic assumptions and failure to reflect actual scientific practices. In her social framework, objectivity is a characteristic of scientific communities, rather than an individual's reasons for favoring a paradigm. A scientific community is objective, that is, free from bias, when its knowledge claims are presented in a public domain and subject to criticism.

Longino assumes that non-epistemic *contextual values* (i.e., socially situated background beliefs and assumptions) are a ubiquitous feature of science, and they determine what counts as scientific evidence in a historical context. Despite the inevitable presence of contextual values, scientific communities are objective if they allow for *transformative criticism*, that is, criticism that can lead to the revision of contextual values: "As long as background beliefs can be articulated and subjected to

---

[7] For criticism of Kuhn's analysis, see Laudan (1984) and McMullin (1993).

criticism from the scientific community, they can be defended, modified, or abandoned in response to such criticism. As long as this kind of response is possible, the incorporation of hypotheses into the canon of scientific knowledge can be independent of any individual subjective preferences" (Longino 1990, 73–74). Scientific communities are objective to the extent that they encourage – and are responsive to – criticism (especially criticism of contextual values).

Longino argues that features of more objective scientific communities include recognized avenues for criticism (e.g., peer-reviewed journals and conferences), the presence of public shared scientific standards (e.g., empirical accuracy) that guide criticism, responsiveness to criticism, and equality of intellectual authority. On this view,

> Objectivity is dependent upon the depth and scope of the transformative interrogation that occurs in any given scientific community. This communitywide process ensures ... that the hypotheses ultimately accepted ... do not reflect a single individual's idiosyncratic assumptions.... To say that a theory ... was accepted on the basis of objective methods [entitles us to say that the theory] reflects the critically achieved consensus of the scientific community.... [I]t's not clear we should hope for anything better. (Longino 1990, 79)

For Longino, objectivity is constituted by social mechanisms that encourage a critical scientific community; by contrast, a community is subjective (i.e., biased) when there are inadequate mechanisms to prevent its theories from reflecting the idiosyncratic assumptions of a few individuals. Whereas Kuhn identifies objectivity with *a shared (or agreed upon) perspective of epistemic values* that provide *clear criteria* (or rules) for choosing among competing paradigms, Longino identifies objectivity with the presence of *a plurality of conflicting perspectives* that facilitate criticism of *contextual values*.

From a methodological perspective, Longino endorses a Kuhn-inspired practice-based methodology, which includes conceptual analysis and the understanding of scientific reasoning "as a practice rather than as the disembodied application of a set of rules" (Longino 1990, 13). Longino utilizes this Kuhnian standard to criticize Kuhn's analysis of objectivity for being artificial and disengaged with actual scientific practices. Longino's methodology is subject to the first horn of Pitt's dilemma insofar as she selectively chooses scientific cases to support normative claims. Longino's strategy involves (1) explicating an operational definition of objectivity and (2) providing contrastive examples, which exemplify objective and subjective aspects of scientific communities. While Hacking opposes (1) as "circular," Longino's definition can be assessed in terms of whether it picks out a desirable epistemic feature of science (viz., a community's encouragement of self-criticism) that captures a useful meaning of "objectivity." While Pitt opposes (2) as a conclusion-driven selection of cases, Longino's examples (e.g., androcentric research on sex differences) can be assessed in terms of whether they accurately illuminate desirable ("objective") and undesirable ("subjective") features of science, and whether there are counterexamples. Longino's strategy for avoiding the question-begging objection is to be conservative in her selection of cases relative to her normative conclusions. She points to relatively *uncontroversial* instances (e.g., peer review, replication of studies) and counterinstances (e.g., politically motivated scientific communities) of her ideal of objectivity. In this regard, Longino's use of history is closer to traditional a priori normative approaches, which utilize cases primarily as anecdotal ("toy") examples that illustrate epistemic virtues and vices of science. Hence, her methodology can be criticized for engaging in the "anecdotal"/Whiggish use of history rejected by Kuhn (cf. Solomon 2001, chapter 3).

12.3.3   *Friedman's Neo-Kantianism*

Michael Friedman argues for anti-Kuhnian conclusions on the basis of the history of science and the history of philosophy of science. Friedman is troubled by the antirationalist and relativist conclusions suggested by Kuhn's analysis of scientific change,[8] and he advances a neo-Kantian account of scientific rationality.

From a methodological perspective, Friedman draws from both the history of science and the history of philosophy of science as resources for supporting normative philosophical claims. One of the main normative arguments that Friedman presents is that the development of modern physics required theoretical principles that have a relativized a priori status. In the history of logical empiricism, the importance of relativized a priori principles was recognized by Reichenbach and Carnap. Friedman writes:

Careful attention to the actual historical development of science … shows that relativized a priori principles of just the kind Carnap was aiming at are central to our scientific theories. Although Carnap may have failed in giving a precise logical characterization or explication of such principles, it does not follow that the *phenomenon* he was attempting to characterize does not exist. On the contrary, everything that we know about the history of science … indicates that precisely

---

[8] The relativist implications of SSR were heartily embraced by sociologists of scientific knowledge (see Wray 2011b).

this phenomenon is an absolutely fundamental feature of science. (Friedman 2001, 41, emphasis in original)

Friedman argues that in the history of mathematical physics, from Newtonian mechanics to the general theory of relativity, physical theories have required the presupposition of relativized a priori principles. The epistemic function of these principles is that they are *constitutive* of empirical laws and tests insofar as they make "the precise mathematical formulation and empirical application of the theories in question first possible" (Friedman 2001, 40). In the history of twentieth-century philosophy of science, Friedman argues that the special status of such principles was recognized not only by Reichenbach and Carnap but also by Kuhn, in his "Kantian" assumption that empirical observations and tests require the presupposition of a paradigm (cf. Hoyningen-Huene 1989/1993; Patton 2021).

Friedman's neo-Kantian account suggests that modern physical theories are distinguishable into three asymmetrically functioning parts:

1. A *mathematical part*: basic mathematical and geometrical principles that describe the spatio-temporal framework.
2. A *mechanical part*: principles that coordinate the mathematical part to the empirical part.
3. An *empirical part*: empirical and physical principles that use the theories in the mathematical part to formulate empirical laws to describe concrete phenomena.

For Friedman, (1) and (2) are relativized a priori in the sense that – *taken together* – they are *constitutive* of the empirical part. In Newtonian mechanics, for example, Euclidean geometry (1) and the laws of motion (2) have a relativized a priori status insofar as they are constitutive of empirical laws such as the law of universal gravitation (3). In special relativity, the geometry of Minkowski space-time (1) and the light principle (2) are constitutive of empirical laws such as Maxwell's equations (3). In general relativity, the theory of semi-Riemannian space-time manifolds (1) and the principle of equivalence (2) are constitutive of empirical laws such as Einstein's equations for the gravitational field (3).

Friedman's argument for the rationality of scientific change is framed in terms of relative a priori principles and draws on Habermas' distinction between instrumental and communicative rationality:

1. Instrumental rationality (subjective): the capacity to engage in effective means-ends reasoning (where an agreed-upon goal is assumed).
2. Communicative rationality (intersubjective): the capacity to engage in deliberative argumentative reasoning (where there is no agreed-upon goal) aimed at bringing about consensus.

In Kuhn's account, communicative rationality is limited to rationality *within* a paradigm (i.e., normal science). Friedman argues that there is an important sense in which communicative rationality, in the sense of achieving a consensus or agreement, is achieved *between* different paradigms (i.e., revolutionary science). For Friedman, the history of mathematical physics illustrates that earlier constitutive principles appear as limiting cases in later theories (which hold under specified conditions), and the concepts of succeeding constitutive frameworks evolve continually, by a series of natural transformations, from previous constitutive frameworks (cf. Worrall 1989). He writes: "we can thus view the evolution of succeeding paradigms ... as a convergent series ... in which we successively refine our constitutive principles in the direction of even greater generality and adequacy" (Friedman 2001, 63).

Friedman's philosophical methodology, like Kuhn's, aims to infer normative conclusions about science by selecting case studies that exemplify an argument. As such, it falls prey to the interpretative bias objection expressed by the first horn of Pitt's dilemma. Conversely, it is notable that Friedman aims to counter Kuhn's view by providing a *more accurate description* of the history of physics. Friedman's analysis also falls prey to the second horn of Pitt's dilemma. While Friedman's analysis may generate normative claims about the rationality of mathematical physics, it is unclear how generalizable this conclusion is to science more generally.[9]

12.3.4 *Chang's Complementary Science*

Hasok Chang is one of the most vocal advocates for pluralism and integrated HPS. Chang's methodological ideal of complementary science is particularly important to examine since it is motivated precisely to address the methodological issues discussed in this chapter.

Chang (2004) presents complementary science as an *ideal function* that HPS should serve. The aim of complementary science is to articulate normative claims about science through historical and philosophical investigation. In this ideal, there is not a strong disciplinary distinction between history, philosophy, and science. Unlike more orthodox methodological approaches in HPS that aim to produce meta-scientific knowledge (e.g., a general account of scientific explanation or scientific

[9] Friedman (2001, 124–129) briefly discusses how his neo-Kantian framework applies to sciences such as chemistry and biology, but his discussion does little to engage with actual chemical or biological science. For a discussion of constitutive principles in biology, see Tsou (2010) and Luchetti (2021).

change), complementary science aims to contribute to scientific knowledge itself:

[Complementary science] contributes to scientific knowledge through historical and philosophical investigations. Complementary science asks scientific questions that are excluded from current specialist science. It begins by re-examining the obvious, by asking why we accept the basic truths of science.... Because many things are protected from questioning and criticism in specialist science, its demonstrated effectiveness is also unavoidably accompanied by a degree of dogmatism and a narrowness of focus that can actually result in a loss of knowledge. History and philosophy of science in its 'complementary' mode can ameliorate the situation. (Chang 2004, 3)

In its complementary function, HPS can recover useful ideas and facts lost in the scientific record ("Kuhn losses"), address foundational questions concerning present science, and explore alternative conceptual systems and lines of experimental inquiry for future science. If these investigations are successful, they will complement and enrich current specialist science. Normative claims (e.g., concerning good or bad science) generated by complementary science are articulated at the level of scientific knowledge itself, rather than at a level removed from science.

As a methodological perspective, complementary science addresses Pitt's dilemma by rejecting the terms in which it is framed. Specifically, Chang rejects the assumption that history deals with the particular and philosophy deals with the general:

[Kuhn] never specified a clear method for the history–philosophy interaction, and without such a method we are condemned to the dilemma between making unwarranted generalizations from historical cases and doing entirely "local" histories with no bearing on an overall understanding of the scientific process.

In attempting to transcend this dilemma, I believe that the first thing we need to do is to see if we can get beyond an inductive view of the history–philosophy relation, which takes history as *particular* and philosophy as *general*. Of course, we cannot get away from inductive thinking entirely, but it is instructive to try seeing the history–philosophy relation as one between the *concrete* and the *abstract*, instead of one between the particular and the general. Abstract [philosophical] ideas are needed for the understanding of *any* concrete [historical] episode.... Any concrete account requires abstract notions [e.g., "confirmed," "observation," "measurement"].... If we extract abstract [normative] insights from the account of a specific concrete episode ... , that is not so much a process of *generalization*, as an *articulation* of what was already put into it. (Chang 2012b, 110, emphasis in original)

Chang challenges Kuhn's stance that – given their divergent aims – history of science and philosophy of science cannot be practiced at the same time (ET 3–20). He also challenges Pitt's assumption that history is concerned with particular cases, whereas philosophy is concerned with general (normative) concepts. Against these views, complementary science advocates an integrated view of HPS, wherein concrete historical episodes and abstract philosophical concepts stand in a mutually dependent relationship (cf. Burian 2001). With respect to the dilemma of case studies, Chang's position recommends resisting an entrenched assumption of HPS, assumed by Kuhn, Pitt, and others, that historical episodes are particular pieces of *evidence* that support general philosophical conclusions. Rather, we should view normative arguments as the articulation of abstract ideas that are embedded in concrete historical episodes.

12.4   Methodological Challenges for HPS

One distinctive feature of the HPS community in the 1980s and 1990s – one generation removed from SSR – is its pluralistic outlook (Cartwright 1983, 1999; Hacking 1983; Galison 1987; Longino 1990; Dupré 1993; Galison and Stump 1996; Giere 1999; Wylie 1999). In advocating scientific pluralism, disunity of science, antireductionism, and methodological engagement with the history of science and scientific practices, post-Kuhnian HPS engaged with a broader range of sciences besides physics (especially biology) and addressed a broader range of philosophical topics (e.g., values in science, models, mechanisms). If the methodological assumptions associated with Stanford and Minnesota schools of pluralism (Solomon and Richardson 2005; Kellert et al. 2006; Cat 2022) provide a fair representation of the methodological commitments of the HPS community, then post-Kuhnian HPS is decidedly more Feyerabendian in character than Kuhnian.[10]

From a metaphilosophical perspective, several methodological challenges for post-Kuhnian HPS – especially the pluralistic tradition that emerged in the 1980s – can be gleaned from the methodological approaches examined in this chapter. While no decisive conclusions can be drawn from such a limited survey, I articulate three related

[10] The pluralist/disunity of science consensus arrived at in the 1980s and 1990s was not self-consciously inspired by Feyerabend. Rather, pluralism was an umbrella concept that post-Kuhnian philosophers of science invoked to articulate *nonpositivist* positions on a variety of issues (Richardson 2006; Ludwig and Ruphy 2021). Despite resonances with Feyerabend's views on pluralism, antireductionism, and disunity of science, the programmatic statements of the Stanford (Suppes 1978) and Minnesota (Kellert et al. 2006) schools of pluralism do not acknowledge Feyerabend as an influence (cf. Hacking 1983, 14; Dupré 1993, 10; Longino 1990, chapter 2). For discussion of Feyerabend's mixed legacy in HPS, see Lloyd (1996), Preston, Munévar, and Lamb (2000), Brown and Kidd (2016), Sankey (2020), and Shaw and Bschir (2021).

methodological challenges for post-Kuhnian HPS: historicism, the relationship between case studies and normative arguments, and the lack of canonical problems.

12.4.1 Historicism

One methodological challenge for HPS concerns the possibility of articulating nonhistoricist and nonrelativist arguments about science. Although Kuhn (SSR-4 203–205) adamantly denied charges of relativism, Popper (1970, 55) correctly diagnosed Kuhn's theory as a species of "historical relativism." Kuhn (and Feyerabend) assumed a form of historicism that eschewed attempts to articulate (stable) transhistorical normative scientific ideals (e.g., "truth," "empirical evidence," "rationality").[11] This chapter suggests that the historicism championed by Kuhn and Feyerabend in the 1960s and 1970s evolved into the pluralistic style of HPS practiced in the 1980s and 1990s.

In the context of twenty-first-century HPS, historicism is most explicitly endorsed in the subfield of "historical epistemology" (Hacking 1975; 1992; 2002; Galison 1987; Daston 1994; Davidson 2001; Daston and Galison 2007).[12] In contrast with traditional normative philosophy of science, which defends *stable normative scientific ideals* (e.g., "probability," "experimentation," "empirical evidence"), historical epistemology assumes that epistemic and scientific ideals are *unstable* and examines the emergence and evolution of such ideals. In the (yet) broader context of twenty-first-century philosophy of science, historicist approaches – such as historical epistemology – are somewhat marginal.[13] At present, philosophy of science embraces a plurality of methodological approaches, including historical, formal, naturalistic, and social approaches. However, historicist considerations are relevant for any HPS analysis that defends *general*

---

[11] For discussion of Kuhn's historicism, see Bird (2015) and Nickles (2017). In contrast with Kuhn's distaste for relativism, Feyerabend (1978a; 1987; 1989) increasingly embraced historicism and relativism in his post–*Against Method* works (Preston 1997a; 1997b; Brown and Kidd 2016; Kusch 2016).

[12] For critical discussion, see Kusch (2010) and Feest and Sturm (2011).

[13] The historicism popularized by Kuhn and Feyerabend in the 1960s and 1970s eventually gave way – especially through the "science wars" – to a realist reaction by philosophers of science in the 1990s and 2000s (Nickles 2017). This realist response also reflects the increased influence of (Quinean) naturalistic methodological approaches to philosophy of science (Tsou forthcoming). It is worth noting that Kuhn's methodology in SSR was naturalistic insofar as it demanded *accurate reconstructions of scientific practices* with the aid of a posteriori sciences (e.g., history, psychology). Bird (2000b; 2002; 2004) argues that Kuhn's naturalism was the most promising feature of SSR; he laments the fact that Kuhn shifted away from naturalism in post-SSR works (e.g., RSS) toward more traditional a priori philosophical methods (cf. Shan 2020).

normative arguments through historical cases. For example, Friedman (2001, 64–65) is acutely aware of historicist considerations in his defense of scientific rationality. His response to historicism involves understanding scientific rationality as a Kantian *regulative ideal of reason*: "We can ... view our present scientific community ... as an approximation to a [Peircean] final, ideal community of inquiry ... that has achieved a universal, trans-historical communicative rationality on the basis of the fully general and adequate constitutive principles reached in the ideal limit of scientific progress" (p. 64). Regardless of its merits as a response to historical relativism (see Richardson 2002; Tsou 2003a), Friedman should be commended for his sensitivity to historicist issues and transparency in responding to them. Few HPS analyses are so explicit.

12.4.2 *How Historical Cases Support Philosophical Arguments*

A related methodological challenge for HPS is how historical cases can support normative positions on the most general philosophy of science issues, such as the demarcation problem.[14] One of the historical legacies of SSR is that it shifted the attention of philosophers of science away from *general* philosophy of science issues toward more qualified topics (e.g., analyses limited to the special sciences) and local normative issues (e.g., structural realism, mechanisms, inductive risk). To the point of this chapter, SSR shifted philosophers away from traditional normative questions concerning the *justification of scientific knowledge*. In my opinion, this state of affairs is odd and regrettable. In the twenty-first century, epistemic questions concerning what distinguishes genuine science from nonscience remain salient in various socially relevant contexts (e.g., debates about climate change models or mRNA vaccines), and philosophers have little to contribute beyond arguments proposed over five decades ago.[15] Others express similar concerns that the disunity of science/pluralist consensus reached in the 1980s and 1990s fragmented HPS and threatened the very possibility of doing general normative philosophy of science (Reisch 1998; Magnus 2013).

[14] Roth (2013) notes the irony that, despite the immense methodological influence of SSR, post-Kuhnian philosophers of science have paid relatively little attention to issues of historical explanation.
[15] For some post-Kuhnian analyses of the demarcation problem, see Pigliucci and Boudry (2013) and Hoyningen-Huene (2015a). Kuhn's historicized answer to the demarcation problem (i.e., science is distinguished by the presence of a paradigm) and normative account of science (i.e., scientists should prefer paradigms that are better at solving puzzles than rivals) are articulated at *too general of a descriptive level* to evaluate the scientific status of individual fields.

The challenge of how to support normative philosophical arguments with cases is brought into focus by Pitt's dilemma. For Pitt, the reliability of the case study approach is limited because one can always question (1) the theoretical neutrality in which cases are selected and interpreted and (2) whether the selected cases provide adequate inductive support for general philosophical conclusions. Pitt's dilemma should be rejected in its implicit presupposition of an unreasonably lofty epistemic standard, namely, the demand for epistemic certainty or infallibility. Pitt's dilemma only has force, *as an argument against using historical cases*, if the goal of philosophical analysis is to infer – *with absolute certainty* – normative conclusions. In this regard, Pitt's dilemma applies equally to science as it does to HPS insofar as scientific conclusions of *a given study* are inferred from particular observations, experiments, or tests. Just as the fallibility of a single scientific study does not speak against inferring general theories from particular observations, the fallibility of generalizations drawn in a single HPS study does not speak against inferring general conclusions from particular case studies. If case studies provide defeasible evidence in support of scientific generalizations, including normative generalizations, and HPS arguments are subject to criticism (e.g., regarding the selection or interpretation of cases), then there is nothing particularly problematic about the case study approach. Following Chang, normative philosophical arguments can be regarded as *abstract generalizations* that are instantiated in and illustrated by concrete historical episodes. In HPS, the challenge is to articulate normative arguments that avoid "naïve abstraction while working at a greater level of generality than the specific sciences" (Magnus 2013, 51).

12.4.3   *Canonical Normative Problems for HPS*

A final challenge for HPS concerns the lack of consensus about canonical normative issues (cf. Galison 2008). In the logical empiricist tradition (e.g., Hempel 1966), philosophers of science worked collectively on a family of interrelated normative issues (e.g., the demarcation problem, induction in science, empirical testability and confirmation, scientific explanation, laws of nature) that focused on the epistemic question of what *justifies* scientific knowledge. SSR introduced ambiguity, not only with respect to the methodologies that philosophers of science should adopt, but for the kinds of normative questions philosophers of science should address. In SSR, Kuhn was preoccupied with what appeared to be a set of descriptive issues (e.g., the continuity and discontinuity in scientific change, science as puzzle-solving, incommensurability). In rejecting the logical empiricists' distinction between the contexts of discovery and justification, Kuhn's historical approach questioned the very assumption that philosophers could address questions about scientific justification, independent from their historical contexts (Hoyningen-Huene 2006a). The pluralism that characterizes much of post-Kuhnian HPS provides little assurance that an emerging consensus on canonical normative issues is forthcoming.

12.5   Concluding Remarks

This chapter examined the legacy of Kuhn's SSR in the history of twentieth-century philosophy of science. From the perspective of normative philosophy of science, the legacy of SSR was primarily destructive: SSR brought into question the very idea of defending ahistorical normative ideals of science and shifted historians and philosophers of science toward more deflationary and localized normative projects. This deflationary impact of SSR need not be viewed as a regressive episode in the history of philosophy of science nor be cause for methodological anxiety. Rather, SSR opened up new methodological possibilities and opportunities for HPS to articulate normative arguments.

From the perspective of contemporary twenty-first-century philosophy of science, historians and philosophers of science face ongoing methodological challenges for articulating and supporting normative arguments about science based on historical (and present) cases of science. Three outstanding methodological challenges for contemporary HPS include (1) articulating nonhistoricist and nonrelativist arguments about science, (2) articulating general ("universal") features of good and bad science, and (3) articulating canonical issues for HPS. These challenges are particularly salient for the pluralistic (quasi-Feyerabendian) tradition of HPS that gained prominence in the generation once removed from SSR. My analysis does not imply that philosophers of science should develop detailed responses to such challenges. Rather, these challenges are highlighted as outstanding methodological issues bestowed on HPS by SSR that provide further opportunities for methodological innovation and evolution.

Acknowledgments

I am grateful to Stephen Biggs, Jamie Shaw, Joseph Pitt, Paul Hoyningen-Huene, Chris Haufe, Vasso Kindi, Markus Seidel, Rob Wilson, Magdalena Malecka, Samuel Schindler, and K. Brad Wray for helpful comments and suggestions. A draft of this chapter was presented at the Kuhn conference sponsored by the Aarhus Institute of Advanced Studies (AIAS) at Aarhus University in June 2022; I thank the participants for feedback and discussion.

13 Thomas Kuhn and the Strong Programme
An Appropriate Appropriation?

Markus Seidel

13.1 Introduction

Thomas Kuhn's *The Structure of Scientific Revolutions* (SSR) has been highly influential not only within the discipline of philosophy of science.[1] In fact, SSR probably is the most influential work in philosophy of science outside the narrow specialty. At least, this claim is well justified when it comes to Kuhn's influence on the social sciences and especially sociology.

In this chapter, I will not try to trace Kuhn's influence on the social sciences and sociology in total but will focus especially on Kuhn's influence on the *sociology of scientific knowledge*. The reason is threefold. First, those who are interested in the larger impact of Kuhn on the social sciences can already consult excellent existing papers (see, e.g., Wray 2015; 2016a; 2021a, chapter 5). Second, arguably but also not very surprisingly, Kuhn's influence was the most thoroughgoing in that part of sociology which investigates the same topic as Kuhn's work, namely, scientific knowledge.[2] Third, my own competence is mainly in philosophically analysing the sociology of scientific knowledge.

Moreover, the chapter will focus especially on the influence of Kuhn's work on the so-called Strong Programme.[3] Why especially the Strong Programme? The main reason is that Kuhn himself scarcely comments

---

[1] Maasen and Weingart (2000, 65–66) provide some quantitative data about the reception. It is noteworthy that the reception remains constant at a high level over time. Furthermore, they show that the vast majority (between 90 and 95%) of all citations to Kuhn are to SSR and most of these are citations such as "Kuhn 1962" and "Kuhn 1970," not substantive references. The data suggest that SSR is "mainly cited for a few catchwords and highly condensed messages" (Maasen and Weingart 2000, 66), thus becoming a kind of "icon."

[2] For example, Collins and Evans see SSR as the foundational work for "Wave Two" of science studies (see Collins and Evans 2017, 10).

[3] The term "Strong Programme" was introduced by Bloor to denote the four tenets of causality, impartiality, symmetry, and reflexivity (see Bloor 1973, 173–174; Bloor 1976/ 1991, 7). I will use the term in a broader way here in that I do not restrict my discussion to just these four tenets but focus on the whole work of the programme's adherents.

on the relation of his account to work in sociology of science.[4] But the
Strong Programme is a notable exception.[5] And Kuhn's reaction seems
to be clearly negative: he is "among those who have found the claims of
the strong program absurd: an example of deconstruction gone mad"
(RSS 110) and aims to "defend notions like truth and knowledge from …
the excesses of postmodernist movements like the strong program" (RSS
91; also Wray 2011b, 91–95). Since this chapter promises to investigate
the appropriateness of appropriation of Kuhn, it will be fruitful to investigate the similarities and differences between Kuhn's thought and those
appropriations in the sociology of scientific knowledge commented on
by Kuhn.[6]

Undoubtedly, Kuhn's ideas have been of major importance to the
Strong Programme. One of its champions, Barry Barnes, who has written
a whole monograph about Kuhn's intellectual influence on the sociology
of scientific knowledge (Barnes 1982a), even goes further by maintaining
that "probably the most important of all work we drew upon initially was
that of Thomas Kuhn, which had been invaluable in the development of
our ideas almost from the start" (Barnes 2011, 27; also Barnes et al.
1996, 111–112).

My questions are: (1) What is that influence? and (2) Is the appropriation by the "Strong Programmers" appropriate?

---

[4] In 1983 Kuhn retrospectively emphasized that SSR surely also is sociological, but that it
involves "an abominable way to do sociology" (Kuhn 1983, 28), because it draws its
sociological conclusions from Kuhn's own "experience with the interpretation of
scientific texts supplemented by my experience as a student of physics" (ibid.).

[5] Other references to work in the sociology of science by Kuhn are often uncommented on
(see e.g., SSR-4 175 n. 5) or are comments unrelated to his work (see e.g., ET 121–122).
At one point, Kuhn suggests that Hagstrom's "work in the sociology of science sometimes
overlaps [his] own" (SSR-4 40 n. 4). The other sociologist of (scientific) knowledge Kuhn
comments on most is Ludwik Fleck (see SSR-4 xli; Kuhn 1979). However, the question
of appropriation would, for obvious historical reasons, be switched in this case.

[6] It should be noted that Kuhn's influence on sociology of science in general shows
considerable national differences. As Ben-David argues, in Great Britain Kuhn's SSR
was received on a theoretical and philosophical level and analysed in contrast to Merton's
sociology of science: "as one of their main innovations, the British sociologists set up two
opposing models: the Mertonian 'model' of a general scientific community acting
according to relatively stable norms and the Kuhnian 'model' of scientific communities
changing their views and rules through periodic revolutions" (Ben-David 1978, 205; see
for the classical source of Merton-criticism by Strong Programmers: Barnes and Dolby
1970). In the United States, rather, Kuhn's ideas were taken up as empirical hypotheses
about scientific behaviour and the development of science without attending to
philosophical conflicts between Merton and Kuhn: "Thus, until about 1970 – that is,
for about a decade – American sociologists of science did not believe that Kuhn's views
competed with those of Merton. They used and quoted both, for different purposes"
(Ben-David 1978, 204).

In the following I will discuss these questions by focussing on similarities and dissimilarities between Kuhn and the Strong Programme with respect to two "isms": relativism and naturalism.[7] Why, some may ask, do I focus on these rather philosophical issues with respect to the Strong Programme in the *sociology* of scientific knowledge? The reason is that – despite their own avowed non- or even anti-philosophical ambitions – philosophical reflections figure large in the writings of the Strong Programme. And it is precisely the underlying philosophical agenda that Kuhn in his critical statements of the Strong Programme objected to.

But before focusing on the "isms," let me make one important remark in advance. I think every philosopher of science has at least an implicit stance about the "isms" mentioned. Thus, the "isms" have strong normative force within debates in philosophy (and sociology) of science. In some quarters, showing that a position is "relativistic" is enough to show that the position is absurd – "this leads to relativism" sometimes seems to function as an argument against the position. And some use the label "naturalism" like a swear word (even if not on the same level as "positivism" or "scientism"). Debates about positions denoted with these labels have been poisoned by such normative commitments. I will try my very best not to use the "isms" in a normative sense. Of course, the question of whether Kuhn and/or the Strong Programme can be described by the "isms" used should be distinguished from the question of whether the position denoted by the "isms" is plausible or convincing.

13.2   Relativism

Amongst philosophers of science the Strong Programme is most (in)famous for its endorsement of relativism. Whereas Strong Programmers openly embrace the relativistic consequences of their account (see Bloor 1976/1991, 158; Barnes 2011), Kuhn was at pains to attenuate the potential relativistic implications of his view (see SSR-4 203–205; RSS 155–162). Nevertheless, the Strong Programmers see Kuhn's work to have "explicit relativistic implications" (Barnes 1982a, 55). In order to assess whether such appropriation is appropriate,

---

[7] It is also possible to consider differences and similarities with respect to a third "ism," namely, *realism*. As far as I see, the similarities especially concern the possibility of interpreting both Kuhn and the Strong Programme along the lines of a Neo-Kantian realism (see Seidel 2014, chapter 2.1). The differences consist in the Strong Programme's endorsement of finitism, which is a radical form of conceptual underdetermination (see Bird 2000b, 218–225; Wray 2010a 316–321; Wray 2011b, 154–160; Wray 2021a, 110–112).

it is undoubtedly requisite to say more clearly what "relativism" is supposed to be.[8]

First, once it comes to cognitive relativism,[9] it is of vital importance to distinguish between alethic and epistemic relativism, the first being relativism with respect to truth itself and the second being relativism with respect to epistemically evaluative notions like "justification," "reason," or "rational." Neither Kuhn nor the Strong Programmers distinguish clearly between alethic and epistemic relativism and much confusion stems from such indifference. Second, both Kuhn and the Strong Programmers discuss the question of alethic relativism in the context of their criticism of the correspondence theory of truth and the question of whether science can be understood as aiming at and approximating truth (see Bloor 1976/1991, 159–160; SSR-4 204–205; RSS 95). This, however, has the consequence that their discussion of alethic relativism collapses into a discussion of (scientific) realism. I do not appreciate such mingling of the relativism and the realism issue. Thus, in what follows I will focus solely on epistemic relativism.

What is epistemic relativism? I propose to define epistemic relativism by four theses:

> **Faultless Disagreement**: People using different epistemic systems (consisting of epistemic standards) can faultlessly disagree over the question of whether a given belief is epistemically justified or not.
>
> **Non-Transcendency**: Beliefs can be justified only with reference to epistemic systems.
>
> **Epistemic Plurality**: There are people using radically different epistemic systems for which *Faultless Disagreement* applies.
>
> **Non-Metajustifiability**: It is impossible to demonstrate by rational argument that one's own epistemic system is superior to all or most of the others (see Seidel 2014, 32).

As I have shown elsewhere (Seidel 2014), the proponents of the Strong Programme are epistemic relativists in the sense just defined. I do not want to repeat my argument here, but let me make just one short comment about one key objection to this interpretation. The objection consists in insisting that the Strong Programme's relativism is just a

---

[8] As far as I see, whenever Kuhn discusses relativism explicitly, he was open to accept the label "relativism" in cases where the position was spelled out adequately (see Kuhn 1974, 508; SSR-4 204–205; Kuhn 2016, 22–23; RSS 76 n. 24, 160, 307).

[9] Probably the most famous non-cognitive forms of relativism are moral and aesthetic relativism.

strictly *methodological* form of relativism (see Bloor 1976/1991, 158; 1983a). Thus, according to the objection, the relativistic implications of the famous four tenets of the Strong Programme (see Bloor 1976/1991, 7) should be treated only as methodological advice for the sociologist to "bracket off" prior evaluative dimensions of belief, such as whether beliefs are rationally or irrationally held (see Bloor 2004, 937). Such a prior evaluation is the basis, so the proponents of the Strong Programme claim, of a sociology of error, invoking sociological explanations only for beliefs regarded as false or irrational (see Bloor 1976/1991, 12, 14, 17). However, such methodological relativism does not imply a more substantial *philosophical* relativism such as epistemic relativism as just defined. Therefore, the objection continues, interpreting the Strong Programme's relativism as epistemic relativism does not take seriously the merely sociological-methodological aspirations of the Strong Programme. Let us call this the "just-methodological objection."

As already said, I do not want to repeat my argument that the Strong Programmers do not *just* propose a methodological relativism here. I only want to emphasize that I am not alone in this judgement. In his recent defence of the Strong Programme, long-time sympathizer Martin Kusch also sees besides methodological relativism a philosophically more substantial form of relativism in the Strong Programme – a form that roughly corresponds to epistemic relativism as I have defined it (see Kusch 2021, 50–51; see Seidel 2021a for discussion). If friend (Kusch) and foe (me) agree in this respect, the bets are good that there is some truth in rejecting the just-methodological objection.

What about Kuhn? Is Kuhn an epistemic relativist as defined above? And in which way is Kuhn's position similar or dissimilar to the position of the Strong Programme? An answer to these questions requires a discussion of Kuhn's invocation of what has been called *methodological incommensurability*. In SSR Kuhn introduces the idea of an "incommensurability of standards" (SSR-4 148). Does such incommensurability of standards imply epistemic relativism as defined?

Let us first focus on the obvious. It seems obvious that Kuhn subscribes to a form of non-transcendency. For Kuhn, paradigms "are the source of the methods, problem-field, and standards of solution accepted by any mature scientific community at any given time" (SSR 4 103). However, as we will see, Kuhn's advocation of non-transcendency should not be seen to imply that trans-paradigmatic justification and persuasion are impossible (see RSS 157). Kuhn's version of non-transcendency should be thought of as the admission that it is impossible to completely escape any perspective and achieve a "God's-eye view" or "view from nowhere" – an admission quite acceptable also for epistemic absolutists (see Siegel 2011, 50–51; Seidel 2014, 33). Only if combined with the theses of epistemic plurality and non-metajustifiability does the thesis of non-transcendency imply the impossibility of trans-paradigmatic justification and persuasion. It is also obvious that Kuhn subscribes to epistemic plurality as a consequence of scientific revolutions: "when paradigms change, there are usually significant shifts in the criteria determining the legitimacy of problems and proposed solutions" (SSR-4 109). Paradigm choice, so Kuhn says, "proves to be a choice between incompatible modes of community life" (SSR-4 94). Furthermore, Kuhn also subscribes to the thesis of faultless disagreement as a consequence of such epistemic plurality. In revolutions we have a second-order disagreement about problems and standards (see SSR-4 147). The obvious consequence of such second-order disagreement of standards is that first-order faultless disagreement about justification is possible. Kuhn famously aims to allow for "rational men to disagree" (ET 332).[10]

What about the thesis of non-metajustifiability? As I will show, Kuhn's account provides us with arguments pro and con with respect to this thesis and this fact reveals similarities and dissimilarities to the Strong Programme. Let us focus on Kuhn's argument for methodological incommensurability in SSR. He gives us a reason for the incommensurability of standards:

To the extent, as significant as it is incomplete, that two scientific schools disagree about what is a problem and what a solution, they will inevitably talk through each other when debating the relative merits of their respective paradigms. In the partially circular arguments that regularly result, each paradigm will be shown to satisfy more or less the criteria that it dictates for itself and to fall short of a few of those dictated by its opponent. (SSR-4 109)

Since criteria for evaluating the merits of paradigms are circularly applied, incommensurability of standards arises. Interestingly, we see that Kuhn applies such a form of circularity argument also in the passages that have subsequently led many to believe that Kuhn subscribes to a form of epistemic relativism, namely, the passages in which Kuhn claims that in paradigm choice "there is no standard higher than the assent of the relevant community" (SSR-4 94). It is worth quoting the passage in full:

---

[10] At this point, I am only pointing to one route to rational disagreement that can be found in Kuhn's writings. There is another route stemming from Kuhn-underdetermination that is significantly different. See Seidel (2021b) for a discussion.

Like the choice between competing political institutions, that between competing paradigms proves to be a choice between incompatible modes of community life. Because it has that character, the choice is not and cannot be determined merely by the evaluative procedures characteristic of normal science, for these depend in part upon a particular paradigm, and that paradigm is at issue. When paradigms enter, as they must, into a debate about paradigm choice, their role is necessarily circular. Each group uses its own paradigm to argue in that paradigm's defense. The resulting circularity does not, of course, make the arguments wrong or even ineffectual. The man who premises a paradigm when arguing in its defense can nonetheless provide a clear exhibit of what scientific practice will be like for those who adopt the new view of nature. That exhibit can be immensely persuasive, often compellingly so. Yet, whatever its force, the status of the circular argument is only that of persuasion. It cannot be made logically or even probabilistically compelling for those who refuse to step into the circle. The premises and values shared by the two parties to a debate over paradigms are not sufficiently extensive for that. As in political revolutions, so in paradigm choice – there is no standard higher than the assent of the relevant community. (SSR-4 94)

This passage reveals two important issues (for a more detailed inter-pretation, see Seidel 2013; 2021b). First, Kuhn claims that the debate about paradigm change is circular and that the argumentative defence of a paradigm already presupposes the very paradigm that is being argued for. This suggests that Kuhn subscribes to the thesis of non-metajustifiability. Since the argumentative defence of a paradigm is *necessarily* circular and since circular arguments are not sufficient to demonstrate the trans-paradigmatic, epistemic superiority of paradigms, it is impossible to demonstrate by rational argument that one's own paradigm is epistemically superior to the other (see also Sankey 2011; 2012; Seidel 2014, 145–150). Second, however, we also see that Kuhn restricts the impact of his circularity-argument in two ways: (1) The exhibit of potential scientific practice "can be immensely persuasive, often compellingly so." (2) If one reads the passage closely, we see that Kuhn restricts the scope of the circularity argument casually: he speaks of "*partially* circular arguments"; claims that "the choice is not and cannot be determined *merely* by the evaluative procedures" of a paradigm, since "these depend *in part* upon a particular paradigm"; and mentions "prem-ises and values *shared*" by the parties in a debate. These two restrictions suggest that Kuhn – contrary to our first impression – does not fully subscribe to the thesis of non-metajustifiability.

As far as I see, Kuhn uses the circularity argument only in SSR – I do not know of any place in his work where he invokes the argument later. In later work, however, he clearly emphasizes the role of shared values in paradigm choice and elaborates on the casual restrictions just men-tioned. Kuhn lists five such values: "accuracy, consistency, scope,

simplicity, and fruitfulness" (ET 322).[11] These values are, according to
Kuhn, not as paradigm-relative as other parts of disciplinary matrices:
"they are more widely shared among different communities" (SSR-4
184); "commitment to them is both deep and constitutive of science"
(SSR-4 185); Kuhn even thinks that they "are necessarily permanent, for
abandoning them would be abandoning science together with the knowledge which scientific development brings" (RSS 252; see also ET 335).
And the existence of these values leads us to expect that Kuhn, especially
in later work, does not subscribe to non-metajustifiability. As Kuhn
explains, "what I am denying then is neither the existence of good
reasons nor that these reasons are of the sort usually described. I am,
however, insisting that such reasons constitute values to be used in
making choices rather than rules of choice" (RSS 157). Therefore,
although the values are part of disciplinary matrices, these values are
also *trans*-paradigmatic.

But if these values are *constitutive* of science and *shared* by proponents
of different paradigms, why are these "not sufficiently extensive" to make
theory choice logically compelling? Why is there underdetermination by
these shared values – "Kuhn-underdetermination," as we may want to
call it (see Carrier 2008, 276)? Kuhn believes that the values *taken in the
abstract* are constitutive of science but that there is room for disagreement
in the *concrete application* of these values in situations of choice. Because
the values "are acquired from the study of examples of past applications
rather than by learning rules about how they are to be applied" (Kuhn
1971, 146), there are two ways in which the shared values underdetermine choice: (1) individual scientists sharing these values may interpret
these values differently – roughly, one's simplicity is another's complexity – and, furthermore, (2) when the values are applied in situations of
choice, they can conflict such that they must be weighed against each
other (see ET 322, SSR-4 185; also Hoyningen-Huene 1989/1993, 150;
Bird 2000b, 241–242). This is why, according to Kuhn, paradigm choice
cannot be determined by shared values.[12]

---

[11] Kuhn himself does not claim that this list is exhaustive (see ET 321). See Hoyningen-
Huene (1989/1993, 149) for more values occasionally mentioned by Kuhn. Kuhn is not
very strict in his nomenclature here: although most of the time, he speaks of "values,"
sometimes he also speaks of "criteria" (see, e.g., ET passim, RSS 251).
[12] As Kuhn emphasizes, such Kuhn-underdetermination does not make theory choice
irrational – quite the contrary: "I do not for a moment believe that science is an
intrinsically irrational enterprise. What I have perhaps not made sufficiently clear,
however, is that I take that assertion not as a matter of fact, but rather of principle.
Scientific behaviour, taken as a whole, is the best example we have of rationality. Our
view of what is to be rational depends in significant ways, though of course not

In sum, we have both a relativistic and an anti-relativistic strain in Kuhn's argument in SSR. The circularity argument – to be found only in SSR – suggests a reading along the lines of the thesis of non-metajustifiability and thus epistemic relativism (see also Bird 2000b, 241). Kuhn's insistence on the role of trans-paradigmatic values instead of algorithmic rules in theory choice – to be found in SSR but especially in later work – suggests a reading contrary to the thesis of non-metajustifiability. Kuhn's stance on epistemic relativism is not really clear, and, despite his insistence that his position is not relativistic, he has not helped much to clear up his position because Kuhn mingles different forms of relativism and does not distinguish between the relativism and the realism debates. At points, it even seems that he contradicts himself once it comes to the question of the permanency of the values shared during paradigmatic change. On the one hand, the values are said to be "necessarily permanent" (RSS 252) and "fixed once and for all" (ET 335); on the other hand, he claims that "there are shared and justifiable, though not necessarily permanent, standards that scientific communities use when choosing between theories" (RSS 76 n. 24). Since the question of the permanency of evaluative standards, values, and criteria just *is* the question around which the debate about epistemic relativism revolves, Kuhn's own seemingly contradictory statements in this respect do forestall an unequivocal classification of him as either an epistemic relativist or an epistemic anti-relativist.

This result is of major importance to addressing the question of whether the appropriation of Kuhn as a relativist by the Strong Programmers is appropriate. And it gives us the opportunity to single out similarities and dissimilarities between their views.

As for the similarities, unsurprisingly, they surface when we focus on the relativist strain in Kuhn. We find the reference to circularity prominently in the Strong Programme's justification for epistemic relativism:

[The relativist] accepts that none of the justifications of his preferences can be formulated in absolute or context-independent terms. In the last analysis, he acknowledges that his justifications will stop at some principle or alleged matter of fact that only has local credibility. The only alternative is that justifications will begin to run in a circle and assume what they were meant to justify. (Barnes and Bloor 1982, 27)

---

exclusively, on what we take to be essential aspects of scientific behavior" (Kuhn 1971). Focusing on values instead of rules, however, explains Kuhn's notorious phrase that in paradigm choice "there is no standard higher than the assent of the relevant community" (SSR-4 94): the values are, according to Kuhn, "irreducibly sociological" (RSS 133).

For Barnes and Bloor, "circularity emerges whenever an attempt is made to ground our most general notions of validity" (p. 41 n. 34).

However, once we focus on Kuhn's invocation of values, we see clear differences. Importantly, these have not gone unnoticed by Kuhn or the Strong Programmers. Kuhn has attacked the Strong Programmers in denying values a decisive role in explaining scientific behaviour: "My own work ... has from the start presupposed [the] existence and role [of scientific values]" (ET xxi–xxii). Barnes has commented on Kuhn's invocation of values: "Needless to say, my own view is that this analysis of the role of values is completely incorrect, and sits most uncomfortably alongside the rest of Kuhn's work. Fortunately, however, it does not figure large in his research.... 'Objectivity, Value-Judgment and Theory Choice' addresses issues rarely touched upon in Kuhn's other work" (Barnes 1982a, 125).[13] The reason is simple: for Barnes, "values must be the products of communal activity, not part of the basis of community" (p. 124; see also Nola 2000). We will see more of the difference between Kuhn and the Strong Programme in this respect once we discuss naturalism.

13.3   Naturalism

There is no doubt that the key to understanding the Strong Programmers is their endorsement of naturalism.[14] For Bloor, "naturalism is the basic perspective which is articulated by the strong programme" (Bloor 2017). However, whether Kuhn can be regarded to be a naturalist is not as easy to figure out – it depends on what "naturalism" means.[15]

The Strong Programmers themselves see Kuhn as entertaining a "naturalistic account" (Bloor 1976/1991, 80; also Shapin 2015, 18–19). Alas, despite its strong commitment to naturalism there is no clear statement by proponents of the Strong Programme about what "naturalism"

---

[13] Barnes sees the reason for Kuhn's remarks on the role of values as his being "both an exponent and advocate of sociological functionalism" (Barnes 1982a, 125; see also Barnes 2010). Barnes is not wrong to count Kuhn in the camp of the functionalists (see RSS 130; SSR-4 185).

[14] The following is a non-exhaustive list of references to that effect: Barnes (1977, 25); Barnes (2011, 26); Barnes et al. (1996, 3); Bloor (1976/1991, 177); Bloor (2004, 919); Bloor (2011, 448–451); Hwang et al. (2010, 602–604). Sometimes Strong Programmers also speak of their programme as "scientistic" (see, e.g., Bloor 1976/1991, 7). As far as I see, they do not make a conceptual difference between their endorsement of naturalism and scientism.

[15] Thus, for example, Bird argues that Kuhn at key points sticks to a Cartesian and empiricist epistemology that contradicts externalist, naturalized epistemology (see Bird 2000b, 141, 266). He sees a development from a more naturalistic approach in SSR to a priori argumentation in later work (see Bird 2002; 2004).

actually is supposed to mean. This also hinders an adequate assessment of the similarities and dissimilarities between Kuhn and the Strong Programme on this issue. For example, Steven Shapin describes the historical sentiments around the publication of SSR by noting that they are naturalistic. Remarking that "naturalism" is a notoriously difficult notion, he settles for "a deflationary sense routinely adopted by such sociologists of scientific knowledge as Barry Barnes and David Bloor … where a naturalistic account of science as it actually proceeds is juxtaposed to its celebration, defense, rational reconstruction, and essentialization" (Shapin 2015, 13 n. 2). Thus, as Shapin understands it, "naturalism is opposed to normativity where the naturalist intention is to describe, interpret, and explain and not to justify, celebrate, or, more rarely, to accuse" (p. 13). I do not want to claim that Shapin is wrong to see the opposition to normativity as key to the naturalism of the Strong Programme (see Bloor 2004, 919). However, such a characterization is too broad and unclear in several respects: it allows for different interpretations of the strength of the naturalistic commitment and it does not explain why the opposition to normativity is called "naturalism" and not "descriptivism" or "empiricism" We need to be more clear about what "naturalism" amounts to in order to investigate the relation between Kuhn and the Strong Programme.

It is common now to distinguish between *ontological* and *methodological* naturalism (see Papineau 2021). Ontological naturalism can be understood as the claim that entities of a certain non-natural brand do not exist, for example, supernatural entities, and that only entities of a specific natural brand do exist, for example, physical entities or entities assumed to exist by mature science. Methodological naturalism, instead, is a thesis about the legitimacy of methods of investigation: only those methods paradigmatically exemplified by the methods of the natural sciences are legitimate. This usually implies a kind of methodological monism, including the rejection of original, genuine philosophical methods like a priori methodology. Beyond doubt, it is methodological naturalism that we need to focus on in order to investigate the similarities and dissimilarities between Kuhn and the Strong Programme. Thus, "naturalism" in the following is assumed to mean "methodological naturalism." Furthermore, I will also restrict the kind of investigations to which methodological naturalism applies. I will concentrate on naturalism with respect to investigations of *science* in order to figure out similarities and dissimilarities between Kuhn and the Strong Programme.

As far as I can see, there does not exist any elaborate, non-critical attempt to categorize the naturalistic commitments by the Strong

Programme.[16] Let me try to make a start by spelling out more clearly what their "opposition to normativity" amounts to. At least, we should distinguish between the Strong Programme's

(a) eliminative naturalism
(b) reductive naturalism
(c) enlightening naturalism.

Let me explain by introducing some more fine-grained distinctions.

The opposition to normativity included in *eliminative* naturalism consists in the elimination of prior normative considerations in investigating science. Such elimination can come in different strengths and forms. Let us distinguish between whether eliminative naturalism is supposed to be (a1) a thesis about the methodology of the *sociology* and *history* of science or (a2) a thesis about the methodology of the *philosophy* of science. The former, (a1), is clearly involved in the naturalism of the Strong Programme. In fact, it is exactly what Bloor's invocation of the four tenets that define the Strong Programme is about. With his causality, impartiality, symmetry, and reflexivity requirements Bloor aims at a sociology of knowledge that eschews any prior normative assessments: "Don't evaluate before you investigate!" should be the sociologists' motto. It is clear that such a stance also applies to the history of science: Bloor sees the four tenets of the Strong Programme prominently and paradigmatically at work in Paul Forman's (in)famous historical study of Weimar physicists.[17] Seen in this way, (a1) does not conflict with a more traditional normative investigation provided by the philosophy of science. In this manner, Barry Barnes has distinguished between the naturalistic and normative orientation of sociologists and philosophers and warns against conflating the two (see Barnes 1982b, 98). Such naturalism in the sociology of scientific knowledge does not aim to attack philosophical conceptions of science and knowledge (see Barnes 1974, 8). However, the Strong Programmers do not just aim to make a claim about the methodology of the sociology and history of science and leave traditional normative-philosophical methodology of science intact. They[18] propose an expansionist project by which the Strong Programme's eliminativist naturalism

---

[16] Nola (2003), Collin (2011), and Bird (2012a) discuss the Strong Programme's naturalism, but always in the context of a critical assessment.

[17] See Bloor (1976/1991, 7). Bird views it as a problem that the Strong Programme cannot clearly distinguish its sociological stance from the history of science (see Bird 2012a, 220–221).

[18] My impression is that Barnes is more reluctant in this respect than Bloor. This might be the result of Bloor's rather strong polemic against philosophers in general. See especially his derogatory remarks in Bloor (2007).

in the sociology of knowledge is considered the only methodologically adequate approach to studying science.[19] Such expansionism is reminiscent of Quine's strong naturalism making philosophy a chapter of empirical science (see Quine 1971, 97). Thus, according to Bloor, the sociology of knowledge is to be the heir of the discipline of philosophy (see Bloor 1983b, 183). The Strong Programme attempts to "reconstruct epistemology along new lines" (Bloor 2004, 919). Normative considerations should be eliminated from *any* theoretical reflection of science – philosophical reflection included. Thus, the Strong Programmers subscribe also to thesis (a2) (see Nola 2003, 257).

But eliminative naturalism is not the only form of opposition to normativity to be found in the Strong Programme. Whereas eliminative naturalism opts for a methodological *elimination* of prior normative considerations in the investigation of science, *reductive* naturalism aims to naturalistically reduce the normative commitments usually made by philosophers as well as scientists. What the Strong Programmers aim at is a genuine sociology of reasons: what is regarded as an epistemic factor, it is claimed, really is a social factor. Thus, "investigators who study the social background of epistemic factors – at least, if they take their enterprise seriously – are not pursuing the social *rather than* the epistemic. They are helping to show the social character *of* the epistemic" (Bloor 1984, 303). In this way, they seek "to make clear the existence of alternative ways of seeing the question at issue – in this case the nature of rationality, objectivity, logical necessity and truth" (Bloor 1976/1991, 83). Far from simply proposing a new non-evaluative methodology for sociology and philosophy of science and being philosophically agnostic about epistemic conceptions, the Strong Programmers aim for a "naturalistic construal of reason" (p. 177). Though distinguishing between eliminative and reductive naturalism is analytically helpful, the two aspects are related.

Third, the opposition to normativity integral in the writings of the Strong Programme also has an *enlightening* aspect. At several places, Bloor uses political and moral notions in order to describe the difference between naturalism and anti-naturalism:

Unless we adopt a scientific approach to the nature of knowledge then our grasp of that nature will be no more than a projection of our ideological concerns. Our theories of knowledge will rise and fall as their corresponding ideology rises and

---

[19] See Collin (2011, 81): "The philosophical project [of the Strong Programme] aims at no less than the *naturalization of the philosophy of science* – that is, the demonstration that sociological explanation will answer all the (legitimate) questions that philosophers have raised concerning science."

declines; they will lack any autonomy or basis for development in their own right. Epistemology will be merely implicit propaganda. (Bloor 1976/1991, 80)[20]

Now, the Strong Programme is not supposed to be anti-ideological in that it is able to *strip off* ideological commitments: the eliminative aspect of naturalism does not mean that the Strong Programme is not itself based on certain values (see ibid., p. 13). But there is a difference between propagandistic epistemology and the Strong Programme: "the strong programme possesses a certain kind of moral neutrality, namely the same kind as we have learned to associate with all the other sciences" (p. 13). Naturalism is meant to explicate, and in this way overcome, unreflective prejudice by adopting an evaluative neutral stance.

Let us consider Kuhn. Does Kuhn subscribe to the thesis (a1)? Does he believe that sociology and history of science should proceed without prior normative assessments? I do not know of any comment by Kuhn pertaining to sociology, but restricting our discussion to Kuhn's treatment of history the answer is simply "yes". Such an answer is justified once we look at Kuhn's comments on Lakatos' idea of writing history as a rational reconstruction and his attack on the latter's distinction between internal and external history. Although Kuhn accepts that rational reconstruction is necessary, that historians must treat the historical material selectively and need philosophy in their work (see RSS 151; ET 10), quoting Lakatos, Kuhn articulates his problems with the idea that "history of *science* ... is a history of events which are selected and interpreted in a normative way", where "philosophy of science ... provides normative methodologies" (Kuhn 1971, 141–142). As far as I can see, Kuhn has two major concerns with such a procedure. His first concern is that such a conception implies that "history could not in principle have the slightest effect on the prior philosophical position which exclusively shaped it" (Kuhn 1971, 143; also RSS 151 n. 32). I will come back to the consequences of this concern for Kuhn's naturalism at the end of this section. The second concern is that Lakatosian rational reconstructions are simply fabricated history. Kuhn comments on Lakatos' suggestion that the idea of electron spin can properly be attributed to Bohr in 1913 despite the fact that "Bohr, in 1913, may not have even thought of the possibility of electron spin" (Lakatos 1971, 106–107): "what Lakatos conceives as history is not history at all but philosophy fabricating examples" (Kuhn 1971, 143; also Kuhn 1980, 181, 183; RSS 151 n. 32). As Kuhn makes clear in the introduction to SSR, history of

---

[20] Of course, such critique of ideology has a Marxian-Mannheimian background in the sociology of knowledge.

science should not be forced to distinguish between, on the one hand, genuine "scientific" components and, on the other hand, error, myth, and superstition in advance. Such a prior distinction is a consequence of an image of science that is nothing more than an "unhistorical stereotype drawn from science texts" (SSR-4 1; ET 11). Kuhn maintains that "the historian is usually well-advised to set expectations aside before beginning research. If science and method, for example, are the subjects, then both should be learned from the people under study not from later scientific and philosophical texts" (Kuhn 1980, 183). "Don't evaluate before you investigate" is also Kuhn's motto.

Is Kuhn an eliminative naturalist with respect to the philosophy of science; does he subscribe to (a2)? And does Kuhn aim to reduce normative commitments in science naturalistically; does he subscribe to (b)? The straightforward answer to both of these questions is "no" – and for the same reason.

Brad Wray has addressed the matter succinctly: "Kuhn believes that *epistemic factors* stabilize belief in the research community.... Moreover, Kuhn claims that belief in science is also *de*stabilized by epistemic factors, not social factors and interests" (Wray 2010a, 324; 2011b, 163–164). Kuhn reproaches the Strong Programmers' "internal sociology of science" for the fact that their treatment of values focused on the wrong sort of values, namely, predominantly on external, socioeconomic interests (see Kuhn 1983, 29–30; also Wray 2021a, 113–114). In this way, Kuhn claims, such a focus loses sight of those internal factors characteristic of science – namely, the cognitive interests of scientists. Kuhn mentions cognitive interests like "love of truth (fear of the unknown, if one prefers); fascination with puzzle-solving (compulsion to exhibit special skills); repudiation of the past (obsession with the law of the excluded middle)" (Kuhn 1983, 30). This difference between Kuhn and the Strong Programmers with respect to naturalism obviously relates to the internalism-externalism debate in philosophy of science (see Kuhn 1983, 29; also Wray 2010a, 321–324; Bird 2012a, 209–210). Whereas Kuhn insists on the decisive role of internal, epistemic factors in the explanation of the development of scientific belief, the Strong Programmers aim to dissolve the distinction between internal and external factors. Despite its sociological appeal, Kuhn's internalism "makes his account a contribution to the *epistemology* of science, rather than the sociology of science" (Wray 2021a, 117; also ET xx).

The difference also becomes clear once we see how Kuhn, like the Strong Programme, aims to "make clear the existence of alternative ways of seeing ... the nature of rationality" (Bloor 1976/1991, 83). Kuhn argues for a rejection of a conception of rationality as an algorithmic application of rules (see SSR-4 Chapter V). As we have already seen, his alternative way of seeing the issue consists in a value-based conception of rationality. In this way, Kuhn is also "trying to change a current notion of what rationality is" (Kuhn 1971, 139). And, to be sure, one important way in which Kuhn's philosophy of science employs social factors is exactly in such a conception of rationality. Thus, according to Kuhn, "there is no neutral algorithm for theory choice, no systematic decision procedure which, properly applied, must lead each individual in the group to the same decision. In this sense it is the community of specialists rather than its individual members that makes the effective decision" (SSR-4 198–199; also RSS 131). This aspect of Kuhn's picture makes it "irreducibly sociological" (RSS 133; also ET xx). However, the values themselves are not *reduced* to social factors (see RSS 115f.). On the contrary, their very existence is presupposed in order to see in which way the effective decision is made by the community (see Kuhn 1983, 30). In fact, Kuhn thinks that "much of what is special about [scientific] communities is ... the shared values of their members" (Kuhn 1971, 146; also Hoyningen-Huene 1992, §3). As we have already seen, this explicitly conflicts with the view of the Strong Programmers, since, for Barnes, "values must be the products of communal activity, not part of the basis of community" (Barnes 1982a, 124). In sum, Kuhn subscribes to neither (a2) nor to (b).

With respect to (c) I do not know of any statements in which it is clear that Kuhn thinks non-naturalized philosophy of science is ideological in principle or that a naturalistic approach is more neutral. However, there is one issue that can give rise to speculation. Kuhn criticizes Popper for providing not a logic of discovery but "an ideology" (ET 283) in supplying procedural maxims for scientists (see RSS 142). That does not imply that Kuhn wants to suggest that normative philosophy of science is ideological *as such*. But, as Kuhn makes clear, what is needed is a better psychological and sociological description of the ideology, that is, the value system of science that Popper merely postulates (see RSS 133–134; ET 290). Kuhn hopes that the empirical investigation of science helps us to enlighten those features actually at work in scientific change rather than – perhaps unwittingly – impose values from the philosophical armchair. According to Kuhn, at points there are passages in Popper's writings that he "can only read as attempts to inculcate moral imperatives in the membership of the scientific group" (ET 292; also RSS 131). Thus, there is a sense in which Kuhn also aims to purge philosophy of science from implicit propaganda. These remarks, though, are highly speculative and surely do not firmly support the interpretation that Kuhn also subscribes to the assumed neutrality of a naturalistic commitment. Most importantly, Kuhn does not aim to eschew ideology by a naturalistic commitment, but urges us to investigate scientific ideology, that is, science's values, in a naturalistic manner. I am reluctant, therefore, to claim that Kuhn also subscribes to (c).

However, now we have the curious situation that Kuhn only appears to subscribe to the restricted naturalism of (a1). That cannot be right, because SSR implies a certain view of the role that descriptive, empirical reflection on science plays for philosophical, normative matters. Let me briefly summarize Kuhn's naturalistic commitment that is considerably weaker than the eliminative and reductionist commitment of the Strong Programme.

We have not taken notice yet of "Kuhn's influence on … philosophy of science['s] turn … from a normative-synchronous orientation to a more descriptive-developmental one" (Hoyningen-Huene 1992, 487). Alexander Bird has called this naturalistic strand "historical-sociological" and correctly notes that it "led to a major trend in science studies" (Bird 2012a, 211). Although in his paper Bird argues that the adoption of Kuhn's historical-sociological naturalism by science studies is inappropriate, nevertheless Kuhn is a key figure of the so-called historical turn in philosophy of science. As Bird says, there is a "characteristic that Kuhn and at least one important strand in science studies do share. Kuhn's approach and that avowed by the Strong Programme are both naturalistic. By a 'naturalistic' approach to some subject matter, I mean an approach that is willing to employ the methods and results of the natural sciences" (Bird 2012a, 219).

We see clearly that Kuhn subscribes to such a naturalism once we have a look at Kuhn's rationale for proposing his alternative conception of rationality. He thinks that "if history or any other empirical discipline leads us to believe that the development of science depends essentially on behavior that we have previously thought to be irrational, then we should conclude not that science is irrational but that our notion of rationality needs adjustment here and there" (Kuhn 1971, 144).

Therefore, though philosophical considerations are not eliminated or reduced to sociological and historical research as in the Strong Programme's eliminative and reductive naturalism, Kuhn nevertheless allows empirical research to have *philosophical consequences* and *change* philosophical conceptions. In fact, this is what a rejection of prior normative, philosophical assessments in the history of science – that is, (a1) – allows in the first place. Since history should be able to change philosophical positions, selecting historical facts in accordance with prior philosophical, normative considerations circularly undermines the justificatory basis of such change. Thus, Kuhn is a naturalist in virtue of the fact that he allows historical-sociological, descriptive considerations about science to have philosophical, normative consequences (see

Kuhn 1971, 142; RSS 130). Of course, Kuhn is aware that he thus may undermine several distinctions or dichotomies that are thought to be substantial in investigating science. He mentions three dichotomies in the introduction to SSR: (1) descriptive versus interpretative/normative, (2) sociology/social psychology versus logic/epistemology, and (3) context of discovery versus context of justification.[21] Though not rejecting them outright, he thinks that these are not "elementary logical or methodological distinctions" (SSR-4 9), but rather are themselves "integral parts" of theories to be investigated with the same scrutiny as other theories. According to Kuhn, there is no principled, methodological difference between philosophy and the empirical sciences: philosophy is a proto-science (see RSS 138).

This kind of naturalism is surely weaker than the Strong Programme's *eliminative* and *reductive* project. Kuhn's naturalism undermines the above-mentioned distinctions by claiming that historical and sociological considerations are *relevant* to the philosophy of science (see ET 4, 12, 120–121).

13.4   Concluding Remarks

In sum, we have seen that the question of whether the Strong Programme's appropriation of Kuhn is appropriate cannot be answered simply by "yes" or "no." Clearly, the Strong Programme goes beyond Kuhn and breaks more radically with philosophical tradition. In particular, it differs from Kuhn in its rejection of shared values as a sociological basis of scientific communities and its eliminativist and reductionist naturalistic commitment. Nevertheless, we have also seen clear philosophical continuities and similarities: both invoke an argument from circularity in order to sustain epistemic relativism as defined, and both reject prior philosophical evaluations in empirical investigations of science as well as several supposedly basic distinctions that aim at restricting disciplinary domains in such investigations. In sum, the question of whether "the Kuhnians"[22] are Kuhnians has the answer: partly yes and partly no.

---

[21] See SSR-4 8–9. See also for distinction (1) RSS 125, 130; for distinction (2) Kuhn (1983, 29), RSS 125 (here as distinction of "logic versus history and social psychology"), and ET Chapter 11; and for distinction (3) ET 326–327 as well as the analysis in Hoyningen-Huene 1989/1993, chapter 7.4.c. Paul Hoyningen-Huene also has tried to separate several issues at play in such distinctions: see Hoyningen-Huene (1987; 2006a).

[22] Merton, expressing his concern about a manufactured dispute between him and Kuhn, referred to the Strong Programmers in this way in a letter to Kuhn (see Wray 2021a, 112).

Acknowledgements

I would like to thank the participants of the Aarhus conference "Thomas Kuhn's Philosophy of Science: In Honour of the 100th Year of His Birth" in June 2022 for their comments on my talk and K. Brad Wray for his invaluable comments on this chapter.

14 Kuhn and the History of Science

Vasso Kindi

14.1 Introduction

The term "history of science" is at least ambiguous: it refers to both the historiography of science and the facts of the past (*res gestae*). In what follows, I will discuss, first, how Kuhn used the facts of the past in his historical philosophy of science and, second, how Kuhn's work impacted developments in the historiography of science and the corresponding discipline. Before I proceed, though, I would like to highlight some oddities and puzzles pertaining to Kuhn's legacy in the history of science.

14.2 Oddities and Puzzles Regarding Kuhn's Relation to the History of Science

Kuhn became famous and influential in various fields including in the history of science, not so much through his historical work, but through his book *The Structure of Scientific Revolutions* (SSR), which, however, had been intended as philosophical in character. In his autobiographical interview, Kuhn said that his ambitions were always philosophical and that he thought of SSR as a book for philosophers (RSS 176; cf. ET 4). I will also argue below that Kuhn's aim and arguments in SSR were philosophical. Yet the book itself was mostly received and interpreted as historiographical. Mary Jo Nye (2019, 10) says that SSR "was mainly intellectual history, or the history of scientific ideas and theories," while Alexander Bird (2000b, viii) calls it "theoretical history." As I will explain later, it is commonly, but wrongly, thought that Kuhn provided in SSR historical evidence for a non-cumulative model of scientific development. The model itself is again treated as a historical generalization or as the bare bones of a historical narrative. The first oddity, then, is that a philosophy book, which is mistakenly taken to be historiographical, influenced the developments in the history of science. Or, so I will argue.

This oddity gives rise to a second one. Kuhn's work had a transformative effect on the historiography of science, but the historians in the field, who acknowledge his impact, claim that their discipline was transformed by moving away from Kuhn. They also distance themselves from his legacy. Lorraine Daston, to take an example, says that most historians of science take Kuhn's account as an attempt to impose a rational order on the history of science and they reject it. They "no longer believe that *any* kind of structure could possibly do justice to their subject matter" (Daston 2016, 117, emphasis in the original).[1] This distancing attitude is strangely also present, if not more pronounced, among Kuhn's own students. Here, there is another, apposite oddity to be noted. Despite the fact that Kuhn, especially after the publication of SSR, increasingly focused his attention on philosophical matters, he had no philosophy students. All his doctoral students were historians of science, but they do not, at least expressly, subscribe to his way of doing history of science. Some of them were present at a symposium in Kuhn's honor, held at the Dibner Institute for the History of Science and Technology at MIT, on November 21–22, 1997, a little over a year after Kuhn's death. In their talks, several made a point of disassociating themselves from Kuhn, being rather critical of, and in some cases, even hostile to, his approach and results. The most critical was Paul Forman, but he was not the only one. Kenneth Caneva, who published a revised version of his talk in 2000, found tensions in the Kuhnian enterprise, incoherence, confusions, anomalies, ambiguities, inconsistencies, and incongruities. He also criticized Kuhn for never providing a "historical exemplar of what the Kuhnian history of science would look like" (Caneva 2000, 92). So, contrary to what Kuhn himself was saying, one might think, Caneva says, that Kuhn believed that he could teach by precept rather than by example. Andrew Pickering, at the same symposium, after stating that he was inspired by Kuhn's work, made a point of noting that his research departed not only from "the letter of the *Structure*," as he says (Pickering 2001, 503), but also from significant tenets of Kuhn's account. For instance, Pickering thought that Kuhn, laying emphasis on the scientists and the scientific community as agents of change, did not take seriously the material world and, so, offered ammunition to that party in the Science Wars who blamed Kuhn for constructivism. Thus, Pickering developed what he called a post-humanist, non-naïve realist approach to science and its history. The humanism of the SSR, Pickering said, "is

---

[1] The historian Michael Gordin (2020, 12) makes the same point more generally: "The revolt of the contributors to the journal [*Historical Studies in the Natural Sciences*] against the supremacy of philosophical structures grew with increasing assertiveness across the decade, as eclectic historians of science drew from other disciplines to find a macroscopic perch."

best regarded as a historical relic, and now is a good time to say goodbye to it, for political as well as scholarly reasons" (ibid., 508). He also said that "the claustrophobic image of the unitary paradigm and its magical self-renewal always felt wrong to [him]" (ibid.).

A third oddity stems from the fact that Kuhn was dissatisfied with the path the history of science had taken, although this path was largely paved by his own work. The contextualist turn in the history of science – a new version of what was formerly called external history – which directed attention to sociological and political factors, the cultural milieu, the institutions, and the rhetoric of science, was indeed inspired by Kuhn's SSR, which did in fact produce a decisive transformation in the image of science that we at the time possessed. History of science was liberated from a certain constricting understanding of science and developed in various new ways. Although Kuhn praised the scholarship of this new history of science as exhibited in books such as *Leviathan and the Air Pump*, he strongly disapproved of it as not being history (RSS 316). What bothered him was that scholars and history of science students did not really care to delve into the science of each case, as he said, concentrating instead on negotiations to the neglect of the role of nature in the advancement of knowledge.

Finally, a puzzle in relation to the history of science emerges from Kuhn's own work: Kuhn is credited with bringing about the so-called historical philosophy of science, but he wanted the two disciplines, history and philosophy, to be kept apart. He criticized attempts to amalgamate them and maintained that his model of science could be derived from first principles.

So, if we are to talk about Kuhn's relation to the history of science, we ought to take these oddities and puzzles into account. In the present chapter, I will try to elucidate and perhaps ease the above tensions as I focus, first, on how Kuhn handled the history of science in SSR and, second, on how the history of science developed under his influence in new directions. I will concentrate on SSR since Kuhn, his historical work notwithstanding, influenced the historiography of science most notably through SSR.[2] Had it not been for SSR, Kuhn's historiography would not have assumed a paradigmatic role. Kuhn's historical work offered a revisionary history of science that gave measured consideration to the so-

---

[2] Stephen G. Brush (2000, 39) was wondering "why did Kuhn's own publications in his primary field, history of science, have so little impact on that field?" Brad Wray (2021a, 131–132) discusses briefly the impact Kuhn's historical work had on the history of science.

called external factors in science;[3] stressed the contingency in scientific development, the diversity of scientific practices, and the conceptual variability of scientific traditions; but did not clearly spell out epistemological and methodological observations as SSR did.[4] SSR, with its extensive impact, could not have been ignored, and it contributed to the expansion and transformation of the discipline.

In the next section, my main contention is that Kuhn did not use history as evidence for his philosophy. I compare his model to Wittgenstein's objects of comparison and argue that this model is used to highlight differences and diversity in the history of science. This feature of his model influenced the practice of the history of science as a discipline despite the fact that historians misinterpreted Kuhn's model. Then, in Section 14.4 I show how new developments in the history of science follow in Kuhn's steps.

14.3   How Did Kuhn Handle the History of Science in *Structure*?

Kuhn's use of historical examples in SSR has been variously interpreted. One interpretation takes Kuhn's account to be a kind of speculative philosophy of history, as an attempt to find a pattern of development in the past. Lorraine Daston, for instance, says that Kuhn's account of science is similar to those that aim to find "overarching regularities in the history of science." She calls this effort a bizarre idea, "a kind of leftover Hegelianism seeking a hidden, inexorable logic in the apparent vagaries of history" (Daston 2016, 117).[5] In her view, Kuhn's was "the

---

[3] Kuhn's consideration of the external factors in science was never as reductionist as the old Marxist approach of Hessen or Bernal or as decisive as the current Science Studies line of research. This is the reason that Kuhn is still interpreted as advancing an internalist history of science. But Kuhn's internalist account of science differed from the standard one at the time in that it made typical external factors, such as the subjective judgments and idiosyncrasies of scientists, vital parts of the hitherto internal scientific process. A better way to see this is to say that Kuhn challenged the distinction itself (cf. SSR-4 8–9).

[4] Kuhn's books on the Copernican Revolution and the black body theory, as well as his articles on specific historical issues and episodes, exemplify the kind of history that he favored: anti-Whiggish and not guided by a philosophical model, not even his own. His discussion of the atomic theory and the chemical revolution (QPT; 1952; 1976) illustrate the qualities that he valued in writing history of science.

[5] Gordon Graham (1997, 33) also compares Kuhn's account to a Hegelian philosophical history, but without Daston's disapproving overtones. It should be noted, however, that the comparison Graham attempts between Hegel and Kuhn on this particular point goes amiss. Hegel's reference to the Owl of Minerva that flies only at dusk does not correspond to Kuhn's use of history in SSR, as Graham surmises. First, Hegel did not mean to say that historical hindsight "gives history some control over philosophical speculation"

last attempt to give Reason (now incarnate in science) a rational history" (ibid.). A Hegelian-type philosophical history is too conjectural and metaphysical to fit Kuhn's purpose. We do not see Kuhn looking for the meaning of history or some underlying logic that governs the unfolding of events. As David Hollinger has put it:

> Connected to no total cosmic scheme, [Kuhn's] theory stands somewhat apart from "speculative philosophy of history." There is no determinate life cycle, no implication that certain changes are "natural" for a cultural unit at certain times, no insistence that traditions will or will not attain a certain degree of stability, no reason, in principle, why a given tradition might not live forever. The theory neither holds that change is always gradual, cataclysmic, or dialectical nor insists that change is generated by elements exclusively within or outside of "social systems." The theory offers a thoroughly nonteleological view of change: no idea of progress is implied, nor one of decline. (Hollinger 1973, 375)

A second interpretation of how Kuhn handles the history of science in SSR says that he is offering us a grand historical narrative, a big picture. But Kuhn does not offer a rich historical developmental plot, involving detailed historical explanation and periodization. Neither does he practice integrated history and philosophy of science in SSR. He has said that he has "resisted attempts to amalgamate history and philosophy of science though simultaneously urging increased interaction between the two" (Kuhn 1980) and that he is never a philosopher and a historian at the same time (RSS 316). Do Kuhnian concepts function as Weber's ideal types in an explanatory project of scientific development (Mladenović 2007)? No. Mladenović herself says that Weberian ideal types are "precisely and unambiguously defined" and are used as heuristic devices in order to form hypotheses that are to be adjusted by comparison to reality (p. 270). Kuhn's concepts, however, are notoriously ambiguous (e.g., the concept of the paradigm) and Kuhn does not put forward or test hypotheses to then refine his views.

The most common interpretation of SSR, and the one I will dwell more upon, says that history of science in this book provides empirical evidence for Kuhn's philosophical model. This interpretation, however, can be criticized in several ways. For one, Kuhn himself denied that he used history as evidence for a philosophical theory. He said to his student, the historian John L. Heilbron (1998, 511), "it is not a theory,

---

(ibid.). Hegel's point was rather that philosophy waits for the things to unfold before it reflects on them. Second, Hegel's thought is not similar to the view that Graham attributes to Kuhn, namely, that "good philosophy of science requires a good knowledge of the history of science" (ibid.). A Hegelian understanding of Kuhn's account can also be found in Bird (2000a, 129–130) and Bird (2015, 27).

and I do not expect it to match the record." Also, in his interview with Skúli Sigurdsson (2016, 27), Kuhn contended that one should not look at history to test the ideas of the book, to find out whether they are true or false. The model is "not meant to be applied this way." Furthermore, Kuhn did not think that historical facts were essential to the development of his model. Long after SSR, he said that "many of the most central conclusions we [Kuhn and his fellow philosophers/historians] drew from the historical record can be derived instead from first principles" (RSS 112). In his view, what emerged as essential from his historical work was not evidential support, but the historical perspective that allowed him to concentrate on the change of beliefs in science rather than the individual beliefs as such. He could, thus, concentrate on a comparative evaluation of beliefs rather than on their individual assessment against the world, by confirmation or falsification (RSS 95).

Kuhn was also aware, already in SSR, of the descriptive/normative gap and other dichotomies current in philosophy at the time, such as the context of discovery and context of justification distinction: "I could scarcely be more aware of their import and force," he said (SSR-4 9). He knew that his "sometimes normative" theses could not be derived from history, "a purely descriptive discipline" (SSR-4 8, 206). His way of dealing with the problem was to say that these distinctions were part of an approach that he would overcome. Equally mindful was Kuhn of the underdetermination of theories by historical data: "If you have a theory you want to confirm, you *can* go and do history so it confirms it, and so forth; it's just not the thing to do" (RSS 313–314). Finally, if the model of science developed in SSR was an empirical generalization from historical evidence, it could be argued that it is very thinly and inadequately supported. Critics have actually challenged Kuhn's interpretation of historical examples and pointed out the model's slim evidential basis. Wes Sharrock and Rupert Read (2002, 107), for instance, found Kuhn's account of science "largely unevidenced," while Janet Kourany (1979) questioned the historical basis of Kuhn's model altogether.[6] Kuhn himself admitted in the Preface of SSR (SSR-3 ix) that his presentation has a schematic character and that he has not exploited all the historical evidence he had. Also, several scholars disputed how Kuhn interpreted

---

[6] According to Caneva (2000, 110), "we find Kuhn looking for answers to his problems in one or another favored analogy, not in the historical record." Caneva also says that "*Structure* provided only snippets of historical evidence in support of its major claims" (116). Brad Wray (2015, 141) responds as follows: "it seems that those historians who have seen *Structure* as an example of poor historical scholarship are mistaken. It is not poor historical scholarship, because it was not historical scholarship at all." For more on the way critics assessed the use of history of science in SSR, see Kindi (2005b).

particular historical episodes to, purportedly, use them to base his model on. Kuhn's understanding of scientific revolutions was particularly contested. Steven Weinberg (1998) maintained that the transition from Newtonian to Einsteinian mechanics, or from classical to quantum physics, was not a revolution in the Kuhnian sense. Dan Garber (2009) makes the same point about "Galileo, Newton and all that": it wasn't a scientific revolution. Ernst Mayr (1972, 988) thinks that the Darwinian revolution that Kuhn appeals to does not conform to the Kuhnian idea of a revolution, while Ursula Klein (2015) thinks that the chemical revolution, which again features prominently in SSR as a paradigmatic example of a Kuhnian revolution, was "[a] Revolution that never happened."[7] So, if we do not want to dismiss Kuhn's model as unevidenced or poorly supported, we need to look for some other use of history in his work, that is, other than as empirical evidence for a philosophical model.[8]

In Kindi (2005b), which discusses the relations between history and philosophy of science in SSR, I advanced a philosophical reading of the text showing that Kuhn's arguments in favor of discontinuity in scientific development are philosophical, not empirical, that is, not based merely on historical evidence. I suggested that Kuhn uses historical examples as anti-essentialist Wittgensteinian "reminders" that expose a variegated

---

[7] To be sure, there are scholars who have argued that Kuhn's model is applicable to the history of science and who do not share the above views about particular historical episodes. See, for example, Hoyningen-Huene (2008) and Chang (2012b) about the Chemical Revolution, or Tanghe et al. (2021) about evolutionary biology. But they would not claim that Kuhn, as a matter of historical fact, has actually based his model on broad and solid historical evidence.

[8] It is true that Kuhn is not always consistent in what he says about his model. In the Postscript to SSR, for instance, responding to his critics, Kuhn speaks of his book as "a viewpoint or theory about the nature of science" and refers to his "descriptive generalizations that are evidence for [his] theory" (SSR-4 206). Taken out of context, these expressions may prove misleading. Kuhn does not mean to say that he formed a philosophical theory about science based on generalizations from experience. The use of the term "theory" for his model is just generic and it is interchanged with "viewpoint." He says of his viewpoint or theory that "it need not be right, anymore than any other theory" (ibid.) and notes that his descriptive generalizations are related circularly with his theory. They provide evidence, but also are derived from it (for other theories of science, they constitute, says Kuhn, anomalous behavior). Kuhn writes that he has used parts of the theory that is presented in SSR as "a useful tool for the exploration of scientific behavior and development" (SSR-4 207). And in the Introduction to SSR, Kuhn speaks of his generalizations that are about the sociology or the social psychology of scientists (SSR-4 8) and of *theories* about knowledge that, presumably, include his own (SSR-4 9). Again, in both cases he does not imply any theory building of the kind we have in science. His generalizations are simply general observations as opposed to singular statements; they have not been reached by any kind of systematic investigation, and they do not exhaust his complete "theory," which also includes "conclusions that belong to logic and epistemology" (SSR-4 8).

landscape in the development of science. My philosophical reading borrowed from the transcendental approach and amounted to the claim that Kuhn in SSR offers the historicized conditions of the possibility of science, that is, exposure to exemplars and dogmatic training. What I wanted to appropriate from a transcendental reading of SSR was *a priori*, so that the reading was in line with what Kuhn himself said about deriving his model from first principles, and *necessity*, so that Kuhn's particular model of science could be singled out from the rest that aspire to capture the nature of science.

I would now like to suggest another possible reading that I find illuminating: I would like to compare Kuhn's model of science, and its relation to history, to Wittgenstein's objects of comparison, an idea that I mention but do not expand on in earlier work (Kindi 2005b). There, I speak of historical examples in Kuhn's work as objects of comparison. Here, I would like to say that the whole model also functions as an object of comparison. Before I present this reading, however, some clarifications are in order: I do not claim that Kuhn deliberately adopted this particular Wittgensteinian idea. Although he read about it in the *Philosophical Investigations* and marked the particular remarks where objects of comparison appear,[9] there is no explicit discussion of this concept in Kuhn's work.[10] In addition, we know that the reference to Wittgenstein's work in SSR does not appear in drafts prior to the published text (Isaac 2012, 105–106; Hoyningen-Huene 2015a, 188). So, Kuhn could not have consciously intended his model as an object of comparison. Why, then, do I propose this interpretation? Because I think that it (1) is exegetically adequate since it offers a coherent account of Kuhn's project, (2) clarifies better than other readings what Kuhn tried to do with his book, (3) allows a better appreciation of Kuhn's contribution, and (4) provides a better evaluation of the critical reception of Kuhn's work. I do not discard the transcendental reading, given especially the fact that Kuhn calls himself a Kantian with moveable categories (RSS 104, 264), but I now think that attempting a transcendental reading may stretch too much, or trivialize, what we understand by transcendental arguments.

What are Wittgenstein's "objects of comparison"? The term appears in two remarks in *Philosophical Investigations* (PI):

---

[9] I have access to Kuhn's own copy of the *Philosophical Investigations*. It is the 1953 bilingual, first edition of Wittgenstein's book.

[10] There was not much discussion in the whole Wittgenstein literature either until rather recently. Oskari Kuusela (2008; 2019) is the one who has most highlighted and discussed this concept in Wittgenstein's work.

PI §130: Our clear and simple language-games are not preliminary studies for a future regimentation of language – as it were, first approximations, ignoring friction and air resistance. Rather, the language games stand there as *objects of comparison* which, through similarities and dissimilarities, are meant to throw light on features of our language.

PI §131: For we can avoid unfairness or vacuity in our assertions only presenting the model as what it is, as an object of comparison – as a sort of yardstick; not as a preconception to which reality *must* correspond. (The dogmatism into which we fall so easily in doing philosophy.) (Wittgenstein 1953/2009, emphases in the original)

In these remarks, Wittgenstein says that language games, that is, real and imaginary examples or models of language used in some particular practice, function as objects of comparison, that is, as measuring rods that we lay against the object that we want to measure. If the measuring rod is a meter, we measure length and we report centimeters; if it is a thermometer, we measure temperature and we report degrees. In the case of language games, we note similarities and dissimilarities between linguistic practices. Wittgenstein warns, however, that we should not proceed to claim that the object of the investigation really has the properties that we project on to it, that it really fits the measuring rod, the mode of representation that we have chosen to discuss it. The features that we project upon the object of the investigation are features of the mode of representation, not intrinsic features of what we measure. If we change the measuring rod and we have a yard instead of a meter, we will be measuring inches and feet. The object of comparison determines only *how* we will conceive the object of the investigation, how we will discuss it; it does not determine what it actually is. If we were to do that, we would be doing arbitrary, speculative metaphysics. And, according to Wittgenstein, we would be falling into dogmatism, thinking that reality must fit our preconceived idea.[11]

The point of the comparisons between a yardstick and a concrete particular is to achieve some end, for instance, precision in measuring or the testing of a hypothesis. In Wittgenstein's case, the point of using, or setting up, language games as objects of comparison is to clarify the function of language, especially by highlighting the different ways language is, and can be, used. He, thus, undermines the essentialist understanding of language, an understanding that lays emphasis on common features and similarities in order to determine necessary and sufficient conditions. The language games that Wittgenstein uses do not carry any

---

[11] Wittgenstein expands on this idea in a comment published in *Culture and Value* (1977/ 1994, 21–22).

ontological import. We are not supposed to ask questions such as: "How many language games are there?" or "Is X a language-game?" Developing and using language games is a technique, a method, to bring about philosophical elucidation.

I want to claim that Kuhn's model of science functions as an object of comparison in order to highlight differences in various scientific traditions and to destroy the ideal image of science that governed both the history and the philosophy of science, an image that projected a seamless uniformity upon scientific practice.[12] Kuhn's examples of paradigms and revolutions are not pointed at as actual instances of his model's concepts, but are themselves used as objects of comparison and as reminders of the different ways science is practiced. Kuhn did not commit himself to there being particular revolutions in history. He discussed the more obvious ones, those that are commonly discussed as revolutions (SSR-4 6), but did not pronounce on which ones are to be recognized as such. He left it to historians to proclaim which of the conceptual changes that they study will be deemed revolutionary. He said: "I am repeatedly asked whether such-and-such a development was 'normal or revolutionary,' and I usually have to answer that I do not know.... Part of the difficulty in answering is that the discrimination of normal from revolutionary episodes demands close historical study" (Kuhn 1970a, 251).

So, the arguments that are supposed to show that certain historical episodes are, or are not really, Kuhnian revolutions, in order to support or discredit Kuhn's model, misunderstand the nature of his proposal and do not affect its validity. One can certainly study a particular historical development and call it a revolution, but there is no point in trying to match it with the schema Kuhn provided. The Kuhnian schema is supposed to help you look at the historical material differently. This is the point of its use.

Kuhn's concept of paradigm has also not been formed by systematic empirical observation. It is a concept that Kuhn used to account for agreement among scientists on a small scale.[13] If scientists were bound by the universal rules of the presumed scientific method, Kuhn would not be able to account for the variability that he had encountered as a historian (RSS 296). All disciplines that aspire to be called scientific

---

[12] I discuss these issues more extensively in Kindi (2015; 2017). Also, in Kindi (2012b), I consider how empirical material, from either history or sociology, features in Kuhn's account of science and criticize attempts in the science studies literature to reduce philosophy to empirical research by appealing to Wittgenstein's philosophy.

[13] Cf. SSR-3 viii. On Kuhn's concept of paradigm, see Kindi (2012a).

would have to follow the same rules. But with paradigms, Kuhn could limit the consensus to particular scientific communities, bound together and shaped by taking particular scientific achievements as models. Kuhn was not committed to the existence of one or more paradigms in each scientific community and did not say whether these remained unchanged or were further articulated during the course of normal science.[14] The use of the concept of paradigm served Kuhn's philosophical purpose, which was to transform the image of science by which we were then possessed (SSR-4 1). He did not derive the concept of paradigm, nor any other concept of his model, for that matter, from experience, even though his coming to use it was informed by his experience as a scientist and historian. Hence, from this perspective, complaints such as Kenneth Caneva's that Kuhn does not tell us "how one actually identifies a paradigm" (Caneva 2000, 116) are off the mark as criticisms of Kuhn's model.[15]

Kuhn's aim in SSR was philosophical, not historiographical. He was interested in "the discovery of essentials" in science (RSS 129) and spoke of his view as "an altered view of the nature of science" (RSS 129, cf. SSR 206–208). The term "structure" in the book's title does not point, as Lorraine Daston seems to think, to overarching regularities in the development of science, but instead points to the logic of science, its skeleton and frame. It is a reference to what makes science what it is (cf. SSR 129, 203). This seems like an essentialist undertaking, but it is not of the kind that seeks once and for all definitional characteristics. It is a logical investigation that explores the essence or the structure of science, similar to Wittgenstein's logical investigation of the essence or structure of language and propositions (Wittgenstein 1953/2009, §§89–92). But instead of trying, with the typical metaphysicians, to penetrate phenomena to find a hidden essence, Kuhn's and Wittgenstein's investigations stay on the surface and survey the different ways science is practiced and language is used, respectively. It can be said that they both look for the essence as expressed in grammar, that is, in the surveyability of different cases (Wittgenstein 1953/2009, §371, 122).[16]

---

[14] Nickles (2012), for instance, criticizes Kuhn for presenting paradigms/exemplars as rigid, fixed, and permanent. Dudley Shapere also raises the criticism that we do not know where "we draw the line between different paradigms and different articulations of the same paradigm" (Shapere 1964, 387).

[15] Kuhn said the following to Skúli Sigurdsson (2016, 27) about his model: "'Can you always locate the paradigm, can you always tell the difference between a revolution and a normal development?' It's not meant to be applied that way."

[16] Wray (2021a, 142–148) advances the view that Kuhn's use of "structure" originates in sociology. Kuhn may have come across the term "structure" in sociological studies with which he was familiar, but he did not try to find the structure of science by conducting

If Kuhn's model of science is an object of comparison that is not assessed empirically, why should we prefer it to other possible objects of comparison? We should prefer it because, first, it fulfils an important objective, set by Kuhn, but also more widely shared, namely, to free the study of science from the ideal image. Second, it gives a more inclusive and coherent picture of science than other models. For instance, what in other models is aberrant behavior of scientists – for example, resorting to values that are subjectively interpreted and assessed – is an essential part of science's practice and success (RSS 129). Finally, Kuhn's model is more fruitful in comparison to other models, say, the logical positivist model, in that it opens up more possibilities of research. This is a consideration that features prominently in Kuhn's discussion of scientific revolutions. A new paradigm needs to carry the promise of fruitful research. As we will see in the next section, where we will discuss the impact of Kuhn's model on the history of science, Kuhn's understanding of science has proven very fruitful indeed.

14.4 Kuhn's Impact on the History of Science and Further Developments

Kuhn reflected at some length on the state of the history of science in the 1980s on two occasions.[17] First, in 1984, in his contribution to the centennial of George Sarton's birth, which was included in a Recollections and Reflections section of *Isis* together with the pieces of other "distinguished statesmen of the history of science," as Arnold Thackray put it in his editorial (Thackray 1984).[18] The second occasion was in 1985 when he addressed the Seventeenth International Congress of History of Science at Berkeley. In the first piece, Kuhn comments on the professionalization and shaping of the field that he relates to the expansion of higher education but, most importantly, to the interest and concern about the power of science as it emerged after the role it had in the Second World War victory of the Allied forces. History of science was thought to be more suitable, in comparison to the more technical philosophy of science, for educating the wider public about

empirical investigations of the type sociologists undertake. Wray rightly criticizes the understanding of "structure" as a Popperian historicist law that underlies scientific development allowing predictions (pp. 148–151).

[17] Kuhn reflected on the history of science before (ET 3–20, 105–126, 127–161) and commented on developments in the discipline in his autobiographical interview (RSS 316–317).

[18] The other distinguished historians were John T. Edsall, I. Bernard Cohen, A. Rupert Hall, and A. C. Crombie.

science and for bridging C. P. Snow's two-culture gap. Yet historians of science, according to Kuhn, concentrated on training future professionals and neglected to create wider interest for the subject.

In the 1985 piece, Kuhn goes over the programs of three international history of science congresses: the sixth at Amsterdam in 1950, the tenth at Ithaca in 1962, and the seventeenth at Berkeley in 1984. He notes (1) the very rapid growth in the volume of papers presented, a growth that led to the transformation of the discipline; (2) a rapid shift of attention from older science (in Antiquity, the Middle Ages, and the Scientific Revolution) to newer and more contemporary topics; (3) a move away from history of ideas to more social and institutional history; and, finally, (4) a greater interest in fields of study and geographical areas that were not sufficiently researched before. He attributes these developments to the same reasons that he discussed in his 1984 comments, educational expansion and the need for scientific literacy. But he also adds that the placement of historians of science in history departments and the task of addressing a lay audience that knew very little science may have led to the turn to social history, away from technical scientific matters. Kuhn says that he welcomes these developments, but he cannot conceal his dissatisfaction with the sidelining of intellectual history and the neglect of the role of reason and experiment in scientific development in favor of socioeconomic interests.[19] He recognizes that the issue is mostly philosophical, but thinks historians of science should also be concerned as they "played a primary role in the destruction of the traditional viewpoint," that is, the viewpoint that takes science to be progressively approaching the real world (Kuhn 1986, 33). Kuhn envisions a greater interaction between intellectual and social-institutional history and finds the prospects for such an enterprise encouraging (ibid.).

Kuhn may not have commented explicitly on the role of his own work in the developments in the history of science, but historians of science have generally acknowledged the great significance of his contribution. To take a few examples: Michael Gordin (2020, 12) has said that the discipline of the history of science emerged "from under Kuhn's overcoat." Jan Golinski was very confident that

there is no way back to the time before Kuhn's intervention. His sonorous words at the opening of *Structure* still hold good: taking history seriously challenges all preconceived ideas of the nature of science. It calls into question any assertions that presuppose the unity of science across time and space or its singularity as a

---

[19] Ken Adler (2013, 94) uses the word "dyspeptic" to characterize Kuhn's reaction to the fact that general historians did not appreciate the technical part of the history of science.

cultural phenomenon. The move cannot be undone, even if the arguments for it do not have quite the force of logical compulsion. (Golinski 2012, 25)

Alex Soojung-Kim Pang (1997, 167) contends that post-constructivist history of science, a quieter approach in comparison to the more pronounced constructivism of science studies and the Edinburgh School, "takes as a starting-point the work of Kuhn, the Edinburgh School, and science studies, but modulates their tone and claims." Finally, the historian Nathan Reingold (1980, 484) calls Kuhn a liberator for freeing history of science "from both practitioners' history and dominant factions in the philosophy of science." Practitioners' history is the old fashioned, anachronistic, triumphant, cumulative history of science, written by scientists and propagated in textbooks. The dominant factions in philosophy of science that Reingold refers to are the various versions of logical positivism, which again see science as achieving a teleological goal (p. 485).

Still, despite the acknowledgment of Kuhn's overall impact on the history of science, many historians of science see his influence as a burden that has been, and needs to be, lifted. What they find particularly bothersome is a certain understanding of Kuhn's model of scientific development. They take it as imposing on the various practices of science universal structures, overarching regularities, or theoretical canons, or as cramming the vast variety and heterogeneity of practices into a few empirical generalizations that are assembled in patterns that repeat themselves in cycles. Peter Galison (1988, 394), for instance, has said that for Kuhn "there was a universal structure for the supplanting of one theory with another in his timeless cycle of epochs: normal science, crisis science, revolutionary science, and the return of normal science."[20] Stephen Brush (2000, 39–40) notes the "hostility and indifference to Kuhn's work displayed by many historians of science," while Mario Biagioli (2012, 480) refers to "evidence of the increasing irrelevance of *Structure* to contemporary history of science (and even doubts about its historical impact on the field)." Biagioli acknowledges that *Structure* helped create the field, but says that it "no longer frames the research agenda" (p. 479).[21]

Historians of science distance themselves from Kuhn's work because they see it as bringing schematic and universalistic *philosophical*

---

[20] Similarly, Bird (2015, 27) thinks that Kuhn finds a "fundamentally cyclical pattern with its alternating phases of normal and extraordinary (revolutionary), science" in the history of science.

[21] A similar view is stated by Wray (2021a, 139): "historians of science generally have a bleak view of the relevance and impact of SSR on the history of science."

considerations to bear on the complicated and heterogeneous manifold of historical reality. Philosophy is seen as the villain from whom they should protect themselves and the discipline. For better or worse, says Michael Gordin (2020, 12), "by the mid- to late 1980s, the historiography of science was emancipated from philosophy." Historians and sociologists, contended the sociologist Trevor Pinch (2017, 128), "fought the common enemy (the philosophers of science who beat up on relativism and felt the sociological approach to be impossible)."

History of science, liberated from philosophy, presumed to be despotic and censorious, turned to other fields and disciplines for inspiration, collaboration, and guidance. Sociology, anthropology, literary theory, political science, political theory, and politics have all provided historians of science with the intellectual and practical resources that helped the discipline grow and develop into a multidisciplinary, expansive field. The "enormous camaraderie" (Pinch 2017, 128) between historians of science and sociologists formed a whole tradition of historical/sociological research – the Strong Programme, SSK, and Science and Technology Studies (STS) – and produced works such as Shapin and Schaffer's *Leviathan and the Air Pump* (1985). Anthropology, with its attention to the local and the particular, and the foregrounding of cultures in their dynamic development (see Dear 1995; Galison 2011), offered a trove of concepts, such as Geertz's notion of thick description, and tools and methods, such as ethnomethodology, that were of lasting significance for the history of science, and led to works such as Latour and Woolgar's *Laboratory Life: The Social Construction of Scientific Facts* (1979). Focus on narrative and texts has invited considerations from rhetoric and literary theory with works such as Mario Biagioli's *Galileo, Courtier* (1993), Jean Dietz Moss' *Novelties in the Heavens: Rhetoric and Science in the Copernican Controversy* (1993), and Peter Dear's anthology *The Literary Structure of Scientific Argument* (1991). Politics in general raised issues that pertain to interests and power relations. A characteristic book in this regard is Naomi Oreskes and Eric Conway's *Merchants of Doubt* (2010).

How did history of science further develop, and did Kuhn have anything to do with it? History of science went, on the one hand, local: it concentrated on case studies, micro-histories, and highly specialized and detailed research topics, focusing considerably on archival material. Daston (2009, 809) spoke of "miniaturizing tendencies." On the other hand, history of science also went global. Studying the circulation of knowledge, historians of science transcended national borders, studied non-Western practices in an inclusive and cosmopolitan spirit, and abandoned metaphors of center and periphery or of unidirectional dissemination.[22] Now these developments interweave with not only negotiations, controversies, and rhetoric that have engaged historians of science since they left behind the purely intellectual history of science, but also "trading zones" and "contact zones," migration and appropriation, translation studies, postcolonial studies, and social, political, and economic relations (Secord 2004). The circulation of knowledge involves diverse groups that need to communicate as they participate in various transactions that are not limited to purely epistemic ones. An opposite development is the demotion of science, from its distinguished and privileged status to its assimilation to other cultural practices, such as art and craft, that yield different, and no less important, types of knowledge (Renn 2015, 242).[23] History of science becomes more history of knowledge. In a parallel line, Ken Adler (2013, 88) also advocates the jettisoning of "moralistic assertions of scientific exceptionalism," enlarging thereby the scope of history of science. Our historical investigations, he says, should be not just science but a set of activities that constitute a certain cultural milieu. Science is not something separate from its environment, but mutually interlocked with and mutually formed by its surround. In his view, historians of science should treat science as any other historical topic. It should be studied

alongside contemporary knowledge-making practices not canonically scientific. And in fact historians of science have repeatedly charted the affinities of emergent scientific practices with cognate ways of knowing and manipulating the world – Renaissance craftwork, early modern natural philosophy, Enlightenment engineering, Victorian quantification, or the twentieth century's discovery of "traditional" healing – the better to identify those aspects of knowledge making that came to be considered scientific. (Adler 2013, 95)[24]

The scope of history of science as a discipline is also broadened through another route: science is no longer understood narrowly and anachronistically. Science as an object of study is variously understood, depending on what the term "science" and the related words, for example, "natural

---

[22] For more on this global trend, see Secord (2004), Fan (2012), and Renn (2015).

[23] SSK's impartiality with respect to "truth and falsity, rationality or irrationality, success or failure," and its symmetry principle, that is, that the same kinds of causes will explain both sides of the above dichotomies (Bloor 1976/1991, 7), have also contributed to the assimilation of science to other practices.

[24] Adler suggests substituting "episcience" for "science" in the "history of science" so that the name of the discipline reflects more accurately what professionals in that discipline do. Episcience, according to Adler, is "the imbrication of science with its 'surround,' or what historians of science generally invoke with terms such as 'environment,' '*milieu*,' or '*Umgebung*'" (2013, 96). One might say that episcience may come close to another science studies concept, "situated knowledge," which requires embedding knowledge in context (Haraway 1988).

philosophy" or "history" signified in different times and places. Exploring affinities and differences between types of knowledge and different genres has led to the transgression of disciplinary boundaries as we understand them today. Art, craft, technology, engineering, data science, theology, law, literature, and humanities are all legitimate objects of research in history of science.

Contemporary history of science not only has transgressed boundaries between disciplines, but also has transgressed and overcome long valued dichotomies, such as the internal/external distinction (e.g., Shapin 1992), the context of discovery/context of justification distinction (e.g., Arabatzis 2006; Schickore and Steinle 2006), and the content/context distinction (e.g., Shapin 1980). Another set of developments relates to the understanding of science as practice. This approach has brought into prominence institutions, experiments, and material culture, while considerations of gender, race, and class have enriched both the perspective and the range of research topics of the history of science. History of science has also given prominence to the study of contingency in scientific development, that is, that science may have developed in an alternative way. Finally, from the focus on ideas, history of science has shifted attention to communities and individual scientists (interest in new, non-idolizing types of biography has surged),[25] reaching, in the contemporary scene, post-humanistic studies.

Now, what is the relation of all these developments in the history of science to Kuhn's work and, more specifically, to SSR? Wray (2021a, 7) says that the impact Kuhn had on the history of science "was comparatively insignificant." Mario Biagioli (2012, 480) believes, as we have seen, that SSR is irrelevant to contemporary history of science and that there are doubts about its historical impact on the field. He also claims that "the 'turn to practice' only sealed ... the increasing drift of *Structure* away from the research agenda of the field."[26] Biagioli notes that "*Structure* was an attractive 'forward-looking statement' about what the history and philosophy of science could be," offering "examples of the exciting work the discipline might produce" (pp. 282–283). Jouni-Matti Kuukkanen (2013, 96) writes that the local turn in the history of science

---

[25] See Nye (2019, 11): "Kuhn shifted many historians' interest away from a heroic history of ideas and individuals toward the history of groups and communities, that is, from histories of eureka moments of discovery to histories of communication, organization, and social context."

[26] In Biagioli's view, Kuhn's work gave the field visibility, but now can function only as a "recruitment tool," "seducing" students into the field of history and philosophy of science by providing a model "that appears to explain everything, or almost everything," leaving at the same time "a lot of the answers open" (Biagioli 2012, 483).

"is incompatible with Kuhn's attempt to draw general conclusions on the nature of science on the basis of historical sources," since it lays emphasis on particular practices and cultures and presents a fragmented image of science. He observes, however, that a localist historiography implies a non-localist theory of historical interpretation (i.e., that "a historical interpretation could be made without any higher-level constructions") and the general idea that science is inherently something local. In that sense, Kuukkanen argues, contemporary localist historiography is not incompatible with Kuhn's approach.

Kuukkanen is right that the localist turn in the history of science is not incompatible with Kuhn's approach, but not for the reasons he cites. We do not need to assume that the localist perspective presupposes some general idea or construction so that it becomes compatible with what Kuhn did. As I have explained in the first part of this chapter, Kuhn's model was not formed by drawing general conclusions from historical evidence. Kuhn did not present a schema that was supposed to fit every developmental case in science.[27] It would take a very insensitive historian to prescribe such a one size-fits-all pattern to science. Far from being incompatible with the localist turn, Kuhn's work made it possible by combating anachronism, the projection of the contemporary conception of science to the past as if it is the only one, and by laying emphasis on the particular practices that develop around a paradigm. The priority of paradigms over rules that Kuhn speaks of in SSR is exactly supposed to highlight that scientists do not follow general rules that allegedly comprise the scientific method. Scientists bond together by using specific achievements as prototypical ways of doing research. This emphasis on diversity paved the way for turning the attention of historians of science to the peculiarities of each historical case.

Biagioli's statement that the turn to practice in the historiography of science increased the drift away from SSR is also questionable. SSR, more than any other book in late 1950s and early 1960s, contributed to the understanding of science as practice rather than as a set of statements. It brought to the fore the elements that build a practice, namely, education and initiation based on paradigms, textbooks, institutions, values and norms, training, research procedures, and the building and communication of communities. History of science took up these newly

[27] Kuhn distances himself from Gaston Bachelard's work. He characterized it as "too much of a constraining [system]." Bachelard, Kuhn said in his autobiographical interview, "had categories, and methodological categories, and moved the thing up the escalator too systematically for me" (RSS 285). Kuhn did not want to form that kind of a model for science, which he regarded as constraining, overly systematic, and too general.

opened fields of study and explored them further. The current research agenda in the history of science may not be framed by SSR (it is already sixty years since its publication), but it owes a lot to Kuhn's pioneering work. SSR's emphasis on practice held the promise of fruitful research and steered the course of development in a new direction, just like Kuhn has said it happens in revolutionary changes in science. Scientists decide between alternative pathways based on the future promise the emerging paradigm holds (SSR 156). This is exactly what Mario Biagioli himself acknowledges when he says that SSR was an attractive "forward-looking statement" and offered "examples of the exciting work the discipline might produce" (2012, 482 and 483)

Kuhn's work, and more particularly his discussion of paradigm shifts and incommensurability, paved the way for many other topics in the historiography of science. Studies of scientific controversies and negotiations relate to Kuhn's showing that the choice of paradigm or theory does not depend on logic and experiment alone, but involves the judgment of individual scientists who may disagree among themselves. Change of allegiance requires persuasion, and rhetoric becomes a legitimate area of study in relation to science. Similarly, Kuhn's rejection of a mechanical understanding of rationality opened the possibility of attributing rational decisions to individuals and groups that differ on how they comparatively assess the merits of the competing paradigms. This means that alternative ways of scientific development should not be excluded or disqualified as irrational, a thought that foreshadows more recent developments in science studies literature. Kuhn has addressed the contingency issue already in his 1951 Lowell Lectures when he discussed the rejection of the phlogiston theory (QPT 57–76). Finally, conceptual incommensurability and problems of communication between scientists defending incommensurable paradigms invite discussions of translation. Translation studies and knowledge transfer research may be seen as involving considerations that were first taken up in SSR in relation to the implications of incommensurability.

Anti-Whiggism and sensitivity to the varying meaning of words, including that of "science," was highlighted in many ways in SSR and, in general, in Kuhn's work (see SSR-4 148; ET 154; RSS 290), as was the blurring of the age-old distinctions, such as the internal/external, the is/ought, and the context of discovery/context of justification distinction.[28] Even the blurring of the language/fact distinction, so crucial for constructivism in the history of science, can be traced back to

---

[28] See, for example, Alexandre Koyré's comment about SSR to Kuhn: "you have brought the internal and external histories of science, which in the past have been very much

Kuhn, who maintained that "the so-called facts proved never to be mere facts, independent of belief and theory. Producing them required apparatus which itself depended on theory, often on the theory the experimenters were supposed to test" (RSS 108). Last, one should not ignore the project "Sources for History of Quantum Physics" that Kuhn conducted between 1962 and 1964, together with his students John Heilbron and Paul Forman and with the assistance of Lini Allen, a project that involved interviews with the leading figures of twentieth-century physics. This project is now considered a predecessor of oral history and new historiography (see Heesen 2020). Similarly, Kuhn's *Black Body Theory and the Quantum Discontinuity, 1894–1912*, despite the criticism that it has received, has been a point of reference in the historiography of quantum theory.

14.5   Conclusion

Kuhn's work decisively influenced, and even transformed, the historiography of science, but historians want nothing to do with it. They keep Kuhn's philosophy at bay, and philosophers treat him as a historian. He was an outcast. At the heart of this state of affairs is a series of misunderstandings that proved, however, as is often the case, very fruitful, at least in the field of the history of science (in philosophy Kuhn was simply ostracized).[29] Historians of science have largely misunderstood Kuhn. Kuhn did not try to impose a philosophical ideal on the history of science; he did not look for laws underlying the scientific development; he did not try to generalize from a few cases; he did not use historical evidence to ground his model. Kuhn's work breached established distinctions and boundaries, and, in consequence, it was difficult to appreciate its revolutionary import. SSR, in particular, did not fit the traditional molds in which it was cast, which meant that tensions and incongruities would, not infrequently, surface. Still, despite the misreadings or the disregard, Kuhn's SSR played a groundbreaking role in how historiography and the discipline of the history of science developed.

---

apart, together." Kuhn said that he had not thought of this as what he was doing, but he felt that Koyré was right in his judgment (RSS 286).

[29] John Searle (1993, 71) says that "a solid and self-confident professorial establishment, committed to traditional intellectual values," blocked the works of authors such as Jacques Derrida, Thomas Kuhn, and Richard Rorty from affecting analytic philosophy, something that did not happen in the rest of the humanities.

Acknowledgments

I would like to thank Brad Wray for his encouragement and very helpful editorial suggestions. I am grateful to Theodore Arabatzis, who is always there to offer generous and incisive comments and continual intellectual support. Finally, I am indebted to the participants of the "Thomas Kuhn's Philosophy of Science" conference, which was organized by Brad Wray, and held in Aarhus, Denmark, on June 1–2, 2022, in honor of the hundredth year of Kuhn's birth. I have profited from their questions and comments.

Bibliography

Abbott, Andrew (2001). *The Chaos of Disciplines.* Chicago: University of Chicago Press.
    (2016). "Structure as Cited, Structure as Read." In Robert J. Richards and Lorraine Daston (eds.), *Kuhn's* Structure of Scientific Revolutions *at Fifty: Reflections on a Science Classic*, pp. 167–181. Chicago: University of Chicago Press.
Abramo, G., D'Angelo, C. A., and Di Costa, F. (2012). "Identifying Interdisciplinarity through the Disciplinary Classification of Coauthors of Scientific Publications." *Journal of the American Society for Information Science and Technology* 63, 11: 2206–2222.
Adler, K. (2013). "The History of Science as Oxymoron: From Scientific Exceptionalism to Episcience." *Isis* 104, 1: 88–101.
Andersen, Hanne (2000). "Learning by Ostension: Thomas Kuhn on Science Education." *Science and Education* 9: 91–106.
    (2001a). "Critical Notice: Kuhn, Conant and Everything – A Full or Fuller Account." *Philosophy of Science* 68, 2: 258–262.
    (2001b). *On Kuhn.* Belmont, CA: Wadsworth/Thomson.
    (2010). "Joint Acceptance and Scientific Change: A Case Study." *Episteme* 7, 3: 248–265.
    (2013). "The Second Essential Tension: On Tradition and Innovation in Interdisciplinary Research." *Topoi* 32, 1: 3–8.
    (2014). "Co-Author Responsibility: Distinguishing between the Moral and Epistemic Aspects of Trust." *EMBO Reports* 15: 915–918.
Andersen, Hanne, Barker, Peter, and Chen, Xiang (2006). *The Cognitive Structure of Scientific Revolutions.* Cambridge: Cambridge University Press.
Andersen, Line E. (2017). "Outsiders Enabling Scientific Change: Learning from the Sociohistory of a Mathematical Proof." *Social Epistemology* 31, 2: 184–191.
Arabatzis, T. (2006). "On the Inextricability of the Context of Discovery and the Context of Justification." In J. Schickore and F. Steinle (eds.), *Revisiting Discovery and Justification*, pp. 215–230. Dordrecht: Springer.
Barber, Bernard (1963). "Review of *The Structure of Scientific Revolutions.*" *American Sociological Review* 28, 2: 298–299.
Barnes, S. Barry (1974). *Scientific Knowledge and Sociological Theory.* London: Routledge and Kegan Paul.
    (1977). *Interests and the Growth of Knowledge.* London: Routledge and Kegan Paul.

(1982a). *T. S. Kuhn and Social Science*. London: Macmillan.

(1982b). "On the Implications of a Body of Knowledge." *Science Communication* 4, 1: 95–110.

(2011). "Relativism as a Completion of the Scientific Project." In R. Schantz and M. Seidel (eds.), *The Problem of Relativism in the Sociology of (Scientific) Knowledge*, pp. 23–39. Frankfurt: Ontos.

Barnes, S. Barry, and Bloor, David (1982). "Relativism, Rationalism and the Sociology of Knowledge." In M. Hollis and S. Lukes (eds.), *Rationality and Relativism*, pp. 21–47. Oxford: Blackwell.

Barnes, S. Barry, Bloor, David, and Henry, J. (1996). *Scientific Knowledge: A Sociological Analysis*. Chicago: University of Chicago Press.

Barnes, S. Barry, and Dolby, R. G. A. (1970). "The Scientific Ethos: A Deviant Viewpoint." *European Journal of Sociology* 11, 1: 3–25.

Bell, Randy, Abd-El-Khalick, Fouad, Lederman, Normal G., McComas, William F., and Matthews, Michael R. (2001). "The Nature of Science and Science Education: A Bibliography." *Science and Education* 10: 187–204.

Bellos, David (2012). *Is That a Fish in Your Ear? Translation and the Meaning of Everything*. New York: Farrar, Straus and Giroux.

Ben-David, J. (1978). "Emergence of National Traditions in the Sociology of Science." *Sociological Inquiry* 48, 3–4: 197–218.

Bernal, John Desmond (1939). *The Social Function of Science*. London: George Routledge and Sons.

Bhopal, Raj. (1999). "Paradigms in Epidemiology Textbooks: In the Footsteps of Thomas Kuhn." *American Journal of Public Health* 89, 8: 1162–1165.

Biagioli, Mario (1990). "Galileo's System of Patronage." *History of Science* 28, 1: 1–62.

(1993). *Galileo, Courtier: The Practice of Science in the Culture of Absolutism*. Chicago: University of Chicago Press.

(2012). "Productive Illusions: Kuhn's *Structure* as a Recruitment Tool." *Historical Studies in the Natural Sciences* 42, 5: 479–484.

Bierman, Paul. (2006) "Reconsidering the Textbook." Science Education Resource Center at Carleton College. Last revised 2006. Accessed November 1, 2022. http://serc.carleton.edu/files/textbook.

Binns, Ian C. (2013) "A Qualitative Method to Determine How Textbooks Portray Scientific Methodology." In Myint Swe Khine (ed.), *Critical Analysis of Science Textbooks: Evaluating Instructional Effectiveness*, pp. 239–258. New York: Springer.

Bird, Alexander J. (2000a). *Thomas Kuhn*. Chesham: Acumen.

(2000b). *Thomas Kuhn*. Princeton, NJ: Princeton University Press.

(2002). "Kuhn's Wrong Turning." *Studies in History and Philosophy of Science Part A* 33, 3: 443–463.

(2004). "Kuhn, Naturalism, and the Positivist Legacy." *Studies in History and Philosophy of Science* 35, 2: 337–356.

(2010). "Social Knowing: The Social Sense of 'Scientific Knowledge.'" *Philosophical Perspectives* 24: 23–56.

(2012a). "Kuhn, Naturalism, and the Social Study of Science." In V. Kindi and T. Arabatzis (eds.), *Kuhn's The Structure of Scientific Revolutions Revisited*, pp. 205–230. London: Routledge.

(2012b). "*The Structure of Scientific Revolutions* and Its Significance: An Essay Review of the Fiftieth Anniversary Edition." *British Journal of the Philosophy of Science* 63: 859–883.

(2014). "When Is There a Group That Knows? Distributed Cognition, Scientific Knowledge, and the Social Epistemic Subject." In Jennifer Lackey (ed.), *Essays in Collective Epistemology*, pp. 42–63. Oxford: Oxford University Press.

(2015). "Kuhn and the Historiography of Science." In W. J. Devlin and A. Bokulich (eds.), *Kuhn's Structure of Scientific Revolutions – 50 Years On*, pp. 23–28. Cham: Springer.

Birge, Raymond Thayer (1929). "Probable Values of the General Physical Constants." *Reviews of Modern Physics* 1, 1: 1–73.

(1932). "The Calculation of Errors by the Method of Least Squares" *Physical Review* 40: 207–227.

(1941). "The General Physical Constants: As of August 1941 with Details on the Velocity of Light Only." *Reports on Progress in Physics* 8, 1: 90–134.

(1943). "Comments on 'The Probable Accuracy of the General Physical Constants.'" *Physical Review* 63, 5–6: 213.

(1957). "A Survey of the Systematic Evaluation of the Universal Physical Constants." *Il Nuovo Cimento* 6, S1: 39–67.

Bloom, Harold (1973). *The Anxiety of Influence*. Oxford: Oxford University Press.

Bloor, David (1973). "Wittgenstein and Mannheim on the Sociology of Mathematics." *Studies in History and Philosophy of Science* 4, 2: 173–191.

(1976/1991). *Knowledge and Social Imagery*. Chicago: University of Chicago Press.

(1983a). "Relativism (methodological)." In W. F. Bynum et al. (eds.), *Macmillan Dictionary of the History of Science*, p. 369. London: Macmillan Press.

(1983b). *Wittgenstein: A Sociological Theory of Knowledge*. London: Macmillan.

(1984). "The Sociology of Reasons: Or Why 'Epistemic Factors' Are Really 'Social Factors.'" In J. R. Brown (ed.), *Scientific Rationality: The Sociological Turn*, pp. 295–324. Dordrecht: Reidel.

(2004). "Sociology of Scientific Knowledge." In I. Niiniluoto et al. (eds.), *Handbook of Epistemology*, pp. 919–962. Dordrecht: Kluwer.

(2007). "Epistemic Grace: Antirelativism as Theology in Disguise." *Common Knowledge* 13, 2–3: 250–280.

(2011). "Relativism and the Sociology of Scientific Knowledge." In D. Hales (ed.), *A Companion to Relativism*, pp. 433–455. Malden, MA: Wiley-Blackwell.

(2017). Interview with David Bloor, by R. McKenna, A.-K. Koch, and N. Ashton. https://emergenceofrelativism.weebly.com/blog/interview-with-david-bloor.

Boas Hall, M. (1963). "Review of *The Structure of Scientific Revolutions*." *The American Historical Review* 68, 3: 700–701.

Bodenmann, S. (2010). "The 18th-Century Battle over Lunar Motion." *Physics Today* 63, 1: 27–32.

Bohm, D. (1964). "Review of *The Structure of Scientific Revolutions*." *The Philosophical Review* 14, 57: 377–379.

Bokulich, Alisa (2020a). "Towards a Taxonomy of the Model-Ladenness of Data." *Philosophy of Science* 87, 5: 793–806.

(2020b). "Calibration, Coherence, and Consilience in Radiometric Measures of Geologic Time." *Philosophy of Science* 87, 3: 425–456.

Bokulich, Alisa, and Parker, William (2021). "Data Models, Representation and Adequacy-for-Purpose." *European Journal for Philosophy of Science* 11, 1: 31.

Boyd, Richard (1979). "Metaphor and Theory Change: What Is 'Metaphor' a Metaphor For?" In Andrew Ortony (ed.), *Metaphor and Thought*, pp. 481–532. Cambridge: Cambridge University Press.

Bratman, M. E. (1992). "Shared Cooperative Activity." *Philosophical Review* 101, 2: 327–341.

Briatte, F. (2007) "Entretien avec David Bloor." *Tracés: Revue de Sciences Humaine* 12: 215–228.

Brown, Matthew J., and Kidd, Ian James (2016). "Introduction: Reappraising Paul Feyerabend." *Studies in History and Philosophy of Science* 57: 1–8.

Brush, Stephen G. (1974). "Should the History of Science Be Rated X?" *Science* 183, 4230 (March 22): 1164–1172.

(2000). "Thomas Kuhn as a Historian of Science." *Science and Education* 9: 39–58.

Buchwald, Jed Z., and Smith, George E. (1997). "Thomas S. Kuhn, 1922–1996." *Philosophy of Science* 64, 2: 361–376.

Burian, Richard M. (2001). "The Dilemma of Case Studies Resolved: The Virtue of Using Case Studies in the History and Philosophy of Science." *Perspectives on Science* 9, 4: 383–404.

Bush, Vannevar (1945). *Science: The Endless Frontier: A Report to the President.* Washington, DC: US Government Printing Office.

Caneva, Kenneth L. (2000). "Possible Kuhns in the History of Science: Anomalies of Incommensurable Paradigms." *Studies in History and Philosophy of Science* 31, 1: 87–124.

Carnap, Rudolf (1932/1987). "On Protocol Sentences," trans. Richard Creath and Richard Nollan. *Noûs* 21: 457–470.

(1934/1937). *The Logical Syntax of Language*, trans. Amethe Smeaton. London: Routledge and Kegan Paul.

(1935). *Philosophy and Logical Syntax*. London: Kegan Paul, Trench, Trubner.

(1936). "Testability and Meaning." *Philosophy of Science* 3, 4: 419–471.

(1937). "Testability and Meaning – Continued." *Philosophy of Science* 4, 1: 1–40.

(1956). "The Methodological Character of Theoretical Concepts." In Herbert Feigl and Michael Scriven (eds.), *The Foundations of Science and the Concepts of Psychology and Psychoanalysis*, pp. 38–78. Minneapolis: University of Minnesota Press.

(1966). *Philosophical Foundations of Physics: An Introduction to the Philosophy of Science*, ed. Martin Gardner. New York: Basic Books.

Carrier, M. (2008). "The Aim and Structure of Methodological Theory." In L. Soler et al. (eds.), *Rethinking Scientific Change and Theory Comparison: Stabilities, Ruptures, Incommensurabilities?*, pp. 273–290. Dordrecht: Springer.

Cartwright, Nancy (1983). *How the Laws of Physics Lie*. Oxford: Oxford University Press.

(1999). *The Dappled World: A Study of the Boundaries of Science*. Cambridge: Cambridge University Press.

Cat, Jordi (2022). "The Unity of Science." In Edward N. Zalta (ed.), *Stanford Encyclopedia of Philosophy* (Spring 2022 Edition). https://plato.stanford.edu/archives/spr2022/entries/scientific-unity.

Cavell, Stanley (2010). *Little Did I Know: Excerpts from Memory*. Palo Alto, CA: Stanford University Press.

Cedarbaum, D. G. (1983). "Paradigms." *Studies in History and Philosophy of Science* 14, 3: 173–213.

Chang, Hasok (2004). *Inventing Temperature: Measurement and Scientific Progress*. Oxford: Oxford University Press.

(2012a). *Is Water H$_2$O? Evidence, Realism and Pluralism*. Dordrecht: Springer.

(2012b). "Incommensurability: Revisiting the Chemical Revolution." In Vasso Kindi and Theodore Arabatzis (eds.), *Kuhn's Structure of Scientific Revolutions Revisited*, pp. 153–176. London: Routledge.

Chu, J. S., and Evans, J. A. (2021). "Slowed Canonical Progress in Large Fields of Science." *PNAS* 118, 41: e2021636118.

Cohen, E. Richard, Crowe, Kenneth M., and DuMond, Jesse W. (1957). *The Fundamental Constants of Physics*, vol. 1. New York: Interscience Publishers.

Cohen, E. Richard, and DuMond, Jesse W. (1957). "The Fundamental Constants of Atomic Physics." In E. Richard Cohen, Kenneth M. Crowe, and Jesse W. M. DuMond (eds.), *Atoms I/Atome I*, pp. 1–87. Berlin: Springer.

(1965). "Our Knowledge of the Fundamental Constants of Physics and Chemistry in 1965." *Reviews of Modern Physics* 37, 4: 537–594.

Cohen, I. B. (1980). *The Newtonian Revolution*. New York: Cambridge University Press.

Collin, F. (2011). *Science Studies as Naturalized Philosophy*. Dordrecht: Springer.

Collins, Harry (2011). "Language and Practice." *Social Studies of Science* 41: 271–300.

(2012). "Comment on Kuhn." *Social Studies of Science* 43, 3: 420–423.

(2018). "Studies of Expertise and Experience." *Topoi* 37, 1: 67–77.

Collins, Harry, and Evans, Robert (2017). *Why Democracies Need Science*. Cambridge: Polity Press.

Collins, Harry, Evans, Robert, Durant, Darrin, and Weinel, Martin (2020). "How Does Science Fit into Society? The Fractal Model." In *Experts and the Will of the People*, pp. 63–88. Cham: Springer.

Collins, Harry, Evans, Robert, and Gorman, Mike (2007). "Trading Zones and Interactional Expertise." *Studies in History and Philosophy of Science* 38: 657–666.

Comte, Auguste (1988). *Introduction to Positive Philosophy*. Indianapolis, IN: Hackett.

Conant, James B. (1947/1952). *On Understanding Science*. Mentor edition. New Haven, CT: Yale University Press.

(1947). *On Understanding Science: An Historical Approach*. New Haven, CT: Yale University Press.

(1950). "Science and Politics in the Twentieth Century." *Foreign Affairs* 28, 2: 189–202.

(1951). "The Impact of Science on Industry and Medicine." *American Scientist* 39, 1: 33–49.

Conant, James B., and Nash, Leonard K. (eds.) (1948/1957). *Harvard Case Histories in Experimental Science.* Cambridge, MA: Harvard University Press.

Cowan, Tyler (2013). "A Profession with an Egalitarian Core." *New York Times*, Section BU (March 17): 4.

Creath, Richard (1990). "Introduction." In W. V. Quine and Rudolf Carnap, *Dear Carnap, Dear Van: The Quine-Carnap Correspondence and Related Work*, ed. Richard Creath, pp. 1–43. Los Angeles: University of California Press.

   (2004). "Quine on the Intelligibility and Relevance of Analyticity." In Roger Gibson (ed.), *Cambridge Companion to Quine*, pp. 47–64. Cambridge: Cambridge University Press.

   (2007). "Quine's Challenge to Carnap." In Michael Friedman and Richard Creath (eds.), *The Cambridge Companion to Carnap*, pp. 316–335. Cambridge: Cambridge University Press.

Csiszar, A. (2018). *The Scientific Journal: Authorship and the Politics of Knowledge in the Nineteenth Century.* Chicago: University of Chicago Press.

Cushing, James T. (1994). *Quantum Mechanics: Historical Contingency and the Copenhagen Hegemony.* Chicago: University of Chicago Press.

D'Agostino, F. (2010). *Naturalizing Epistemology.* Basingstoke: Palgrave Macmillan.

Daston, Lorraine (1994). "Historical Epistemology." In James Chandler, Arnold I. Davidson, and Harry Harootunian (eds.), *Questions of Evidence: Proof, Practice, and Persuasion across the Disciplines*, pp. 282–289. Chicago: University of Chicago Press.

   (2009). "Science Studies and the History of Science." *Critical Inquiry* 35: 798–813.

   (2016). "History of Science without *Structure*." In R. J. Richards and L. Daston (eds.), *Kuhn's Structure of Scientific Revolutions at Fifty*, pp. 115–132. Chicago: University of Chicago Press.

Daston, Lorraine, and Galison, Peter (2007). *Objectivity.* New York: Zone Books.

Davidson, Arnold I. (2001). *The Emergence of Sexuality: Historical Epistemology and the Formation of Concepts.* Cambridge, MA: Harvard University Press.

Davidson, Donald (1974/1984). "On the Very Idea of a Conceptual Science." In *Inquiries into Truth and Interpretation*, pp. 183–198. Oxford: Clarendon Press.

   (1997). "Gadamer and Plato's *Philebus*." In L. E. Hahn (ed.), *The Philosophy of Hans-Georg Gadamer.* Chicago: Open Court.

Dear, Peter (ed.) (1991). *The Literary Structure of Scientific Argument.* Philadelphia: University of Pennsylvania Press.

   (1995). "Cultural History of Science: An Overview with Reflections." *Science, Technology, and Human Values* 20, 2: 150–170.

   (2012). "Fifty Years of *Structure*." *Social Studies of Science* 43, 3: 424–428.

Del Vicario, M., Vivaldo, G., Bessi, A., Zollo, F., Scala, A., Caldarelli, G., and Quattrociocchi, W. (2016). "Echo Chambers: Emotional Contagion and Group Polarization on Facebook." *Scientific Reports* 6, 1: 1–12.

Dewey, John (1916). *Democracy and Education: An Introduction to the Philosophy of Education*. New York: Macmillan.

Dimopoulos, Kostos, and Karamanidou, Christina (2013). "Towards a More Epistemologically Valid Image of School Science: Revealing the Textuality of School Science Textbooks." In Myint Swe Khine (ed.), *Critical Analysis of Science Textbooks: Evaluating Instructional Effectiveness*, pp. 61–78. New York: Springer.

Donner, Wendy (1998). "Mill's Utilitarianism." In John Skorupski (ed.), *The Cambridge Companion to Mill*, pp. 255–292. Cambridge: Cambridge University Press.

Duhem, Pierre (1914/1954). *The Aim and Structure of Physical Theory*, 2nd ed., trans. Marcel Rivière. Princeton, NJ: Princeton University Press.

Dupré, John (1993). *The Disorder of Things: Metaphysical Foundations of the Disunity of Science*. Cambridge, MA: Harvard University Press.

Earman, John (1993). "Carnap, Kuhn, and the Philosophy of Scientific Methodology." In Paul Horwich (ed.), *World Changes: Thomas Kuhn and the Nature of Science*, pp. 9–36. Cambridge, MA: MIT Press.

Edgerton, D. (2004). "The Linear Model Did Not Exist: Reflections on the History and Historiography of Science and Research in Industry in the Twentieth Century." In K. Grandin, N. Wormbs, and S. Widmalm (eds.), *The Science-Industry Nexus. History, Policy, Implications*. New York: Science History Publications, pp. 31–57.

Fagan, Melinda Bonnie (2011). "Is There Collective Scientific Knowledge? Arguments from Explanation." *The Philosophical Quarterly* 61, 243: 247–269.

   (2012a). "The Joint Account of Mechanistic Explanation." *Philosophy of Science* 79, 4: 448–472.

   (2012b). "Collective Scientific Knowledge." *Philosophy Compass* 7, 12: 821–831.

Fan, F. (2012). "The Global Turn in the History of Science." *East Asian Science, Technology and Society: An International Journal* 6: 249–258.

Feest, Uljana, and Sturm, Thomas (2011). "What (Good) Is Historical Epistemology? Editors' Introduction." *Erkenntnis* 75, 3: 285–302.

Feyerabend, Paul K. (1970). "Consolations for the Specialist." In Imre Lakatos and Alan Musgrave (eds.), *Criticism and the Growth of Knowledge*, pp. 197–230. Cambridge: Cambridge University Press.

   (1975). *Against Method: Outline of an Anarchistic Theory of Knowledge*. London: New Left Books.

   (1978a). *Science in a Free Society*. London: New-Left Books.

   (1978b). *Science in a Free Society*. London: Verso Books.

   (1987). *Farewell to Reason*. London: Verso.

   (1989). "Realism and the Historicity of Knowledge." *Journal of Philosophy* 86, 8: 393–406.

Fiske, Edward B. (1987). "Colleges Prodded to Prove Worth." *New York Times*, Section 1 (January 18): 1.

Fleck, Ludwik (1935/1979). *Genesis and Development of a Scientific Fact*, trans. Fred Bradley and Thaddeus J. Trenn. Chicago: University of Chicago Press.

Friedman, Michael (1987). "Carnap's *Aufbau* Reconsidered." *Nous* 21: 521–545.
(2001). *Dynamics of Reason: The 1999 Kant Lectures at Stanford University*. Stanford, CA: CSLI Publications.

Fuller, Steve (2000a). "From Conant's Education Strategy to Kuhn's Research Strategy." *Science and Education* 9: 21–37.
(2000b). *Thomas Kuhn: A Philosophical History for Our Times*. Chicago: University of Chicago Press.
(2003). *Kuhn vs. Popper: The Struggle for the Soul of Science*. Duxford: Icon.
(2004). *Kuhn vs. Popper: The Struggle for the Soul of Science*. New York: Columbia University Press.

Galison, Peter (1987). *How Experiments End*. Chicago: University of Chicago Press.
(1988). "History, Philosophy, and the Central Metaphor." *Science in Context* 2: 197–212.
(1997). *Image and Logic. A Material Culture of Microphysics*. Chicago: University of Chicago Press.
(1999). "Trading Zone: Coordinating Action and Belief." In M. Biagioli (ed.), *The Science Studies Reader*, pp. 137–160. New York: Routledge.
(2008). "Ten Problems for History and Philosophy of Science." *Isis* 99, 1: 111–114.
(2011). "Scientific Cultures." In J. C. Alexander, P. Smith, and M. Norton (eds.), *Interpreting Clifford Geertz: Cultural Investigation in the Social Sciences*. New York: Palgrave Macmillan.
(2016). "Practice All the Way Down." In R. Richards and L. Daston (eds.), *Kuhn's "Structure of Scientific Revolutions" at Fifty: Reflections on a Science Classic*, pp. 42–69. Chicago: University of Chicago Press.

Galison, Peter, and Stump, David J. (eds.) (1996). *The Disunity of Science: Boundaries, Contexts, and Power*. Stanford, CA: Stanford University Press.

Garber, Daniel (2009). "Galileo, Newton and All That: If It Wasn't a Revolution What Was It?" *Circumscribere* 7: 9–18.

Geertz, C. (1997/2000). "The Legacy of Thomas Kuhn: The Right Text at the Right Time." In *Available Light*, pp. 160–166. Princeton, NJ: Princeton University Press.

Giere, Ronald N. (1973). "History and Philosophy of Science: Intimate Relationship or Marriage of Convenience?" *British Journal for the Philosophy of Science* 24, 3: 282–297.
(1999). *Science without Laws*. Chicago: University of Chicago Press.

Gilbert, Margaret (1987). "Modelling Collective Belief." *Synthese* 73, 1: 185–204.
(1990). "Walking Together: A Paradigmatic Social Phenomenon." *Midwest Studies in Philosophy* 15: 1–14.
(1992). *On Social Facts*. Princeton, NJ: Princeton University Press.
(2000). "Collective Belief and Scientific Change." In *Sociality and Responsibility: New Essays in Plural Subject Theory*, pp. 37–49. Lanham, MD: Rowman and Littlefield.

Gillispie, C. C. (1962). "The Nature of Science." *Science* 138, 3546 (December 14): 1251–1253.

Giri, Leandro, and Melongo, Pablo (2023). "Towards a Genealogy of Thomas Kuhn's Semantics." *Perspectives on Science* 31: 385–404.

Godfrey-Smith, P. (2003). *Theory and Reality: An Introduction to the Philosophy of Science*. Chicago: University of Chicago Press.

Golinski, J. (2012). "Thomas Kuhn and Interdisciplinary Conversation: Why Historians and Philosophers of Science Stopped Talking to One Another." In S. Mauskopf and T. Schmaltz (eds.), *Integrating History and Philosophy of Science*, pp. 13–28. Dordrecht: Springer.

Goodwin, W. (2021). "Mop-Up Work." In K. Brad Wray (ed.), *Interpreting Kuhn: Critical Essays*, pp. 85–104. Cambridge: Cambridge University Press.

Gordin, M. D. (2020). "When National Styles Were Stylish." *Historical Studies in the Natural Sciences* 50, 1–2: 11–16.

Graham, G. (1997). *The Shape of the Past: A Philosophical Approach to History*. Oxford: Oxford University Press.

Grégis, Fabien (2019a). "Assessing Accuracy in Measurement: The Dilemma of Safety versus Precision in the Adjustment of the Fundamental Physical Constants." *Studies in History and Philosophy of Science Part A* 74: 42–55.

(2019b). "On the Meaning of Measurement Uncertainty." *Measurement* 133, 41–46.

Grendler, P. F. (2002). *The Universities of the Italian Renaissance*. Baltimore: Johns Hopkins University Press.

Guarnieri, M. (2010). "The Early History of Radar." *IEEE Industrial Electronics Magazine* 4, 3: 36–42.

Hacking, Ian (1975). *The Emergence of Probability: A Philosophical Study of Early Ideas about Probability, Induction and Statistical Inference*. Cambridge: Cambridge University Press.

(1982). "Experimentation and Scientific Realism." *Philosophical Topics* 13, 1: 71–87.

(1983). *Representing and Intervening: Introductory Topics in the Philosophy of Natural Science*. Cambridge: Cambridge University Press.

(1992). "'Style' for Historians and Philosophers." *Studies in History and Philosophy of Science* 23, 1: 1–20.

(1999). *The Social Construction of What?* Cambridge, MA: Harvard University Press.

(2002). *Historical Ontology*. Cambridge, MA: Harvard University Press.

(2012). "Introductory Essay." In T. S. Kuhn (ed.), *The Structure of Scientific Revolutions: 50th Anniversary Edition*, pp. vii–xxxvii. Chicago: University of Chicago Press.

(2015). "Let's Not Talk about Objectivity." In Padovani Flavia, Alan Richardson, and Jonathan Y. Tsou (eds.), *Objectivity in Science: New Perspectives from Science and Technology Studies*, pp. 19–33. Cham: Springer.

(2016). "Paradigms." In R. J. Richards and L. Daston (eds.), *Kuhn's Structure of Scientific Revolutions at Fifty: Reflections on a Science Classic*, pp. 96–112. Chicago: University of Chicago Press.

Hagstrom, W. O. (1974). "Competition in Science." *American Sociological Review* 39, 1: 1–18.

Hakli, R. (2006). "Group Beliefs and the Distinction between Belief and Acceptance." *Cognitive Systems Research* 7, 2–3: 286–297.

Handelsman, Jo, Ebert-May, Diane, Beichner, Robert, Bruns, Peter, Chang, Amy, DeHaan, Robert, Gentile, Jim, et al. (2004). "Scientific Teaching." *Science* 304, 5670: 521–522. Accessed November 2, 2022. www.jstor.org/stable/3836701.

Hankins, T. L. (1985). *Science and the Enlightenment.* Cambridge: Cambridge University Press.

Hannaway, O., Kargon, R., and Davis, A. B. (1983). "Annual Meeting of the History of Science Society, 28–31 October 1982." *Isis* 74, 2 (June): 243–248.

Hanson, Norwood Russell (1958). *Patterns of Discovery.* Cambridge: Cambridge University Press.

Haraway, D. (1988). "Situated Knowledges: The Science Question in Feminism and the Privilege of Partial Perspective." *Feminist Studies* 14, 3: 575–599.

Hardwig, John (1985). "Epistemic Dependence." *Journal of Philosophy* 82, 7: 335–349.

   (1991). "The Role of Trust in Knowledge." *Journal of Philosophy* 88, 12: 693–708.

Harman, G. (1986). *Change in View: Principles of Reasoning.* Cambridge, MA: MIT Press.

Haufe, C. (2024). *Fruitfulness.* New York: Oxford University Press.

Heesen, A. te (2020). "Thomas S. Kuhn, Earwitness: Interviewing and the Making of a New History of Science." *Isis* 111, 1: 86–97.

Heilbron, John L. (1998). "Thomas Samuel Kuhn, 18 July 1922–17 June 1996." *Isis* 1998, 89: 505–515.

Helmholz, A. C. (1980). "Raymond Thayer Birge." *Physics Today* 33, 8: 68–70.

Hempel, Carl G. (1950). "Problems and Changes in the Empiricist Criterion of Meaning." *Revue Internationale de Philosophie* 41, 11: 41–63.

   (1966). *Philosophy of Natural Science.* Englewood Cliffs, NJ: Prentice-Hall.

Henrion, M., and Fischhoff, B. (1986). "Assessing Uncertainty in Physical Constants." *American Journal of Physics*, 54, 9: 791–798.

Hershberg, James (1993). *James B. Conant: Harvard to Hiroshima and the Making of the Nuclear Age.* New York: Alfred A. Knopf.

Hesse, Mary (1963). "Review of *The Structure of Scientific Revolutions* by Thomas S. Kuhn." *Isis* 54, 2: 286–287.

   (1980). "The Hunt for Scientific Reason." In Peter D. Asquith and Ronald N. Giere (eds.), *PSA 1980: Proceedings of the 1980 Biennial Meeting of the Philosophy of Science Association*, vol. 2, pp. 3–22. East Lansing, MI: Philosophy of Science Association.

Holbrook, J. B. (2005). "Assessing the Science–Society Relation: The Case of the US National Science Foundation's Second Merit Review Criterion." *Technology in Society* 27, 4: 437–451.

Hollinger, A. D. (1973). "T. S. Kuhn's Theory of Science and Its Implications for History." *The American Historical Review* 78, 2: 370–393.

Hooker, Brad (2015). "Rule Consequentialism," bibliography. *The Stanford Encyclopedia of Philosophy*, Stanford University, accessed November 4, 2022. https://plato.stanford.edu/entries/consequentialism-rule/.

Horwich, Paul (ed.) (1993). *World Changes: Thomas Kuhn and the Nature of Science*. Cambridge, MA: MIT Press.

Hoyningen-Huene, Paul (1987). "Context of Discovery and Context of Justification." *Studies in History and Philosophy of Science Part A* 18, 4: 501–515.

(1989/1993). *Reconstructing Scientific Revolutions: Thomas S. Kuhn's Philosophy of Science*, trans. Alex Levine. Chicago: University of Chicago Press.

(1992). "The Interrelations between the Philosophy, History and Sociology of Science in Thomas Kuhn's Theory of Scientific Development." *British Journal for the Philosophy of Science* 43: 487–501.

(1995). "Two Letters of Paul Feyerabend to Thomas S. Kuhn on a Draft of the Structure of Scientific Revolutions." *Studies in History and Philosophy of Science Part A* 26, 3: 353–387.

(2006a). "Context of Discovery versus Context of Justification and Thomas Kuhn." In Jutta Schickore and Friedrich Steinle (eds.), *Revisiting Discovery and Justification: Historical and Philosophical Perspectives on the Context Distinction*, pp. 119–131. Dordrecht: Springer.

(2006b). "More Letters by Paul Feyerabend to Thomas S. Kuhn on *Proto-Structure*." *Studies in History and Philosophy of Science* 37, 4: 610–632.

(2008). "Thomas Kuhn and the Chemical Revolution." *Foundations of Chemistry* 10: 101–115.

(2012). "Philosophical Elements in Thomas Kuhn's Historiography of Science." *Theoria* 75: 281–292.

(2015a). "Kuhn's Development before and after *Structure*." In W. J. Devlin and A. Bokulich (eds.), *Kuhn's Structure of Scientific Revolutions – 50 Years On*, pp. 185–195. Cham: Springer.

(2015b). *Systematicity: The Nature of Science*. Oxford: Oxford University Press.

Hufbauer, Karl (2012). "From Student of Physics to Historian of Science: T. S. Kuhn's Education and Early Career, 1940–1958." *Physics in Perspective* 14: 421–470.

Hull, David L. (1989). *Science as a Process*. Chicago: Chicago University Press.

(1996). "A Revolutionary Philosopher of Science." *Nature* 382 (July 18): 203–204.

Hwang, R.-C., et al. (2010). "Dropping the Brand of Edinburgh School: An Interview with Barry Barnes." *East Asian Science, Technology and Society: An International Journal* 4: 601–617.

Irzik, Gürol, and Grünberg, Teo (1995). "Carnap and Kuhn: Arch Enemies or Close Allies?" *British Journal for the Philosophy of Science* 46: 285–307.

Isaac, J. (2012). "Kuhn's Education: Wittgenstein, Pedagogy, and the Road to *Structure*." *Modern Intellectual History* 9: 89–107.

(2013). "Review of Thomas S Kuhn, *The Structure of Scientific Revolutions*, Fiftieth Anniversary Edition." *Isis* 104, 3: 658–659.

James, R. J. (1989). "A History of Radar." *IEE Review* 35, 9: 343–349.

Jamieson, Annie, and Radick, Gregory (2017). "Genetic Determinism in the Genetics Curriculum: An Exploratory Study of the Effects of Mendelian and Weldonian Emphases." *Science and Education* 26: 1261–1290.

Josephson, P. R. (1985). "Soviet Historians and *The Structure of Scientific Revolutions*." *Isis* 76: 551–559.

Kellert, Stephen H., Longino, Helen E., and Waters, C. Kenneth (2006). "Introduction: The Pluralist Stance." In Stephen H. Kellert, Helen E. Longino, and C. Kenneth Waters (eds.), *Scientific Pluralism*, pp. vii–xxix. Minneapolis: University of Minnesota Press.

Kelves, D. (1977). "The National Science Foundation and the Debate over Postwar Research Policy, 1942–1945." *Isis* 68: 5–26.

Khine, Myint Swe (ed.) (2013). *Critical Analysis of Science Textbooks: Evaluating Instructional Effectiveness*. New York: Springer.

Kindi, Vasso (2005a). "Should Science Teaching Involve the History of Science? An Assessment of Kuhn's View." *Science and Education* 14: 721–731.

(2005b). "The Relation of History of Science to Philosophy of Science in the *Structure of Scientific Revolutions* and Kuhn's Later Philosophical Work." *Perspectives on Science* 13, 4: 495–530.

(2012a). "Kuhn's Paradigms." In V. Kindi and T. Arabatzis (eds.), *Kuhn's The Structure of Scientific Revolutions Revisited*, pp. 91–111. London: Routledge.

(2012b). "The *Structure*'s Legacy: Not from Philosophy to Description." *Topoi* 32, 1: 81–89.

(2015). "The Role of Evidence in Judging Kuhn's Model: On the Mizrahi, Patton, Marcum Exchange." *Social Epistemology Review and Reply Collective* 4, 11: 25–33. http://wp.me/p1Bfg0-2sQ.

(2017). "The Kuhnian Straw Man." In M. Mizrahi (ed.), *The Kuhnian Image of Science: Time for a Decisive Transformation?*, pp. 95–112. London: Rowman and Littlefield.

Kindi, Vasso, and Arabatzis, Theodore (eds.) (2012). *Kuhn's* The Structure of Scientific Revolutions *Revisited*. New York: Routledge.

Kitcher, P. (1990). "The Division of Cognitive Labor." *The Journal of Philosophy* 87, 1: 5–22.

Klein, U. (2015). "The Revolution That Never Happened." *Studies in History and Philosophy of Science* 49: 80–90.

Klug, Anastasia, Müller, Olaf, Reinacher, Anna, Vine, Troy, and Yürüyen, Derya (eds.) (2021). *Goethe, Ritter und die Polarität: Geschichte und Kontroversen*. Leiden: Brill.

Kopec, M., and Miller, S. (2018). "Shared Intention Is Not Joint Commitment." *Journal of Ethics and Social Philosophy* 13, 2: 179–189.

Kourany, J. (1979). "The Nonhistorical Basis of Kuhn's Theory of Science." *Nature and System* 1: 46–59.

Kuhn, Thomas S. (1952). "Robert Boyle and Structural Chemistry in the Seventeenth Century." *Isis* 43, 1: 12–36.

(1957). *The Copernican Revolution: Planetary Astronomy in the Development of Western Thought*. Cambridge MA: Harvard University Press.

(1959). Thomas S. Kuhn Papers, MC240, Box 3: Folder 12, Lectures, General, 1957–1959.

(1959/1977). "The Essential Tension: Tradition and Innovation in Scientific Research." Reprinted in T. S. Kuhn (ed.), *The Essential Tension: Selected Studies in Scientific Tradition and Change*, pp. 225–239. Chicago: University of Chicago Press.

(1961/1977). "The Function of Measurement in Modern Physical Science."
In T. S. Kuhn's (ed.), *Essential Tension: Selected Studies in Scientific Tradition
and Change*, pp. 178–224. Chicago: University of Chicago Press. Originally
published in *Isis* 52, 2: 161–193.

(1961a). Letter from Thomas S. Kuhn to James B. Conant. Berkeley, CA,
June 29, 1961. Harvard University Archives.

(1961b). *Proto-Structure.* Unpublished early manuscript of *The Structure of
Scientific Revolutions.* Thomas S. Kuhn Archives, MIT.

(1962). *The Structure of Scientific Revolutions.* Chicago: University of Chicago
Press.

(1962/1970). *The Structure of Scientific Revolutions*, 2nd ed. Chicago: University
of Chicago Press.

(1962/1996). *The Structure of Scientific Revolutions.* 3rd ed. Chicago: University
of Chicago Press.

(1962/2012). *The Structure of Scientific Revolutions*, 4th ed., 50th Anniversary
Edition, with an Introductory Essay by Ian Hacking. Chicago: University of
Chicago Press.

(1963). "The Function of Dogma in Scientific Research." In A. C. Crombie
(ed.), *Scientific Change: Historical Studies in the Intellectual, Social and
Technical Conditions for Scientific Discovery and Technical Invention, from
Antiquity to the Present*, pp. 347–369. London: Heinemann; New York:
Basic Books.

(1964). Thomas S. Kuhn Papers, MC240, Box 3: Folder 13, Lectures,
General, 1960–1964.

(1968/1977). "The History of Science." In T. S. Kuhn (ed.), *The Essential
Tension: Selected Studies in Scientific Tradition and Change*, pp. 105–126.
Chicago: University of Chicago Press.

(1970/2012). "Postscript – 1969." In T. S. Kuhn (ed.), *The Structure of
Scientific Revolutions*, 2nd–4th ed., pp. 173–208. Chicago: University of
Chicago Press.

(1970a). "Logic of Discovery or Psychology of Research?" In I. Lakatos and A.
Musgrave (eds.), *Criticism and the Growth of Knowledge*, pp. 1–23. London:
Cambridge University Press.

(1970b). "Reflections on My Critics." In Imre Lakatos and Alan Musgrave
(eds.), *Criticism and the Growth of Knowledge*, pp. 231–278. Cambridge:
Cambridge University Press.

(1971). "Notes on Lakatos." In R. C. Buck and R. S. Cohen (eds.), *PSA 1970:
In Memory of Rudolf Carnap*, pp. 137–146. Dordrecht: Reidel.

(1971/1977). "The Relations between History and History of Science." In T.
S. Kuhn (ed.), *The Essential Tension: Selected Studies in Scientific Tradition and
Change*, pp. 127–161. Chicago: University of Chicago Press.

(1972). "Scientific Growth: Reflection on Ben-David's Scientific Role."
Minerva 10: 166–178.

(1973/1977). "Objectivity, Value Judgment, and Theory Choice." In T. S.
Kuhn (ed.), *The Essential Tension: Selected Studies in Scientific Tradition and
Change*, pp. 320–339. Chicago: University of Chicago Press.

(1974). "Discussion." In F. Suppe (ed.), *The Structure of Scientific Theories*, pp. 500–517. Urbana: University of Illinois Press.

(1974/1977). "Second Thoughts on Paradigms." In T. S. Kuhn (ed.), *The Essential Tension: Selected Studies in Scientific Tradition and Change*, pp. 293–319. Chicago: University of Chicago Press.

(1976). "Mathematical versus Experimental Traditions in the Development of Physical Science." *Journal of Interdisciplinary History* 7, 1: 1–31.

(1976/1977a). "Mathematical versus Experimental Traditions in the Development of Physical Science." In T. S. Kuhn (ed.), *The Essential Tension: Selected Studies in Scientific Tradition and Change*, pp. 31–65. Chicago: University of Chicago Press.

(1976/1977b). "The Relations between the History and the Philosophy of Science." In T. S. Kuhn (ed.), *The Essential Tension: Selected Studies in Scientific Tradition and Change*, pp. 3–20. Chicago: University of Chicago Press.

(ed.) (1977). *The Essential Tension: Selected Studies in Scientific Tradition and Change*. Chicago: University of Chicago Press.

(1977a). "Preface." In T. S. Kuhn (ed.), *The Essential Tension: Selected Studies in Scientific Tradition and Change*, pp. ix–xxiii. Chicago: University of Chicago Press.

(1978). *Black-Body Theory and the Quantum Discontinuity, 1894–1912*. Oxford: Oxford University Press.

(1979). "Foreword." In Ludwik Fleck, T. J. Trenn, and R. K. Merton, trans. F. Bradley and T. J. Trenn, *Genesis and Development of a Scientific Fact*, pp. vii–xi. Chicago: University of Chicago Press.

(1980). "The Halt and the Blind: Philosophy and History of Science." *British Journal for Philosophy of Science* 31: 181–192.

(1983). "Reflections on Receiving the John Desmond Bernal Award." *4S Review* 1, 4 (Winter): 26–30.

(1984). "Professionalization Recollected in Tranquility." *Isis* 75, 1: 29–32.

(1986). "The Histories of Science: Diverse Worlds for Diverse Audiences." *Academe* 72, 4: 29–33.

(1989/1993). "Foreword." In Paul Hoyningen-Huene (ed.), *Reconstructing Scientific Revolutions: Thomas S. Kuhn's Philosophy of Science*, pp. xi–xiii. Chicago: University of Chicago Press.

(1991/2000). "The Road since *Structure*." In J. Conant and J. Haugeland (eds.), *The Road since Structure: Philosophical Essays, 1970–1993, with an Autobiographical Interview*, pp. 90–104. Chicago: University of Chicago Press.

(1992/2000). "The Trouble with the Historical Philosophy of Science." In J. Conant and J. Haugeland (eds.), *The Road since Structure: Philosophical Essays, 1970–1993, with an Autobiographical Interview*, pp. 105–120. Chicago: University of Chicago Press.

(1993/2000). "Afterwords." In J. Conant and J. Haugeland (eds.), *The Road since Structure: Philosophical Essays, 1970–1993, with an Autobiographical Interview*, pp. 224–252. Chicago: University of Chicago Press.

(1997/2000). "A Discussion with Thomas S. Kuhn." In *The Road since Structure: Philosophical Essays, 1970–1993, with an Autobiographical*

Interview, ed. J. Conant and J. Haugeland, pp. 253–323. Chicago: University of Chicago Press.

(2000). *The Road since Structure: Philosophical Essays, 1970–1993, with an Autobiographical Interview*, ed. J. Conant and J. Haugeland. Chicago: University of Chicago Press.

(2016). "The Nature of Scientific Knowledge: An Interview with Thomas Kuhn (conducted by Skúli Sigurdsson)." In A. Blum et al. (eds.), *Shifting Paradigms. Thomas S. Kuhn and the History of Science*, pp. 17–30. Berlin: Max-Planck-Institute for the History of Science.

(2021). *The Quest for Physical Theory. Problems in the Methodology of Scientific Research*, ed. George A. Reisch. Boston: MIT Libraries.

(2022). *The Last Writings of Thomas S. Kuhn: Incommensurability in Science*, edited by Bojana Mladenovic. Chicago: University of Chicago Press.

Kusch, Martin (2010). "Hacking's Historical Epistemology: A Critique of Styles of Reasoning." *Studies in History and Philosophy of Science* 41, 2: 158–173.

(2016). "Relativism in Feyerabend's Later Writings." *Studies in History and Philosophy of Science* 57: 106–113.

(2021). *Relativism in the Philosophy of Science*. Cambridge: Cambridge University Press.

Kuukkanen, Jouni-Matti (2013). "Kuhn's Legacy: Theoretical and Philosophical Study of History." *Topoi* 32: 91–99.

Kuusela, O. (2008). *The Struggle against Dogmatism*. Cambridge, MA: Harvard University Press.

(2019). *Wittgenstein on Logic as the Method of Philosophy*. Oxford: Oxford University Press.

Lakatos, Imre (1970). "Falsification and the Methodology of Scientific Research Programmes." In Imre Lakatos and Alan Musgrave (eds.), *Criticism and the Growth of Knowledge: Proceedings of the International Colloquium in the Philosophy of Science, London, 1965*, vol. 4, pp. 91–196. Cambridge: Cambridge University Press.

(1971). "History of Science and Its Rational Reconstructions." In R. C. Buck and R. S. Cohen (eds.), *PSA: Proceedings of the Biennial Meeting of the Philosophy of Science Association 1970*, pp. 91–136. Dordrecht: Reidel.

Lakatos, Imre, and Musgrave, Alan (eds.) (1970). *Criticism and the Growth of Knowledge*. Cambridge: Cambridge University Press.

Latour, B. (1987). *Science in Action: How to Follow Scientists and Engineers through Society*. Cambridge, MA: Harvard University Press.

Latour, Bruno, and Woolgar, Steve (1979). *Laboratory Life: The Social Construction of Scientific Facts*. Princeton, NJ: Princeton University Press.

Laudan, Larry (1984). *Science and Values: The Aims of Science and Their Role in Scientific Debate*. Berkeley: University of California Press.

Levine, Alex (1999). "Scientific Progress and the Fregean Legacy." *Mind and Language* 14, 3: 263–290.

(2000). "Which Way Is Up? Thomas S. Kuhn's Analogy to Conceptual Development in Childhood." *Science and Education* 9: 107–122.

Levine, Alex, and Adriana Novoa (2012). *¡Darwinistas! The Construction of Evolutionary Thought in Nineteenth Century Argentina*, ed. Mordechai Feingold. Leiden: Brill.

List, C., and Pettit, P. (2011). *Group Agency: The Possibility, Design, and Status of Corporate Agents*. Oxford: Oxford University Press.

Lloyd, Elisabeth A. (1996). "The Anachronistic Anarchist." *Philosophical Studies* 81, 2–3: 247–261.

Longino, Helen E. (1990). *Science as Social Knowledge: Values and Objectivity in Scientific Inquiry*. Princeton, NJ: Princeton University Press.

Loving, Cathleen C., and Cobern, William W. (2000). "Invoking Thomas Kuhn: What Citation Analysis Reveals about Science Education." *Science and Education* 9: 187–206.

Luchetti, Michele (2021). "Coordination in Theory Extension: How Reichenbach Can Help Us Understand Endogenization in Evolutionary Biology." *Synthese* 1999, 3–4: 9855–9880.

Ludwig, David, and Ruphy, Stéphanie (2021). "Scientific Pluralism." In Edward N. Zalta (ed.), *Stanford Encyclopedia of Philosophy* (Winter 2021 edition). https://plato.stanford.edu/archives/win2021/entries/scientific-pluralism.

Lynch, M. (2012). "Self-Exemplifying Revolutions? Notes on Kuhn and Latour." *Social Studies of Science* 42, 3: 449–455.

Maasen, S., and Weingart, P. (2000). *Metaphors and the Dynamics of Knowledge*. London: Routledge.

Mach, Ernst (1911). *History and Root of the Principle of the Conservation of Energy*, trans. P. E. B. Jourdain. Chicago: Open Court.

Magnus, P. D. (2013). "Philosophy of Science in the 21st Century." *Metaphilosophy* 44, 1–2: 48–52.

Malpas, Jeff (2021). "Donald Davidson." In Edward N. Zalta (ed.), *Stanford Encyclopedia of Philosophy* (Fall 2021 Edition).

Mandelbaum, Maurice (1969). "Two Moot Issues in Mill's Utilitarianism." In J. B. Schneewind (ed.), *Mill: A Collection of Critical Essays*, pp. 206–233. Notre Dame, IN: Notre Dame University Press.

Marcum, James A. (2015). *Thomas Kuhn's Revolutions: A Historical and an Evolutionary Philosophy of Science?* London: Bloomsbury.

Masterman, Margaret (1970). "The Nature of Paradigms." In I. Lakatos and A. Musgrave (eds.), *Criticism and the Growth of Knowledge: Proceedings of the International Colloquium in the Philosophy of Science, London, 1965*, vol. 4, pp. 59–89. Cambridge: Cambridge University Press.

Mathiesen, K. (2006). "The Epistemic Features of Group Belief." *Episteme* 2, 3: 161–175.

Matthews, Michael R. (2000). "Editorial." *Science and Education* 9: 1–10.

(2003). "Thomas Kuhn's Impact on Science Education: What Lessons Can Be Learned?" *Science Education* 88, 1: 90–118.

(2022). "Thomas Kuhn and Science Education: A Troubled Connection." *HP&ST Newsletter*, April: 6–24.

Mayoral, Juan Vicente (2009). "Intensions, Belief and Science: Kuhn's Early Philosophical Outlook (1940–1945)." *Studies in History and Philosophy of Science Part A* 40, 2: 175–184.

(2011). "Hacia una reinterpretación de la ciencia normal: Kuhn y la física de su tiempo (1940–1951)." *Asclepio* 63, 1: 221–248.

(2017). *Thomas S. Kuhn: La búsqueda de La Estructura*. Zaragoza: Prensas de la Universidad de Zaragoza.

Mayr, E. (1972). "The Nature of the Darwinian Revolution." *Science* 176: 981–989.

McMullin, Ernan (1984). "A Case for Scientific Realism." In J. Leplin (ed.), *Scientific Realism*, pp. 8–40. Berkeley: University of California Press.
  (1993). "Rationality and Paradigm Change in Science." In Paul Horwich (ed.), *World Changes: Thomas Kuhn and the Nature of Science*, pp. 55–78. Cambridge, MA: MIT Press.

Melogno, Pablo (2022). "From Externalism to Internalism: The Historiographical Development of Thomas Kuhn." *Foundations of Science* 27, 2: 371–385.

Mercier, J. (1924). "De la synchronisation harmonique et multiple." *Journal de Physique et le Radium* 5: 168–179.

Michelson, A. A. (1927). "Measurement of the Velocity of Light between Mount Wilson and Mount San Antonio." *The Astrophysical Journal* 65, 1: 1–14.

Mizrahi, M. (ed.) (2018). *The Kuhnian Image of Science: Time for a Decisive Transformation?* London: Rowman and Littlefield International.

Mladenović, B. (2007). "'Muckracking in History': The Role of the History of Science in Kuhn's Philosophy." *Perspectives on Science* 15, 3: 261–294.

Morris, E. (2018). *The Ashtray; or, The Man Who Denied Reality.* Chicago: University of Chicago Press.

Moss, J. D. (1993). *Novelties in the Heavens: Rhetoric and Science in the Copernican Controversy.* Chicago: University of Chicago Press.

Nash, L. K. (1956). "The Origin of Dalton's Chemical Atomic Theory." *Isis* 47, 2: 101–116.

National Science Board (2007). *Enhancing Support of Transformative Research at the National Science Foundation.* Arlington, VA: National Science Foundation.

NGSS Lead States (2013). "Appendix H, Understanding the Scientific Enterprise: The Nature of Science in the Next Generation Science Standards." In *Next Generation Science Standards for States by States*, pp. 1–10. Washington, DC: National Academies Press.

Niaz, Mansoor (2001). "A Rational Reconstruction of the Origin of the Covalent Bond and Its Implications for General Chemistry Textbooks." *International Journal of Science Education* 23, 6: 623–645.
  (2016). *Chemistry Education and Contributions from History and Philosophy of Science.* Dordrecht: Springer.

Niaz, Mansoor, and Coştu, Bayram (2013). "Analysis of Turkish General Chemistry Textbooks Based on a History and Philosophy of Science Perspective." In Myint Swe Khine (ed.), *Critical Analysis of Science Textbooks: Evaluating Instructional Effectiveness*, pp. 199–218. New York: Springer.

Niaz, Mansoor, and Maza, Arelys (2011). *Nature of Science in General Chemistry Textbooks.* Springer Briefs in Education. Accessed November 6, 2022. DOI: 10.1007/978-94-007-1920-0_1.

Nickles, Thomas (ed.) (2003a). *Thomas Kuhn.* Cambridge: Cambridge University Press.
  (2003b). "Normal Science: From Logic to Case-Based and Model-Based Reasoning." In Thomas Nickles (ed.), *Thomas Kuhn*, pp. 142–177. Cambridge: Cambridge University Press.

(2012). "Some Puzzles about Kuhn's Exemplars." In V. Kindi and T. Arabatzis (eds.), *Kuhn's The Structure of Scientific Revolutions Revisited*, pp. 112–133. London: Routledge.

(2017). "Historicist Theories of Scientific Rationality." In Edward N. Zalta (ed.), *Stanford Encyclopedia of Philosophy* (Spring 2021 Edition). https://plato .stanford.edu/entries/rationality-historicist/.

(2021). "Kuhn on Scientific Discovery as Endogenous." In K. Brad Wray (ed.), *Interpreting Kuhn: Critical Essays*, pp. 185–201. Cambridge: Cambridge University Press.

Nola, Robert (2000). "Saving Kuhn from the Sociologists of Science." *Science and Education* 9, 1: 77–90.

(2003). *Rescuing Reason: A Critique of Anti-Rationalist Views of Science and Knowledge*. Dordrecht: Kluwer.

North, J. (1992). "The Quadrivium." In H. de Ridder-Symoens (ed.), *A History of the University in Europe, vol. 1: Universities in the Middle Ages*, pp. 337–359. Cambridge: Cambridge University Press.

Novoa, Adriana, and Levine, Alex (2010). *From Man to Ape: Darwinism in Argentina, 1870–1920*. Chicago: University of Chicago Press.

Nye, Mary Jo (2014). "Science and Politics in the Philosophy of Science: Popper, Kuhn, and Polanyi." In M. Epple and C. Zittel (eds.), *Science as Cultural Practice*, pp. 201–216. Berlin: De Gruyter.

(2019). "Shifting Trends in Modern Physics, Nobel Recognition, and the Histories That We Write." *Physics in Perspective* 21: 3–22.

O'Donohue, William, and Willis, Brendan (2018). "Problematic Images of Science in Undergraduate Psychology Textbooks: How Well Is Science Understood and Depicted?" *Archives of Scientific Psychology* 6, 1: 51–62. Accessed October 10, 2022. doi:10.1037/arc0000040.

Olby, Robert C. (1966/1985). *Origins of Mendelism*, 2nd ed. Chicago: University of Chicago Press.

Olesko, Kathryn M. (2006). "Science Pedagogy as a Category of Historical Analysis: Past, Present, and Future." *Science and Education* 15: 863–880.

Oliveira, J.C. de Pinto (2021). "Kuhn and Logical Positivism: On the Image of Science and the Image of Philosophy." In K. Brad Wray (ed.), *Interpreting Kuhn: Critical Essays*, pp. 65–82. Cambridge: Cambridge University Press.

Oreskes, N., and Conway, E. M. (2010). *Merchants of Doubt: How a Handful of Scientists Obscured the Truth on Issues from Tobacco Smoke to Climate Change*. New York: Bloomsbury.

Pang, A. S. K. (1997). "Visual Representation and Post-Constructivist History of Science." *Historical Studies in the Physical and Biological Sciences* 28, 1: 139–171.

Papineau, D. (2021). "Naturalism." In E. Zalta (ed.), *Stanford Encyclopedia of Philosophy*. https://plato.stanford.edu/archives/sum2021/entries/naturalism/.

Patton, Lydia (2018). "Kuhn, Pedagogy, and Practice: A Local Reading of *Structure*." In M. Mizrahi (ed.), *The Kuhnian Image of Science: Time for a Decisive Transformation?*, pp. 113–130. London: Rowman and Littlefield.

(2021). "Kuhn's Kantian Dimensions." In K. Brad Wray (ed.), *Interpreting Kuhn: Critical Essays*, pp. 27–44. Cambridge: Cambridge University Press.

Pettigrove, G. (2016). "Changing Our Mind." In M. Brady and M. Fricker (eds.), *The Epistemic Life of Groups*, pp. 111–130. Oxford: Oxford University Press.

Pickering, Andrew (2001). "Reading the *Structure*." *Perspectives on Science* 9, 4: 499–510.

(2012). "The World since Kuhn," *Social Studies of Science* 42, 3: 467–473.

Pigliucci, Massimo, and Boudry, Maarten (eds.) (2013). *Philosophy of Pseudoscience: Reconsidering the Demarcation Problem*. Chicago: University of Chicago Press.

Pinch, Trevor (2017). "All Pumped Up about the Sociology of Scientific Knowledge." *Isis* 108, 1: 127–129.

Pirozelli, P. (2021). "The Structure of Scientific Controversies: Thomas Kuhn's Social Epistemology." *Filosofia Unisinos: Unisinos Journal of Philosophy* 22, 3: 1–17.

Pitt, Joseph C. (2001). "The Dilemma of Case Studies: Toward a Heraclitian Philosophy of Science." *Perspectives on Science* 9, 4: 372–383.

Popper, Karl R. (1935/1959). *The Logic of Scientific Discovery*. London: Hutchinson.

(1963). *Conjectures and Refutations: The Growth of Scientific Knowledge*. London: Routledge.

(1970). "Normal Science and Its Dangers." In Imre Lakatos and Alan Musgrave (eds.), *Criticism and the Growth of Knowledge: Proceedings of the International Colloquium in the Philosophy of Science, London, 1965*, vol. 4, pp. 51–58. Cambridge: Cambridge University Press.

Porter, A., Cohen, A., David Roessner, J., and Perreault, M. J. S. (2007). "Measuring Researcher Interdisciplinarity." *Scientometrics* 72, 1: 117–147.

Porter, A., and Rafols, I. (2009). "Is Science Becoming More Interdisciplinary? Measuring and Mapping Six Research Fields over Time." *Scientometrics* 81, 3: 719–745.

Porter, A., and Rossini, F. (1985). "Peer Review of Interdisciplinary Research Proposals." *Science, Technology, and Human Values* 10, 3: 33–38.

Poulsen, M.-B. J. (2001). "Competition and Cooperation: What Roles in Science Dynamics?" *International Journal of Technology, Policy and Management* 22, 7–8: 782–793.

Preston, John (1997a). *Feyerabend: Philosophy, Science and Society*. Cambridge: Polity Press.

(1997b). "Feyerabend's Retreat from Realism." *Philosophy of Science* 64 (Proceedings): S421–S431.

Preston, John, Munévar, Gonzalo, and Lamb, David (eds.) (2000). *The Worst Enemy of Science? Essays in Memory of Paul Feyerabend*. New York: Oxford University Press.

Price, Derek de Solla (1963). *Little Science, Big Science*. New York: Columbia University Press.

Putnam, Hilary (1975). "The Meaning of 'Meaning.'" *Minnesota Studies in the Philosophy of Science* 7: 131–193.

Quine, W. V. O. (1951). "Two Dogmas of Empiricism." *Philosophical Review* 60: 20–43.

(1960). *Word and Object*. Cambridge, MA: MIT Press.

(1969). "Epistemology Naturalized." In *Ontological Relativity and Other Essays*, pp. 69–90. New York: Columbia University Press.

(1971). "Epistemology Naturalized." In L. Gabriel (ed.), *Akten des XIV. Internationalen Kongresses für Philosophie Wien, 2.–9. September 1968*, vol. 6, pp. 87–103. Vienna: Herder.

Quinn, T. (2011). "Time, the SI and the Metre Convention." *Metrologia* 48, 4: S121.

Radick, Gregory (2016). "Teach Students the Biology of Their Time." *Nature* 533 (May 19): 293.

Rasmussen, S. B. (2014). *Potentialeledelse: Om strategisk ledelse i fagprofessionelle organisationer*. Copenhagen: Barlebo Forlag.

Reingold, Nathan (1980). "Through Paradigm-Land to a Normal History of Science." *Social Studies of Science* 10, 4: 475–496.

(1987). "Vannevar Bush's New Deal for Research: Or the Triumph of the Old Order." *Historical Studies in the Physical and Biological Sciences* 17, 2: 299–344.

Reisch, George A. (1991). "Did Kuhn Kill Logical Positivism?" *Philosophy of Science* 58: 264–277.

(1998). "Pluralism, Logical Empiricism, and the Problem of Pseudoscience." *Philosophy of Science* 65, 2: 333–348.

(2016). "Aristotle in the Cold War: On the Origins of Thomas Kuhn's *Structure of Scientific Revolutions*." In R. J. Richards and L. Daston (eds.), *Kuhn's Structure of Scientific Revolutions at Fifty: Reflections on a Science Classic*, pp. 12–29. Chicago: University of Chicago Press.

(2019). *The Politics of Paradigms: Thomas S. Kuhn, James B. Conant, and the Cold War "Struggle for Men's Minds."* Albany: State University of New York Press.

(2021a). "A Public Intellectual and a Private Scholar: On Thomas Kuhn, James B. Conant, and the Place of History and Philosophy of Science in Postwar America." In K. Brad Wray (ed.), *Interpreting Kuhn: Critical Essays*, pp. 45–64. Cambridge: Cambridge University Press.

(2021b). "Thomas Kuhn's Quest for Physical Theory: Editor's Introduction." In George Reisch (ed.), *The Quest for Physical Theory: Problems in the Methodology of Scientific Research*. Cambridge, MA: MIT Libraries Digital Collections.

Renn, J. (2015). "The History of Science and the Globalization of Knowledge." In T. Arabatzis, J. Renn, and A. Simões (eds.), *Relocating the History of Science*, pp. 241–252. Cham: Springer.

Renzi, B. G. (2009). "Kuhn's Evolutionary Epistemology and Its Being Undermined by Inadequate Biological Concepts." *Philosophy of Science* 76: 143–159.

Reydon, T. A. C., and Hoyningen-Huene, P. (2010). "Discussion: Kuhn's Evolutionary Analogy in *The Structure of Scientific Revolutions* and 'The Road since *Structure*.'" *Philosophy of Science* 77: 468–476.

Richards, Robert J., and Daston, Lorraine (2016). *Kuhn's* Structure of Scientific Revolutions *at Fifty: Reflections on a Science Classic*. Chicago: University of Chicago Press.

Richardson, Alan W. (1998). *Carnap's Construction of the World: The* Aufbau *and the Emergence of Logical Empiricism*. Cambridge: Cambridge University Press.

(2002). "Narrating the History of Reason Itself: Friedman, Kuhn, and a Constitutive A Priori for the Twenty-First Century." *Perspectives on Science* 10, 3: 253–274.

(2006). "The Many Unities of Science: Politics, Semantics, and Ontology." In Stephen H. Keller, Helen E. Longino, and C. Kenneth Waters (eds.), *Scientific Pluralism*, pp. 1–25. Minneapolis: University of Minnesota Press.

Rolin, K. (2008). "Science as Collective Knowledge." *Cognitive Systems Research* 9, 1–2: 115–124.

Rorty, R. (1997/1999). "Thomas Kuhn, Rocks and the Laws of Physics." In *Philosophy and Social Hope*, pp. 175–189. London: Penguin Books.

Rosa, E. B., and Dorsey, N. E. (1907). "A New Determination of the Ratio of the Electromagnetic to the Electrostatic Unit of Electricity." *Bulletin of the Bureau of Standards* 3, 3: 433–540.

Roth, Paul A. (2013). "The Silence of the Norms: The Missing Historiography of *The Structure of Scientific Revolutions*." *Studies in History and Philosophy of Science* 44, 4: 545–552.

Rothleitner, C., and Schlamminger, S. (2017). "Invited Review Article: Measurements of the Newtonian Constant of Gravitation, G." *Review of Scientific Instruments* 88, 11: 111101.

Rowley, W. R. C. (1984). "The Definition of the Metre: From Polar Quadrant to the Speed of Light." *Physics Bulletin* 35, 7: 282.

Sahlins, M. D. (1964). "Review of *The Structure of Scientific Revolutions*." *Scientific American*, 210, 5 (May): 142–144.

Sankey, Howard (2011). "Epistemic Relativism and the Problem of the Criterion." *Studies in History and Philosophy of Science A* 42, 4: 562–570.

(2012). "Scepticism, Relativism and the Argument from the Criterion." *Studies in History and Philosophy of Science A* 43, 1: 182–190.

(2020). "The Relativistic Legacy of Kuhn and Feyerabend." In Martin Kusch (ed.), *The Routledge Handbook of the Philosophy of Relativism*, pp. 379–387. London: Routledge.

Saul, Stephanie (2022). "N.Y.U. Students Were Failing Class. The Professor Lost His Job." *New York Times*, Section A (October 3): 1.

Scheffler, Israel (1967). *Science and Subjectivity*. Indianapolis: Bobbs-Merrill.

Schickore, Jutta (2011). "More Thoughts on HPS: Another 20 Years Later." *Perspectives on Science* 19, 3: 453–481.

Schickore, Jutta, and Steinle, Friedrich (2006). "Introduction: Revisiting the Context Distinction." In J. Schickore and F. Steinle (eds.), *Revisiting Discovery and Justification*, pp. 7–19. Dordrecht: Springer.

Schindler, Samuel (2013). "The Kuhnian Mode of HPS." *Synthese* 190, 18: 4137–4154.

Schwab, Joseph J. (1962). *The Teaching of Science as Enquiry*. Cambridge, MA: Harvard University Press.

Searle, John R. (1993). "Rationality and Realism, What Is at Stake?" *Daedalus* 122, 4: 55–83.

Secord, James A. (2004). "Knowledge in Transit." *Isis* 95, 4: 654–672.

Seidel, Markus (2013). "Between Relativism and Absolutism? – The Failure of Kuhn's Moderate Relativism." In M. Hoeltje, T. Spitzley, and W. Spohn (eds.), *Was dürfen wir glauben? Was sollen wir tun? Sektionsbeiträge des achten internationalen Kongresses der Gesellschaft für Analytische Philosophie e.V.*, pp. 172–185. Online Publication of the University of Duisburg–Essen (DuEPublico), https://duepublico2.uni-due.de/servlets/MCRFileNodeServlet/duepublico_derivate_00033085/GAP8_Proceedings.pdf.

(2014). *Epistemic Relativism: A Constructive Critique.* Basingstoke: Palgrave Macmillan.

(2021a). "Portraying the Relativist Spectrum." *Metascience* 30: 357–360.

(2021b). "Kuhn's Two Accounts of Scientific Disagreement in Science: An Interpretation and Critique." *Synthese* 198: 6023–6051.

Šešelja, D., and Straßer, C. (2013). "Kuhn and the Question of Pursuit Worthiness." *Topoi* 32, 1: 9–19.

Sewell, W. H., Jr. (2005). *Logics of History: Social Theory and Social Transformation.* Chicago: University of Chicago Press.

Shan, Yafeng (2020). "Kuhn's 'Wrong Turning' and Legacy Today." *Synthese* 197, 1: 381–406.

Shapere, Dudley (1964). "Review of *The Structure of Scientific Revolutions.*" *The Philosophical Review* 73, 3: 383–394.

Shapin, Steven (1980). "Social Uses of Science." In G. S. Rousseau and R. Porter (eds.), *The Ferment of Knowledge: Studies in the Historiography of Eighteenth-Century Science*, pp. 93–139. Cambridge: Cambridge University Press.

(1992). "Discipline and Bounding: The History and Sociology of Science as Seen through the Externalism-Internalism Debate." *History of Science* 30: 333–369.

(2008). *The Scientific Life: A Moral History of a Late Modern Vocation.* Chicago: University of Chicago Press.

(2015). "Kuhn's Structure: A Moment in Modern Naturalism." In W. J. Devlin and A. Bokulich (eds.), *Kuhn's Structure of Scientific Revolutions – 50 Years On*, pp. 11–21. Cham: Springer.

Shapin, Steven, and Schaffer, Simon (1985). *Leviathan and the Air Pump.* Princeton, NJ: Princeton University Press.

Shapiro, S. J. (2002). "Law, Plans, and Practical Reason." *Legal Theory* 8, 4: 387–441.

Sharrock, W., and Read, R. (2002). *Kuhn: Philosopher of Scientific Revolutions.* Cambridge: Polity Press.

Shaw, Jamie, and Bschir, Karim (2021). "Introduction: Paul Feyerabend's Philosophy in the Twenty-First Century." In Karim Bschir and Jamie Shaw (eds.), *Interpreting Feyerabend: Critical Essays*, pp. 1–10. Cambridge: Cambridge University Press.

Shipman, Harry L. (2000). "Thomas Kuhn's Influence on Astronomers." *Science and Education* 9: 161–171.

Siegel, Harvey (1990). *Educating Reason: Rationality, Critical Thinking and Education.* New York: Routledge.

(2011). "Relativism, Incoherence, and the Strong Programme." In R. Schantz and M. Seidel (eds.), *The Problem of Relativism in the Sociology of (Scientific) Knowledge*, pp. 41–64. Frankfurt: Ontos.

Sigurdsson, S. (2016). "The Nature of Scientific Knowledge: An Interview with Thomas S. Kuhn." In A. Blum, K. Gavroglu, C. Joas, and J. Renn (eds.), *Shifting Paradigms: Thomas S. Kuhn and the History of Science*, pp. 17–30. Berlin: Edition Open Access.

Smart, J. J. C. (1956). "Extreme and Restricted Utilitarianism." *Philosophical Quarterly* 6: 344–354.

Smith, George E. (2010). "Revisiting Accepted Science: The Indispensability of the History of Science," ed. Sherwood J. B. Sugden. *Monist* 93, 4: 545–579.

   (2014). "Closing the Loop: Testing Newtonian Gravity, Then and Now." In Z. Biener and E. Schliesser (eds.), *Newton and Empiricism*, pp. 262–351. Oxford: Oxford University Press.

Sober, Elliott (2009). *Core Questions in Philosophy: A Text with Readings*, 5th ed. Englewood Cliffs, NJ: Prentice Hall.

Soler, Léna (2022). "Incompatible Judgments about the Operational Coherence and Success of Competing Epistemic Systems of Practice." Presentation at the conference "Can Realism Allow Pluralism and Contingency," Université de Lorraine, October 19.

Solomon, Miriam (2001). *Social Empiricism*. Cambridge, MA: MIT Press.

Solomon, Miriam, and Richardson, Alan (2005). "A Critical Context for Longino's Critical Contextual Empiricism." *Studies in History and Philosophy of Science* 36, 1: 211–222.

"Stem Futures: The Future Substance of STEM Education Project." STEM Futures. Science Education Resource Center at Carleton College. Last revised October 7, 2022. Accessed October 31, 2022. https://serc.carleton.edu/stemfutures/index.html.

Stewart, David, Blocker, H. Gene, and Petrik, James (2013). *Fundamentals of Philosophy*, 8th ed. Boston: Pearson.

Stopes-Roe, H. V. (1964). "Review of *The Structure of Scientific Revolutions*." *British Journal for the Philosophy of Science* 15, 58: 158–161.

Struik, D. (1967). *A Concise History of Mathematics*. London: Bell and Sons.

Sunstein, C. R. (2002). "The Law of Group Polarization." *Journal of Political Philosophy* 10, 2: 175–195.

Suppe, F. (ed.) (1974). *The Structure of Scientific Theories*, pp. 459–482. Urbana: University of Illinois Press.

Suppes, Patrick (1978). "The Plurality of Science." In Peter Asquith and Ian Hacking (eds.), *PSA 1978: Proceedings of the 1978 Biennial Meeting of the Philosophy of Science Association*, vol. 2, pp. 3–16. East Lansing, MI: Philosophy of Science Association.

Tal, E. (2011). "How Accurate Is the Standard Second?" *Philosophy of Science* 78, 5: 1082–1096.

Tanghe. K. B., Pauwels, L., De Tiège, A., and Braeckman, J. (2021). "Interpreting the History of Evolutionary Biology through a Kuhnian Prism: Sense or Nonsense?" *Perspectives on Science* 29, 1: 1–35.

Tarski, Alfred (1956). "The Concept of Truth in Formalized Languages." In John Corcoran (ed.) and Translated by J. W. Woodger, *Logic, Semantics, Metamathematics: Papers from 1923 to 1938*. Oxford: Oxford University Press.

Taylor, B. N., Parker, W. H. and Langenberg, D. N. (1969). "Determination of e/h, Using Macroscopic Quantum Phase Coherence in Superconductors:

Implications for Quantum Electrodynamics and the Fundamental Physical Constants." *Reviews of Modern Physics* 41, 3: 375–496.

Taylor, Barry N. (1971). "Comments on Least-Squares Adjustments of the Constants." In D. N. Langenberg and B. N. Taylor (eds.), *Precision Measurement and Fundamental Constants*, pp. 495–498. Washington, DC: National Bureau of Standards.

Thackray, A. (1984). "Sarton, Science, and History." *Isis* 75, 1: 7–9.

Tiesinga, E., Mohr, P. J., Newell, D. B., and Taylor, B. (2021). "CODATA Recommended Values of the Fundamental Physical Constants: 2018." *Reviews of Modern Physics* 93, 2: 025010.

Toon, Ernest R., and Ellis, George L. (1978). *Foundations of Chemistry*. New York: Holt, Rinehart and Winston.

Toulmin, Stephen (1970). "Does the Distinction between Normal and Revolutionary Science Hold Water?" In I. Lakatos and A. Musgrave (eds.), *Criticism and the Growth of Knowledge*, pp. 29–48. London: Cambridge University Press.

Tsou, Jonathan Y. (2003a). "A Role for Reason in Science." *Dialogue: Canadian Philosophical Review* 42, 3: 573–598.

(2003b). "Reconsidering Feyerabend's 'Anarchism.'" *Perspectives on Science* 11, 2: 208–235.

(2010). "Putnam's Account of Apriority and Scientific Change: Its Historical and Contemporary Interest." *Synthese* 176, 3: 429–445.

(2015). "Reconsidering the Carnap-Kuhn Connection." In W. J. Devlin and A. Bokulich (eds.), *Kuhn's Structure of Scientific Revolutions – 50 Years On*, pp. 51–69. Cham: Springer.

(forthcoming). "Philosophical Naturalism and Empirical Approaches to Philosophy." In Marcus Rossberg (ed.), *The Cambridge Handbook of Analytic Philosophy*. Cambridge: Cambridge University Press.

Tuomela, Raimo (1992). "Group Beliefs." *Synthese* 91, 3: 285–318.

Urmson, J. O. (1953). "The Interpretation of the Moral Philosophy of J. S. Mill." *Philosophical Quarterly* 3: 33–39.

van Berkel, Berry, de Vos, Wobbe, Verdonk, Adri H., and Pilot, Albert (2000). "Normal Science Education and Its Dangers: The Case of School Chemistry." *Science and Education* 9: 123–159.

Wagenknecht, Susann (2016). *A Social Epistemology of Research Groups: Collaboration in Scientific Practice*. London: Palgrave Macmillan.

Walker, T. C. (2010). "The Perils of Paradigm Mentalities: Revisiting Kuhn, Lakatos, and Popper." *Perspectives on Politics* 8, 2: 433–451.

Warwick, Andrew (2003). *Masters of Theory: Cambridge and the Rise of Mathematical Physics*. Chicago: University of Chicago Press.

Watkins, John (1970). "Against 'Normal Science.'" In Imre Lakatos and Alan Musgrave (eds.), *Criticism and the Growth of Knowledge*, pp. 25–37. Cambridge: Cambridge University Press.

Weatherall, James O., and Gilbert, Margaret (2016). "Collective Belief, Kuhn, and the String Theory Community." In M. Brady and M. Fricker (eds.), *The Epistemic Life of Groups*, pp. 191–218. Oxford: Oxford University Press.

Weinberg, S. (1998). "The Revolution That Didn't Happen." *New York Review of Books* (October 8): 48–52.

Williams, David Lay (2020). "Was Slavery a 'Necessary Evil'? Here's What John
    Stuart Mill Would Say." *Washington Post* (July 30). Accessed November 4,
    2022. www.washingtonpost.com/politics/2020/07/30/was-slavery-necessary-
    evil-heres-what-john-stuart-mill-would-say/.
Wittgenstein, Ludwig (1953/2009). *Philosophical Investigations*, 4th ed., ed. E.
    Anscombe. Oxford: Blackwell.
    (1977/1994). *Culture and Value*, ed. G. H. von Wright, H. Nyman, and A.
    Pichler. Oxford: Blackwell.
Woolgar, Steve (1981). "Discovery: Logic and Sequence in a Scientific Text."
    In Karin D. Knorr, Roger Krohn, and Richard Whitley (eds.), *The Social
    Process of Scientific Investigation*, pp. 239–268. Boston: D. Reidel.
Worrall, John (1989). "Structural Realism: The Best of Both Worlds?" *Dialectica*
    43, 1–2: 99–124.
    (2003). "Normal Science and Dogmatism, Paradigms and Progress: Kuhn
    'versus' Popper and Lakatos." In Thomas Nickles (ed.), *Thomas Kuhn*,
    pp. 65–100. Cambridge: Cambridge University Press.
Wray, K. Brad (2001). "Collective Belief and Acceptance." *Synthese* 129: 319–333.
    (2003). "Is Science Really a Young Man's Game?" *Social Studies of Science* 33,
    1: 137–149.
    (2004). "An Examination of the Contributions of Young Scientists in New
    Fields." *Scientometrics* 61, 1: 117–128.
    (2007). Who Has Scientific Knowledge?" *Social Epistemology* 21, 3: 337–347.
    (2010a). "Kuhn's Constructionism." *Perspectives on Science* 18, 3: 311–327.
    (2010b). "Rethinking the Size of Scientific Specialties: Correcting Price's
    Estimate." *Scientometrics* 83, 2: 471–476.
    (2011a). "Kuhn and the Discovery of Paradigms." *Philosophy of the Social
    Sciences* 41, 3: 380–397.
    (2011b). *Kuhn's Evolutionary Social Epistemology*. Cambridge: Cambridge
    University Press.
    (2015). "Kuhn's Social Epistemology and the Sociology of Science." In W. J.
    Devlin and A. Bokulich (eds.), *Kuhn's Structure of Scientific Revolutions –
    50 Years On*, pp. 167–183. Dordrecht: Springer.
    (2016a). "Kuhn's Influence on the Social Sciences." In L. McIntyre and A.
    Rosenberg (eds.), *The Routledge Companion to Philosophy of Social Science*,
    pp. 65–75. London: Routledge.
    (2016b). "The Influence of James B. Conant on Kuhn's *Structure of Scientific
    Revolutions*." *HOPOS: The Journal of the International Society for the History of
    Philosophy of Science* 6, 1: 1–23.
    (2018). "Thomas Kuhn and the T. S. Kuhn Archives at MIT." *OUPBlog*,
    May 27. Accessed September 10, 2021. https://blog.oup.com/2018/05/
    thomas-kuhn-archives-mit/.
    (2021a). *Kuhn's Intellectual Path: Charting* The Structure of Scientific
    Revolutions. Cambridge: Cambridge University Press.
    (ed.) (2021b). *Interpreting Kuhn: Critical Essays*. Cambridge: Cambridge
    University Press.
    (2021c). "Reassessing Kuhn's Theoretical Monism: Addressing the Pluralists'
    Challenge." In K. Brad Wray (ed.), *Interpreting Kuhn: Critical Essays*,
    pp. 222–237. Cambridge: Cambridge University Press.

Wray, K. Brad, and Andersen, Line E. (2020). "Reporting the Discovery of New Chemical Elements: Working in Different Worlds, Only 25 Years Apart." *Foundations of Chemistry* 22, 2: 137–146.

Wright, G., van der Heijden, K., Bradfield, R., Burt, G., and Cairns, G. (2004). "Why Organizations Are Slow to Adapt and Change – And What Can Be Done about It." *Journal of General Management* 29, 4: 20–35.

Wu, L., Wang, D., and Evans, J. A. (2019). "Large Teams Develop and Small Teams Disrupt Science and Technology." *Nature* 566, 7744: 378–382.

Wuchty, S., Jones, B. F., and Uzzi, B. (2007). "The Increasing Dominance of Teams in Production of Knowledge." *Science* 316: 1036–1039.

Wylie, Alison (1999). "Rethinking Unity as a 'Working Hypothesis' for Philosophy of Science: How Archaeologists Exploit the Disunities of Science." *Perspectives on Science* 7, 3: 293–317.

Xu, F., Wu, L., and Evans, J. (2022). "Flat Teams Drive Scientific Innovation." *Proceedings of the National Academy of Sciences* 119, 23: e2200927119.

Zachary, G. (1997). *Endless Frontier: Vannevar Bush, Engineer of the American Century*. New York: Simon and Schuster.

Index

Abbott, Andrew, 197
Adler, Ken, 269
analytic-synthetic distinction, 23–24, 26, 28–30, 32
Andersen, Hanne, 49, 121
anomalies, 5, 67, 71–74, 77, 80, 85, 88–93, 94, 109, 114, 129, 132, 149, 154–155, 160–164, 199, 219
  quantitative, 55–56, 60–61, 71, 73, 77
anti-rationality, 226
antirealism, 9, 72–73
applied science, 39, 40, 41, 42–44, 49, 50, 197
arationality, 149, 163, 167, 169–170, 175, 178
atomic bomb, 39, 43, 51

bandwagon effect, 66–69, 77
Barnes, Barry, 195, 236, 237, 243–245, 246, 250
basic/pure sciences, 37, 40, 42–45, 46, 50–51, 123–124, 129, 199
behavioral criteria, 29, 33, 35–36
Berkeley, 43, 55, 66
  Kuhn at, 56–57, 62, 70, 77–78, 265–266
  physics department of, 57, 62, 70
Bernal, John Desmond, 37, 195
Biagioli, Mario, 267–268, 270–272
Bird, Alexander, 7, 16, 48, 148, 242–243, 249, 251, 254
Birge, Raymond, 55–57, 62–70
Bloor, David, 195, 237–239, 243–248, 249
Bohmian mechanics, 133
Boyd, Richard, 141–142, 145
Brush, Stephen, 267
Buchwald, Jed, 1, 8–9
Bush, Vannevar, 39–40, 42–46, 50, 51

Caneva, Kenneth, 255, 264
Carnap, Rudolf, 17, 21–33, 35–36, 217, 226
case studies, dilemma of, 218, 220–221, 222, 224, 225, 228, 229–230, 233

case study method, 32, 96–97, 101, 104, 110–115, 218, 219–226, 228, 229–233, 234, 256, 268, 271, 273
Chang, Hasok, 218, 222, 228–230
circularity, philosophical, 219–220, 223, 225, 240–244, 251, 252
Clairaut, Alexis Claude, 3, 161–163
Cohen, E. Richard, 65, 70–73
collaboration, 43, 50, 112, 119, 141, 157, 198, 202–204, 206–209, 268
collective belief, 149–158, 160–161, 163
collective knowledge, 147
Collins, Harry, 11–12, 206
commitments, scientific, 23, 24, 27, 28–29, 32, 35
  shared fundamental, 81–82, 83, 84–89, 90, 91, 92, 94
Committee on Data for Science and Technology (CODATA), 56, 71–76
communities
  expert, 123
  scientific/research, 13, 22–23, 33–35, 47, 84–94, 98–99, 103, 104–105, 128, 131, 147–148, 153–155, 158, 160–161, 163, 166, 173, 184–185, 195, 198, 200–202, 224–226, 244, 249, 250, 252, 264, 270, 271
community-based activity, 80
complementary science, 218, 222, 228–230
Comte, Auguste, 97–98, 112–113, 115, 141
Conant, James B., 16, 17, 38–46, 47, 50–51, 95–98, 104–105, 110–112, 114–116, 119, 126–127, 183
conceptual change, 8, 24–25, 100, 263
consensus, 13, 15, 68, 80, 81, 83, 86–89, 93–94, 95, 113–114, 117, 124, 137, 148, 153–155, 157, 163, 195, 199–200, 225, 227–228, 232–234, 264
conservative, normal science as, 79–80, 82–85, 88–89, 90, 92, 94, 148, 155, 157, 159, 199

301

Printed in the United States
by Baker & Taylor Publisher Services